AF179774

Sustainable Development

Environment, Energy and Water Resources

Sustainable Development

Environment, Energy and Water Resources

Prof. M. K. Ghosh Roy

Taylor & Francis
Taylor & Francis Group
Boca Raton London New York

CRC is an imprint of the Taylor & Francis Group,
an informa business

Ane Books Pvt. Ltd.

Sustainable Development

M.K. Ghosh Roy

© Author

First Published in 2011

Ane Books Pvt. Ltd.

4821, Parwana Bhawan, 1st Floor
24 Ansari Road, Darya Ganj, New Delhi -110 002, India
Tel: +91 (011) 2327 6843-44
Fax: +91 (011) 2327 6863
e-mail: kapoor@anebooks.com, anebooks@vsnl.net
Website: www.anebooks.com

For

CRC Press
Taylor & Francis Group
6000 Broken Sound Parkway, NW, Suite 300
Boca Raton, FL 33487 U.S.A.
Tel : 561 998 2541
Fax : 561 997 7249 or 561 998 2559
Web : www.taylorandfrancis.com

For distribution in rest of the world other than the Indian sub-continent

ISBN : 978 1 43988 8254

All rights reserved. No part of this publication may be reproduced, stored in a retrieval system, or transmitted in any form or by means, electronic, mechanical, photocopying, recording and/or otherwise, without the prior written permission of the publishers. This book may not be lent, resold, hired out or otherwise disposed of by way of trade in any form, binding or cover other than that in which it is published, without the prior consent of the publishers.

British Library Cataloguing in Publication Data
A catalogue record for this book is available from the British Library

Printed at : Thomson Press (India) Ltd., Delhi

To

my wife Sulekha

whose constant support saw the book through three arduous years of its birth

Foreword

Contemporary times can hardly disown any facet of globalization; indeed, globalization is in a state of continual flux. Even its economic parenthood, as it were, has kept it alive through a variety of generations. Globalization, often revealingly or unwittingly has crept into sectors, which in turn, keep on acquiring different complexions. Perhaps, seeking a global identity may go on receding largely because of globality coupling itself with, and bringing in its trail, new forms that couldn't be made discernible earlier. Identification on this score is really an onerous task and becomes all the more complicated and worthwhile too, when such intermingling pose as challenges. Global Challenges, as the main title of the book, may seem to be transitory occurences. Hence, the book cannot but be legitimately fascinating. The title makes the job a fairly challenging proposition. The author has ably met it.

There can always be desirable proclivity from the post-Kyoto framework, but that the Protocol did unleash a genre of climate fix, is not often realized. Any post-Kyoto Protocol framework has to reckon discourses on energy security and climate change. The transitory aspect, as mentioned in the above lines on globalization, is currently taken care of by the most recent COP-14 in Poznan, Poland and so the timely presentation through this book, of the elements or areas that can hardly be decoupled. The concern for mitigating the impact on humanity of the climate change is becoming unbearingly difficult. The detail, however monotonous they may appear, of how to go about limiting carbon emissions and often simultaneously putting a price on activities harmful to environment, continue to be intriguingly deliberated.

Professor Ghosh Roy seems to be favouring a new sort of hierarchy in his list of the challenges segment. Sustainable development, historically speaking, emerged as a concept because of the World Commission of Environment and Development, under the aegis of the United Nation, way back in 1982. While the environment as a vital concept came up a decade ago in the well-known Stockholm Conference, CSD (Commission on Sustainable Development) had the follow-ups as CSD-1 in 1993 and CSD-12 in 2004. Climate change, global warming etc., came well within the purview of such get-togethers. The interface between energy and environment is well fathomed in this treatise. UN Millennium Development Goals can hardly be missing in the semi-textual presentation in the book. Policy matters, economic or sociological, can in no way, be dispensed with, as borne out by some of its adroitly well put up sections. That appallingly ubiquitous scarcity of water resources can occur, which is likely to be ignored, has been excellently posed as challenges. Energy needs have to have an overriding priority among the categories of four concerns considered by the Author.

Dr. Bjorn Lomberg in his marvellous book *The Sceptical Environmentalist* opts for 10 global challenges: air pollution, conflict, disease, global warming, hunger and malnutrition, lack of education, gender inequity, lack of water and sanitation, terrorism and trade barriers. Coming back to Kyoto Protocol, following the arguments of Dr. Lomberg, all its likely successors ought to be seriously concerned how to take the challenges on. Nuclear energies, nuclear efforts etc., should in no way take a back seat if the country like India has to acquire energy self-sufficiency. Millennium Development Goals (MDGs) have overlaps with challenges mentioned earlier. That this needs to be attended to through massive efforts has been looked by the author. Hopefully, the book will be great facilitator for leapfrogging to clean and cleaner technologies. While one can set forth a seemingly magnificently array of energy options because of pressing actionable features of the climate plan, in the ultimate analysis, we have too have a form of sustainable development that is stable from the economic standpoint, equitable because of social imperatives and above all, environmentally accountable. I commend this book to readers who are not bogged down to ironies of climate and climate change as well, but receptive to identifying capably strong strategies.

D.K. Sinha
Former Rashbehari Ghosh Professor of
Applied Mathematics, Cal. University
Former Pro-Vice Chancellor, C.U.

Former Vice Chancellor, Vishva Bharati U.

Preface

India's great savant leader and national conscience-keeper Mahatma Gandhi had forewarned quite some time ago " the world has enough for everybody's needs, but not enough for everybody's greed." How a small sentence succinctly epitomizes a global truth. On one side, we have the high opulence, wasteful splurging of the affluent societies of the Western world juxtaposing the abject poverty and unmet hunger of vast humanities in Africa and much of Asia. This apparently unsustainable model has provoked the enunciation of the UN. The most important environmental challenges staring the earth are the unacceptably high pollution of the biosphere in the recent decades, manifested in many ways by the global warming, change of weather, unpredictable floods and draughts, shrinking biodiversity from deforestation and desertification, soil erosion affecting agriculture etc. After long and arduous research, investigation and deliberation at the national level and UN level, it has now been inferred that the pernicious global warming is due to high concentration of carbon dioxide and the other so-called Greenhouse gases (GHG) in the atmosphere. Excessive GHG emission has been caused by heavy consumption of fossil fuels (coal, oil, and gas). The high consumption of the fossil fuel has been ascribed to the abnormal energy-intensity of the current global technological-cum-industrial infrastructure.

Engineering the industrialized world thus throws another challenge. Admittedly, the modern life thrives on energy. When we drive a car on roads, rides on railways or fly in the sky, the motive engine consumes energy all the time. From the kitchen to the office and the factories, the modern civilization needs energy. Even after adjusting for the rationalization of consumption and increasing energy efficiency, it may not be feasible to much reduce the consumption of fossil fuel, albeit its high carbon-pollution. Then the fossil fuels are exhaustible resources of the earth and may not last even till the end of the 21st Century. The challenge of alternative energies is two-fold, to hit upon a carbon-neutral energy source immediately to replace or reduce the fossil fuel consumption and secondly to develop a large source of clean energy for the coming decades or centuries. The development trend is already visible towards renewable energy source of the earth—hydropower, wind power, and solar energy. As carbon-neutral energy source, nuclear energy is also finding large-scale revival. Hydrogen energy and Fuel cell technology are the incoming energy alternatives too, which are highly potential.

Greek philosopher Thales Miletus propounded "the best of everything is water". Down the ages, Leonardo da Vinci too reaffirmed water as the 'driver of nature'. Water is indeed the driver of life, the human body constituting 70 % of water. But because of its easy availability in the past, water has traditionally been regarded as the inexhaustible free gift of nature; this has led to extravagant and wasteful use; thus this life-saving resource has become scarce threatening human life, livelihood, agriculture, development and progress. A UN document (Mordel Plata Conference 1977) concluded that though potentially there may be enough water to meet the forthcoming global demands, there may be serious shortages of water at the right place, right time and right quality. The World Bank, the various UN agencies and the World Water Council have all published nearly similar world water maps showing more and more countries of the world will become water-stressed by 2050, because of its increasing scarcities. Thus there is a deep global challenge to plan, augment and rationalize demand and supply of water resources to carry the world forward in the 21st Century.

So much is the raison d'etre for the title of this book: *"Sustainable Development: Challenges of Environment, Energy and Water Resources"*. For each challenge, we have devoted a Part and there are Three Parts for the three major challenges and one introducing part on overview on Sustainable Development.

In Part I, we have first introduced concepts of sustainable development in Chapter 1.In the humanist-cum-environmentalist approach of the Brundtland Report, it connotes development that includes preservation of ecology and equity for all, that is inclusive development and growth. In the economist's parlance (Chapter 4), according to Robert Solow and others, sustainable development is to ensure inter-generational equity *i.e.,* natural resources shall be consumed keeping the long-term well-being and sustainability of the future generation in view. In case, the exhaustible natural resource must be used up faster, consumption must also lead to creation of capital stock for future generation in terms of technology and physical infrastructure. In Chapter 2, we have attempted a limited review and audit of various natural resources, which include agriculture, fisheries and aquaculture, wild life and forest resource and also just touched energy sources (Mongillo 2004). In Chapter 3, we developed the theme that application of science and technology for solutions to sustainability is essential. In Chapter 4, besides the subject of economic sustainability, we also submit that special governance model for sustainable development is necessary at the individual, corporate and the national levels (Lafferty 2001). Last but not the least, we should mention that we have included some aspects of the International Conventions including Earth Summit at Rio de Janeiro 1992 in Chapter 1. and UN Millennium Development and Goals in Chapter 2.

In Part II, we have presented an outline of the Challenges of Environment. The basic theme is that clean air which is vital for life is threatened by the general atmospheric pollution (Chapter 5) and automobile emission (Chapter 6). The categories of pollutants are principally carbon dioxide, carbon monoxide, sulphur dioxide, nitrous oxide and the SSP-10 (suspended solid particulates below 10 micron size). Solid particulates often give rise to smog acted upon by solar photolytic action. The more serious challenge comes from formation of large voids (known as Ozone Holes) in the upper atmosphere (stratosphere) allowing carcinogenous ultraviolet rays to pass through to the earth. Not so serious is the occurrence of acid rain (Chapter 8); the atmospheric Sox and Nox get hydrolyzed into their respective sulphuric and nitric acids and fall with the rain retarding or destroying the aquatic lives, plants and vegetation, buildings and other infrastructure. In the next three Chapters (9, 10 and 11), we break into the most burning issues of environmental challenges, firstly the increasing atmospheric concentrations of carbon dioxide, methane, nitrous oxide and chlorofluorocarbon, the so-called GHGs (glasshouse gases), and secondly, consequent global warming and climate change; all these may lead to glacial ice melting, unacceptable rise of sea level, drying up of rivers, erratic rainfall, floods and droughts. Such devastating challenges demand most urgent solutions.

In Part III on Energy, we have tried to present a long list of alternative energy options. Chapter 12 is really a preamble to the status of global stock of fossil energy, its global consumption and the pollution hazards thereof. In the subsequent seven chapters 13 to 19, we covered the availability and feasibility of several energy resources—nuclear energy, bio-energy (biomass, ethanol and bio-diesel), solar energy (thermal and Fuel Cells in Chapter 15, Wind Energy (Chapter 16), Hydrogen Energy & Fuel Cells (Chapter 17), Hydro-Energy (Chapter 18), and Geothermal, Wave Energy & Ocean Thermal Energy (Chapter 19). The obvious conclusion is there is still life and hope beyond the petroleum fossil fuel. The renewable Bio-fuels are able to bring down carbon emission substantially. But the other renewable solar and wind energies are totally carbon-neutral; so is the hydrogen-fed Fuel Cell, though there is possibility of carbon emission in producing hydrogen fuel. All these three forms of energy promise virtually limitless energy, provided continued research is devoted for upgrading their application technology and engineering.

The third dimension of our global challenge rest with development of water resources for the mankind, described in Part IV. There are two reviews, one carried out for global water resources (Chapter 20) and the other for the Indian water resources (Chapter 21 and 22). India, being of near continental size in land area and population, we have examined its rainfalls, rivers, river basins and potential inter-basin transference of water. Globally, total quantity of water available is more or less constant. The problems here are basically infrastructure development and water management, eliminating extravagance and waste, economizing uses, treating water for quality and recycle water where supply falls short of demand.

Next, I like to identify the readers for whom this book is intended for. This book is not structured for any specific university courses but shall lend insights and identify milestones for Graduate students and Researchers working in the fields of environment, energy and sustainable development. It touches thoughts and provide numerous supporting data. However, despite the broad discourses in these subjects, the book makes occasional forays into some mathematical formulations, which the scientific readers may pursue, others may skip over. For a wider range of today's environmentally conscious public, the Part I of this book will indicate some guidelines of a socially equitable and responsible living and Part III (Energy) and Part IV (Water Resources) throw some enlightenment for the future.

In the call of social responsibility, I have now undertaken to write this book. In this task I have drawn heavily on the original work of many scientists, researchers, professors and social activists, to all of whom I am greatly indebted. Some of these sources I have acknowledged in the book and I regret, if I have missed to acknowledge many others. I shall be very gratified if the book is able to serve its objectives for the benefit of general public, researchers and students.

M.K. Ghosh Roy

Postscript

Copenhagen Accord 2009

In the process of publication of this book, the United Nations Framework Convention on Climate Change (UNFCCC) Conference took place at Copenhagen, 7–18 December 2009 which was attended by 192 Nations including the US. Let us hope that the Copenhagen Accord will usher in a new regime of equitable sustainable development through the instrumentality of mitigation and adaptation *vis-à-vis* global warming.

M.K. Ghosh Roy

Contents

Part I: Overview of Sustainable Development

Part II: Challenges of Environment

Part III: Challenges of Energy

Part IV: Challenges of Water Resources

Part I
Overview of Sustainable Development

CHAPTER 1

Global Sustainability–An Overview

1.1 General Review

Over-exploitation of earth's resources and resulting pollution of biosphere in meeting the demands of the high consuming societies of the western world has been causing growing concern among the world community for the past several decades. India's great savant leader and national conscience-keeper Mahatma Gandhi had forewarned quite early "That the world has enough for everybody's needs, but not enough for everybody's greed".

High Consumerism

How a small sentence succinctly epitomizes a global truth. On one side, we have the high opulence, wasteful splurging of the affluent societies of the America and the Europe juxtaposing the abject poverty and unmet hunger of vast humanities in Africa and much of Asia. Mercifully, the whole of the world of six billion plus people is not addicted to the feverish consumerism of the West, which could have created cataclysmic havoc to the world economic order and the earth's ecology. But if the energy guzzling western civilization structure could be moderated, restructured and redesigned to a lower but optimum level of energy and resource consumption and the stagnant or archaic economy of the less developed world could be modernized to higher productivity and common well-being, it will then be possible to create an equitable world order, which will be sustainable within the earth's overall eco-system.

Earth's Carrying Capacity

Now comes the problem of unacceptably high pollution of the biosphere in the recent decades, manifested in many ways by global warming, change of weather of unpredictable floods and droughts, shrinking biodiversity from deforestation and desertification, soil erosion affecting agriculture, dwindling availability of water affecting agriculture and hence human health with plethora of air-borne water-borne diseases.

It has now come to be widely recognized that global warming is being accentuated by the steadily increasing concentration of carbon dioxide and other greenhouse gases (GHG) in the stratosphere environment. It is also now known that the GHG emissions are being mostly caused by the combustion of fossil fuels coal and petroleum in houses, industries, and transportation. Apart from the pernicious properties of fossilized hydrocarbons during combustion, there is another important dimension, defined by the so-called carrying capacity of the upper atmosphere up to certain extent. The excess carbon of the atmosphere is recyclable into forest plants and vegetation. The amount carbon fixation rate of the biomass is estimated at about 2 gigatons per annum. As long as the carbon emission rate does not exceed the above limit, the emission level will be sustainable. The above cycle thus represents the sustainable limit of GHG emissions.

Energy Sustainability

The global sustainability problems throw other challenges. It is evident that the present high-energy demands will not allow much reduction in the consumption of fossil fuel, even after adjusting for increasing efficiencies in the energy production, and the rationalization of consumption. That means that if the scale of use of fossil fuel has to come down, it has to be proportionately replaced by other benign non-GHG-emitting fuel and better still other combustion technologies. This opens the vast possibilities of bio-fuels, solar and wind energies in the short and medium term and the hydrogen fuel in the long term.

Resource Sustainability

Depending upon the availability of natural resources, there is another aspect of sustainability to consider. Some of the earth's resources like minerals, e.g. iron, aluminum etc. are exhaustible i.e. their stock is limited. To ensure sustainability, there are two possibilities, firstly to economize their use to highest essentiality and secondly, to recycle the used products, e.g. scrap steel or scrap aluminum is reprocessed to newer uses. This process can continue till new substitutable materials are discovered or developed. This will ensure continuing sustainability.

There is a class of exhaustible resources, which cannot be reprocessed and recycled. Energy resources like coal and petroleum fuel once consumed cannot be renewed. Here, the best option is to gradually switch over to some renewable fuel resources like wind energy or bio-fuels. We have the Ethanol (fermented from crop sugar) in use already in Brazil and secondly the Bio-diesel (abstracted from corn/vegetable edible oil in US but from non-edible oil Jatropha plants in India). Originating from plants, such products are renewable up to the earth's carrying capacity. Both these bio-fuels will emit 25 to 30 per cent less CO_2.

The most benign non-GHG-emitting renewable energy forms are Solar energy and Wind energies. Both these energies are thinly distributed over the earth; but given today's technologies to concentrate, collect and store these energies, they can be made available in abundant quantities for electrical power generation or direct heating.

Looming on the horizon is the very bright prospect of Hydrogen fuel. This is one fuel, which can completely eliminate the global concern of the atmospheric carbon once for all. There are two technological challenges, firstly of the hydrogen production by using solar energy and secondly, the commercial production of Fuel cells customized to automotives at the lower end and powers at the higher end. The 'Hydrogen Economy' can really solve the energy sustainability of this world. Limitless oceans are the infinite sources of hydrogen.

Nuclear energy from fission, breeding or fusion has definitely a stellar role to play. But we are desisting to open up discussion on this subject in the introductory overview, because of some strong reservation on the nuclear hazards to human life. However, we have included a separate chapter on Nuclear energy in this book later.

In Fig F1.1 below we present a sustainable energy spectrum from early time till 2150 adapted from Dunn (2004).

Environmental Economy

Environmental economists have played a key role in developing the concept of the sustainability. The Brundtland Report (1987) of the World Commission on Environment and Development defines sustainable development as meeting the needs of the present generation without compromising the needs of the future generation. Work of Nobel Laureate Robert M Solow (Solow 1991) is mentioned here in this respect. Solow and others believe that the rent from the use of the non-renewable resources shall be redirected to the creation of technological and other infrastructure capital to keep the total stock of resource capital intact over time extending intergenerationally.

Of late, environmental sociologists have also been raising issues on the impact of environmental degradation and inequitability of wealth and poverty on the human behavior and well being of the disadvantaged.

From the above general consideration of the concept of sustainability, we now try to overview international initiatives of the United Nation and other organizations towards environmental amelioration and sustainability.

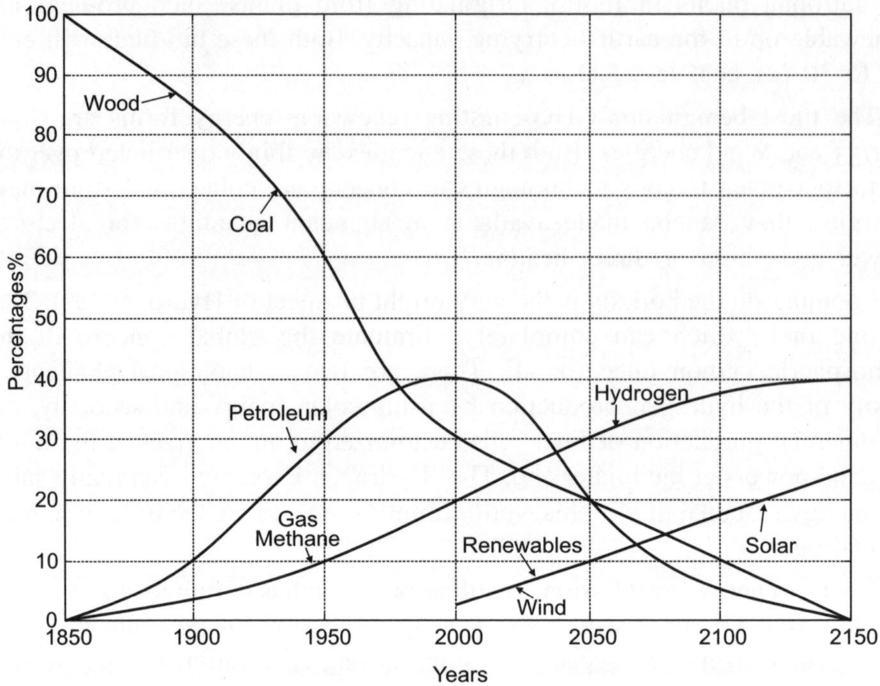

Fig. F1.1: Long-term Energy Supply Trend

1.2 Overview of International Environmental Initiatives

A series of initiatives have been undertaken under the auspices of the United Nations in view of the deteriorating global environment, the landmarks being the followings:

1972 Stockholm Conference under United Nation Environment Programme (UNEP);

1985 Montreal Protocol on depletion of Ozone layer under the Vienna Convention;

1987 Brundtland Report on "Our Common Future" under the World Commission on Environment and Development under the U.N.;

1988 Establishment of Intergovernmental Panel on Climate Change (IPCC) jointly by UNEP and the World Meteorological Office;

1992 Earth Summit at Rio de Janeiro enunciating the famous 'Agenda 21' for sustained development;

1997 Kyoto Protocol under the Framework Convention on Climate Change 1992;

2002 World Summit, Johannesburg, proposing Millennium Declarations on improving the quality of life.

We will now elaborate on some aspects of the International Conventions, Protocols, and Reports towards environmental protections and sustainability.

Stockholm Conference (5-16 June, 1972)

In 1972 UN sponsored the first major Conference on Human environment under the aegis of the United Nation Environment Programme, which Agency was designed to be the 'environmental conscience of the United Nation' and work as its central coordinating secretariat. It adopted a Declaration consisting a Preamble and 26 Principles.

By the Preamble, the Conference, among other things, declared that in the developing counties, environmental problems generally arose from poverty, whereas in the affluent societies of the industrialized countries, it arose from excessive energy dissipation, waste and wanton consumption. The Principle 17 exhorted the creation of the appropriate national institutions for care of environmental resources and the Principle 25 encouraged international cooperation towards protection and improvement of the environment.

The Stockholm Conference waked up, for the first time, the global responsibility towards environment and more than eighty conventions, protocols, and treaties including the Law of the Sea (1982), the Convention on Trade in Endangered Species, Vienna Convention on Protection of the Ozone Layer (1985) followed.

Montreal Protocol (1985/1987)

This Protocol on substances that deplete ozone was signed under the Vienna Convention (1985) and came into force in Jan. 1987. The ozone layer in the stratosphere absorbs carcinogenic ultraviolet radiation protecting humans on earth. CFCs (Chlorofluorocarbons), used to be the input in the refrigeration and as solvents for air sprays are the worst offender in destroying ozone. These substances were to be totally banned from 2010.

World Commission on Environment and Development (1983) and Brundtland Report "Our Common Future" (1987)

In 1983, being conscious of the accelerating global environmental degradation and the international responsibility for affirmative action, UN set up the World Commission on Environment and Development (WCED). The Commission, headed by Gro Harlem Brundtland of Norway, after arduous work, submitted a report "Our Common Future" in 1987, popularly known as Brundtland Report, which boldly presented the concept of sustainable development as an alternative to the policy of only economic growth.

"Sustainable development is the development that meets the needs of the present without compromising the ability of the future generation to meet their needs. It contains within it two key concepts:

- the concept of needs in particular the essential; needs of the world's poor, to which the overriding priority should be given;
- the idea of limitations, imposed by the state of technology and social organization on the environment's ability to meet the present and future needs (WCED 1987:43)"

In effect, the Bruntland Report affirming that equity and growth with the due preservation of the environment are simultaneously possible. But it also emphasizes that achieving equity and sustainable growth would involve technological and social change.

Earth Summit, Rio de Janeiro under UNCED

After considering the 1987 Brundtland Repot, the UN General Assembly called for the UN Conference on Environment and Development at Rio de Janeiro in 1992.

This 1992 conference at Rio is the second UN conference after the first UN conference at Stockholm in 1972 and sometimes referred to as STOCKHOLM-II.

This Rio meet is also known as the 'Earth Summit', which was unprecedented in its size of participation by 108 Heads of State or representatives and many thousands of ordinary delegates, journalists and other participants. It aroused global response and awareness that nothing less than the total transformation of various national growth, development and governance strategies and people's attitude as responsible earth citizens will save the earth and the environment towards the long-term sustainability and well being of the mankind at large.

The summit arrived, inter alia, at the following major international agreements:

- Rio Declaration of a series of 27 Principles defining the rights and responsibilities of individual State countries towards sustainability of environment and Developments
- Famous Agenda 21 – a very comprehensive set of action programmes towards sustainable developments;

The Declaration of Principles included, *inter alia*, the following salient issues:

- That human beings are at the center of concern for sustainable development. They are entitled to a healthy and productive life in harmony with the nature;
- That right to development must be fulfilled so as to meet the development and environmental needs of the present and the future generation;
- That states have a sovereign right to exploit their own resources pursuant to their own development and environmental needs of the present and the future generation;

- That eradicating poverty and reducing economic disparities of living standards among people of the world are indispensable requirements for sustainable development;
- That women having a vital role in the environmental management, their full participation is essential for achieving sustainable development.

Because of their importance, the above twenty-seven principles are presented in Box B 1.1.

Box B 1.1

Directive Principles, Earth Summit, 1992

Principle 1 Human beings are at the center of concerns for sustainable development. They are entitled to a healthy and productive life in harmony with the nature.

Principle 2 States have, in accordance with the Charter of the United nations and the principles of international law, the sovereign right to exploit their own resources pursuant to their own environmental and developmental policies, and the responsibility to ensure that activities within their jurisdiction or control do not cause damage to the environment of other States or of areas beyond the limits of national jurisdiction.

Principle 3 The right to development must be fulfilled so as to equitably meet developmental and environmental needs of present and future generations.

Principle 4 In order to achieve sustainable development, environmental protection shall constitute an integral part of development process and cannot be considered in isolation from it.

Principle 5 All states and all people shall cooperate in the essential task of eradicating poverty as an indispensable requirement for sustainable development, in order to decrease the disparities in standards of living and meet the needs of the majority of the people of the world.

Principle 6 The special situation and needs of developing countries, particularly the least developed and those most environmentally vulnerable, shall be given special priority. International actions in the field of environment and development should also address the interests and needs of all countries.

Principle 7 States shall cooperate in a spirit of global partnership to conserve, protect and restore the health and integrity of contributions to global environmental degradation. States have common but differentiated responsibilities. The developed countries acknowledge the responsibility that they bear in the international pursuit of sustainable development in view of the

Contd...

pressures their societies place on the global environment and of the technologies and financial resources they command.

Principle 8 To achieve sustainable development and a higher quality of life for all people, the State should reduce and eliminate unsustainable patterns of production and consumption and promote appropriate demographic policies.

Principle 9 States should cooperate to strengthen endogenous capacity, building for sustainable development by improving understanding through exchanges of scientific and technological knowledge, and by enhancing the development, adaptation and transfer of technologies, including new and innovative technologies.

Principle 10 Environmental issues are best handled with the participation of all concerned citizens, at the relevant level. At the national level, each individual should have appropriate access to information concerning environment that is held by public authorities, including information on hazardous materials and activities in their communities and the opportunity to participate in the decision-making processes. States shall facilitate and encourage public awareness and participation by making information widely available. Effective access to judicial and administrative proceedings, including redress and remedy, shall be provided.

Principle 11 States shall enact effective environmental legislation. Environmental standards, management objectives and priorities should reflect the environmental and developmental context to which they apply. Standards applied by some countries may be inappropriate and of unwarranted economic and social cost to other countries, in particular, the developing countries.

Principle 12 States should cooperate to promote a supportive and open international economic system that would lead to economic growth and sustainable development in all countries, to better address the problems of environmental degradation. Trade policy measures for environmental purposes should not constitute a means of arbitrary or unjustified discrimination or a disguised restriction on international trade. Unilateral actions to deal with environmental challenges outside the jurisdiction of the importing country should be avoided.Environmental measures addressing trans-boundary or global environmental problems should as far as possible be based on international consensus.

Principle 13 States should develop national law regarding liability and compensation for the victims of pollution and other environmental damage. States shall also cooperate in an expeditious and determined manner to develop further

Contd...

	international law regarding liability and compensation for adverse effects of environmental damage caused by activities within their jurisdiction.
Principle 14	States should effectively cooperate to discourage or prevent the relocation and transfer to other States of any activities and substances that cause severe environmental degradation or are found to be harmful to human health.
Principle 15	In order to protect the environment, the precautionary approach shall be widely applied by States according to their capabilities. Where there are threats of serious irreversible damage, lack of full scientific certainty shall not be used as a reason for postponing cost-effective measures to prevent environmental degradation.
Principle 16	National authorities should endeavour to promote the internationalization of environmental costs and the use of economic instruments, taking into account the approach that the polluter should, in principle, bear the cost of pollution, with due regard to the public interest and without distorting international trade and investment.
Principle 17	Environmental impact assessment, as a national instrument, shall be undertaken for proposed activities that are likely to have a significant adverse impact on the environment and are subject to a decision of a competent national authority.
Principle 18	States shall immediately notify other States of any natural disasters or other emergencies that are likely to produce sudden harmful effects on the environment of those States. Every effort shall be made by the international community to help States so affected.
Principle 19	States shall provide prior and timely notification and relevant information to potentially affected States on activities that may have a significant adverse trans-boundary environmental effect and shall consult with those States an early stage and in good faith.
Principle 20	Women have a vital role in environmental management and development. Their full participation is therefore essential to achieve sustainable development.
Principle 21	The creativity, ideals and courage of the youth of the world should be mobilized to forge a global partnership in order to achieve sustainable development and ensure a better future for all.
Principle 22	Indigenous people and their communities and other local communities have a vital role in environment management and development because of their knowledge and traditional practices. States should recognize and duly support their

Contd...

	identity, culture and interests and enable their effective participation in the achievement of sustainable development.
Principle 23	The environment and natural resources of people under oppression, domination and occupation shall be protected.
Principle 24	Warfare is inherently destructive of sustainable development. States shall therefore respect international law providing protection for the environment in times of armed conflict and cooperate in its further development, as necessary.
Principle 25	Peace, development and environmental protection are interdependent and indivisible.
Principle 26	States shall resolve all their environmental disputes peacefully and use appropriate means in accordance with the Charter of the United Nations.
Principle 27	States and people shall cooperate in good faith and in a spirit of partnership in fulfillment of the principles embodied in the Declaration and in further development of international law in the field of sustainable development.

Agenda 21 of the Rio Earth Summit encompasses comprehensive action plans for the social and economic development such as poverty alleviation, eco-efficient production system, energy conservation, and preservation of natural resources, demographic planning, protecting the atmosphere, oceans, and biodiversity; preventing deforestation; promoting sustainable agriculture. The gist of all comprehensive action plans under Agenda 21 is listed in Box B1.2.

Box B 1.2

Themes of Agenda 21
Earth Summit 1992

1. Combating Poverty;
2. Changing Consumption Pattern;
3. Demographic Dynamics and Sustainability;
4. Protection and Promotion of Human Health;
5. Promoting Sustainable Human settlement development;
6. Integrating Environment and Development in decision making;
7. Protecting the Atmosphere;
8. Integrated Approach to planning and management;
9. Combating Deforestation;
10. Managing Fragile Ecosystem: Combating Desertification and Drought;
11. Managing Fragile Ecosystem: Sustainable Mountain Development;
12. Promoting Sustainable Agriculture and Rural development;
13. Conservation and Biological Diversity;
14. Environmentally sound Management and Biotechnology;
15. Protection of the Oceans, all kinds of seas, including enclosed and

Contd...

semi enclosed seas and coastal areas and the protection, rational use and development of their living resources;

16. Protection of the quality and supply of fresh water resources: application of integrated approaches to the development, management and use of water resources;
17. Environmentally sound management of toxic chemicals including prevention of illegal international traffic in toxic and dangerous products; Environmentally sound management of hazardous wastes including prevention of illegal international traffic in hazardous wastes; Safe and environmentally sound management of radioactive wastes;
18. Global Action for Women towards sustainable and equitable development;
19. Children and youth in sustainable development;
20. Recognizing and strengthening the role of indigenous people and their communities;.
21. Strengthening the role of Non-Government Organizations as partners for sustainable development;
22. Local authorities' initiatives in support of Agenda 21;
23. Strengthening the role of workers and their trade union;
24. Strengthening the business and industry, scientific and technological community;
25. Strengthening the role of farmers;
26. Financial resources and mechanisms;
27. Transfer of environmentally sound technology, cooperation and capacity building;
28. Science for sustainable development;
29. Promoting Education, Public awareness;
30. National mechanisms and international cooperation for capacity building;
31. International Institutional arrangements; Legal Instruments and mechanism;
32. Information for decision-making.

The following two Conventions were also opened at the Rio Summit Convention –

(*i*) The United Nation Framework Convention on Climate Change;

(*ii*) Convention on Biological Diversity.

Convention on Biological Diversity

This convention signed at the time of the Rio Earth Summit 1992, had the overall aim of the sustainable use of biological diversity for the benefit of the present and future generations. It comprised, inter alia, following main provisions:

- Biological Diversity: It is defined in Article 2, meaning all the variability among the living organisms from all sources terrestrial and marine.
- Measures for Conservation of all Biological resources (Article 6) shall be included in all natural strategies, plans or programmes duly integrated.
- In-situ and Ex-situ Conservation (Articles 8 & 9) shall include conservation of eco-systems, natural habitats and viable populations of species in their natural surroundings and similar conservation outside their natural habitats.
- Sustainable uses of the bio-resources by integrated consideration, which does not adversely impact the biological diversity.
- Implementation of the Convention includes an establishment of a Secretariat under Article 24 and a subsidiary scientific body under the Article 25 and settlement of any disputes under the same Article.

UN Commission on Sustainable Development (UNCSD) 1993

The above UNCSD was set up on 16.02.1993 as a part of UNECO under Article 68 of the UN Charter mainly to ensure follow-up action on the Rio Earth Summit, 1992 and for the implementation of Agenda 21 of the same Summit.

UN Conference on Human Settlement (HABITAT II), 1996

This Conference is a follow-up of the first Human Settlement Conference in 1976 at Vancouver, Canada. This Second Conference on Human Settlement was held at Istanbul, Turkey in June 1996 and also known as 'City Summit' or Habitat II. The Habitat Agenda of the Conference included a Preamble affirming,

"Everyone has the right to adequate standard of living for themselves and their families, including adequate food, clothing, housing, water and sanitation and to continuous improvement of living conditions."

Besides the preamble, the Agenda includes a statement of Goals and Principles and six Commitments for Governmental support of its objectives and strategies for the implementation of global plan of action.

The habitat II aimed at the achievement of sustainable development of the world's cities, towns and villages within the next two decades of the next century.

Framework Convention on Climate Change, 1992 and Kyoto protocol, 1997 on Global Warming

FCCC, 1992

It has been known for sometime that Greenhouse gases (GHG) consisting mainly of carbon dioxide, CFC (Chlorofluorocarbons), Nitrogen oxide and methane cause global warming creating a thermal shield over the atmosphere. Global warming, in turn, affects the global climate changing the pattern of rain, draught, storm, and floods, glacial melt and river flow, cycle of weather, agriculture etc. IPPCC (International panel on Climate Change), since its formation by the UN in 1988, has been doing very valuable investigative research and reporting work. At the UNCED Summit, 1992 at Rio' after much negotiation, the Framework Convention on Climate Change (FCCC) was signed and adopted with a view to achieve reduction and stabilization of greenhouse gases emission and consequent concentration in the atmosphere at a level that will prevent very serious anthropogenic interference with the world global climatic system. The FCCC laid down, inter alia, following important guidelines:

(*i*) the need to protect the climate system on the basis of equity and in accordance with the States' common but differentiated responsibilities and respective capabilities; accordingly, the developed countries' parties should take the lead;

(*ii*) the specific need and special circumstances of the developing countries' parties, especially those are particularly vulnerable;

(*iii*) the need for precautionary measures in the absence of full scientific certainties, qualified by the need to be "cost effective" and comprehensive, that is, address all sources and sinks, adoption of all economic sectors;

(*iv*) the parties have a right to and should promote sustainable development integrated with the national development programme; and

(*v*) the parties should cooperate to promote a supportive and open international system; measures taken should not constitute a means of arbitrary or unjustifiable discrimination or a disguised restriction on international trade.

Taking cognizance of the higher carbon emission of the developed countries and at the same time their income and capabilities, the countries were divided into two categories:

- Annexure I Countries (OECD countries, Eastern Europe, & former Soviet Union)
- Non-Annexure I Countries (all other countries deemed developing)

It will be seen from The Table T1.1 that Annex countries having population of 21.7%, releases 63.9% of global carbon emission, while having a per capita income five times that of the Non-Annex countries.

Table T1.1: Global Carbon Emission Trend 1996

Countries	Population Million tons (%)	Per Capita Income PPP $	Per Capit Carbon Emission (Tones)	Share in Carbon Emission (%)
Annex I	1,299 (21.7)	15,986	3.1	63.9
USA			5.6	24.4
EU			2.5	15.1
Non-Annex I	4,438 (78.3)	2,904	0.5	36.1
China			0.6	13.1
India			0.8	12.6

Kyoto Protocol 1997

The signatories to the Framework Convention on Climate change met at a conference from December 1 to 11, 1997 at Kyoto to draw up a Protocol of implementation of the recommendations of FCCC 1992 within an international Climate Change regime. The key elements of the Kyoto Protocols are:

(*i*) Agreements of quantified emission limitation or reduction commitments by Annex I countries of FCCC, which will reduce the aggregate GHG emission by an average of 5.2 per cent below 1990 carbon dioxide level by the period 2008 – 12 with the leading Industrialized countries reducing by the following figures:

European Union 8%

USA 7%

Japan ...... 6%, by 2012;

(*ii*) Undertaking by the Annex countries to demonstrable progress in this regard by 2005;

(*iii*) Undertaking by Annex countries to the preparation of inventories of reduction of GHG emission and reporting national plans to the FCCC;

(*iv*) No requirements of any reduction of GHG emission by the developing countries of Non-Annex categories;

(*v*) Introduction of several instruments of carbon trading, which allows Annex I countries to fulfill their quota of GHG reduction by financing carbon reduction in other countries, *e.g.*,:

- Clean Development Mechanism (CDM);
- Joint Implementation (JI);
- International Emission Trading (IET).

As mentioned, the main reason for project-based mechanism is to offer credit in the carbon reduction quota of Annex I countries. CDM allows a Annex I country to finance a valid GHG emission reduction project in non-Annex country and the actual reduction gets credited to the committed quota of the sponsoring-cum-financing country. This way, GHG emission reduction also gets accomplished in the developing country.

Similarly, a JI allows an Annex country to finance a GHG reduction project in another Annex country and to receive credit equal to part or all of emission reduction.

IET is a tradable quota regime, which establishes a national emission limit for each participating country and each country to hold quota equal to part or full of the emission reduction achieved.

Tradable carbon reduction is now a growing business in India, China and Philippines. But so far, USA and Australia did not sign the Kyoto Protocol.

Johannesburg World Summit, 2002

The Johannesburg is a follow-up of the Rio Earth Summit of 1992 but was of smaller scale with less than 100 national representations. India and USA did not participate. The main objectives of the summit were emphasized as under:

(*i*) Good Governance for sustainable development;

(*ii*) Poverty Eradication is the greatest challenge facing the world today and an indispensable requirement for sustainable development;

(*iii*) Changing unsustainable pattern of production and consumption.

Among the plethora of Johannesburg Declarations, we present some of the important ones as follow:

From our origin to the future

- We, the representatives of the peoples of the world, assembled at the World Summit on sustainable Development in Johannesburg, South Africa from 2 – 4 September 2002, reaffirm our commitment to sustainable development;
- We commit ourselves to build a humane, equitable and caring global society cognizant of the need for human dignity for all.

- From this Continent, the cradle of Humanity, we declare, through the plan of Implementation and this Declaration, our responsibility to one another, to the greater community of life and to our children.

From Stockholm to Rio de Janeiro to Johannesburg

- Thirty years ago, in Stockholm, we agreed on the urgent need to respond to the problem of environmental deterioration. Ten years ago, at the United Nation Conference on Environment and development, held at Rio de Janeiro, we agreed that the protection of the environment, and social and economic development are fundamental to the sustainable development, based on the Rio Principles. To achieve such development, we adopted global programme, Agenda 21 and the Rio Declaration, to which we affirm our commitment.

The challenges we face

- We recognize that poverty eradication, changing consumption and production pattern, and protecting and managing the natural resource bases for economic and social development are overarching objectives of and essential requirements for sustainable development
- Globalization has added new dimension to these challenges. The rapid integration of the markets, mobility of capital and significant increase in investment flows around the world have opened up new challenges and opportunities for pursuit of sustainable development.

Our commitment to sustainable development

- We are determined to ensure that our rich diversity, which is our collective strength, will be used for constructive partnership for change and for achievement of the common goal of sustainable development.
- We welcome the Johannesburg Summit focus on indivisibility of human dignity and are resolved through decisions on targets, timetables and partnerships to speedily increase access to basic requirements such as clean water, sanitation, adequate shelter, energy, health care, food security and the protection of biodiversity.
- We are committed to ensure that women's empowerment and emancipation, and gender equality are integrated in all activities encompassed within Agenda 21, the Millennium Development Goals and the Johannesburg plan of implementation.

Delhi Ministerial Declaration, October 2002

(COP-8 of FCCC, UN)

The Eighth Conference of Parties (COP-8) to the UN Framework Convention on climate Change (FCCC), was held on 28th October 2002 in New Delhi. The conference at the Ministerial level was attended by 170 countries. The Delhi Declarations urged the developed countries (Annex I) to meet their commitments on emission reduction, technology transfer and financial assistance under FCCC and Montreal Protocol. Some of the important Declarations are presented here:

- Parties have a right to, and should, promote sustainable development;
- All parties, taking into account their common but differentiated responsibilities and their specific national and regional development priorities, objectives and circumstances, shall continue to advance the implementation of their commitments to address climate change and its adverse effects in order to achieve sustainable development;
- Annex I Parties to Kyoto Protocol should implement their commitments under the Convention, including those relating to financial resources and technology transfer and demonstrate that they are taking lead in modifying long-term trends in anthropogenic emissions.

1.3 Overview – Indian Environmental Initiatives

No doubt, the United Nation Conference on Human Environment held on 5-16, June 1972 had a profound impact on the Indian consciousness and awareness on the environmental challenges. Just prior to the conference, the then Prime Minister Mrs. Indira Gandhi of India established a National Committee on Environmental Planning and Coordination (NCEPC), under the aegis of the Department of Science and Technology with the objectives:

"Our emphasis should be on the improvement in the quality of life as a whole rather than the quantitative growth of various sectors in the economy."

Arising out of the spirit of the Stockholm Conference, followed the 42nd Amendment of the Indian Constitution and several updating of existing Environment-related laws and enactments of several new laws *e.g.,* Laws on Air (1981) and Environment (1986), to which we will refer to later.

Indian Forest Act, 1927

Even before India gained independence in 1947, there was the existing Indian Forest Act 1927, which categorized four categories of forests, namely, Reserve Forests, Village forests, protected Forests and non-Government Private Forests. A State may declare forestlands or waste lands as reserved forests and may sell the produce from these forests.

In the aftermath of the Stockholm Summit, in order to arrest the rapid deforestation, the Government of India enacted the Forest (Conservation) Act, 1980, as amended 1988, which required the consent of the Central Government before a State dereserves a 'Reserved' forest or use a reserved forest land for non-forest purposes.

The Factories Act, 1948

Much before the Stockholm meet, came the Factories Act, 1948 as amended 1988, introduced special provisions to control hazardous industrial activities.

The Wildlife Protection Act, 1972

Similarly, the Wildlife Protection Act of 1972 is an example of national initiative without any input from the Stockholm. Enacted upon enabling Resolutions of 11 States under Article 252(1) of the Indian Constitution, this Act provides for State Wildlife Advisory Boards, establishments of Sanctuaries and national parks and other regulatory measures to protect wildlife. In any case, India had a long tradition in Animal protectionism exemplified by the famous Corbett Park in North India.

The Water Prevention and Control of Pollution Act, 1974

Another instance of environmental legislation on national initiative is the Water (Prevention and Control of Pollution) Act, 1974. Enacted after a decade of national debate and deliberation and supported by the enabling Resolutions of twelve States under the Article 252(1) of the Indian Constitution, the Act as amended 1988 led to the constitution of Pollution Control Boards as the regulatory authorities at the central and the state levels.

42nd Amendment of Indian Constitution

Inspired and encouraged by the 1972 UNEP Summit at Stockholm, India amended the Constitution incorporating some very important environment concerns by the Constitution (42nd Amendment) Act of 1976. The salient features of these amendments are as follows:

- Article 48A were inserted in the Directive Principles of State Policy, which states "State shall endeavour to protect and improve the environment and safeguard the forests and wildlife of the country."
- Article 51A(g) in a new Chapter entitled "Fundamental Duties" enjoins a similar responsibility on every citizen in the words, "to protect and improve the natural environment including the forests, lakes, rivers and wildlife and to have compassion for living creatures."
- Expansion of "Concurrent List" by inserting the following:
- Population Control and Family Panning;
- Forests (by transfer from the State List);
- Protection of wild animals and birds (by transfer from the State List).

Explanatory Note: The Seventh Schedule of the Indian Constitution contains three 'Lists', *e.g.,*

- List I: Central List (concerning matters, which require legislation by the Parliament);

- List II: State List (concerning matters which require enactment by State Legislature);
- List III: Concurrent List (concerning common matter which can be enacted by either the central or state Governments).
- Article 253 of the Constitution: This Article empowers the Indian parliament to legislate on matters concerning any Indian obligations under any international Conference, Association or Body like U.N. In the wake of the Stockholm Conference, India used Article 253 to enact the following two laws:

  Air (Prevention and Control of Pollution) Act of 1981;

  Environment (Protection) Act, 1986.

  It now appears that in the light of the various provisions of the United Nation Conference Human environment, 1972 at Stockholm and many other subsequent UN conventions, India can enact laws under this Article 253.

Other Provisions of Indian Constitution related to Environment and Sustainability

- **Part III: Fundamental Rights**

  The Part III of the Indian Constitution confers some fundamental rights to every citizen under Article 14 and 21 as briefly elaborated below:

  Article 14: The Right to Equality

  The right to equality strikes at arbitrariness, because an action that is arbitrary must necessarily evolve a negation of equality. The urban environmental groups often resort to Article 14 to quash arbitrary municipal provisions for construction that are contrary to development regulations. Similarly, Article 14 has also been invoked to challenge governmental solutions for wanton mining impacting environment.

  Article 21: The Right to Protection to Life and Personal Liability

  This Article provides that "no person shall be deprived of his Life or personal Liberty except according to procedure established by law". Over the years, the scope of these two fundamental rights have been enhanced by Indian Supreme Court and the High Courts by means of judicial reviews and a measure of judicial activism to ensure the right to pollution-free environment.

- **Part IV: Directive Principles of State Policy**

  The Directive Principles are Policy Recommendations to the Government and are really unenforceable in a Court of Law. But, fortunately arising out of judicial activism, the directive principles are being frequently cited by the Superior Courts as being complimentary to the Fundamental Rights (AIR, 1981).

The following Articles of the Part IV are conspicuously supportive of the Environmental Protection.

Article 47: Duty of State to improve Public Health;

Article 48: State to take steps for prohibiting the slaughter of Cows etc.

Article 48A: Duty of the State to protect and improve environment, safeguard forests and wildlife (This Article has been incorporated under the 42nd Amendment of the Constitution).

- **Part IVA: Fundamental Duties**

  Article 51A(g): It imposes duty on every citizen of India to protect and improve the natural environment including forests, lakes, rivers, and wildlife, and to have compassion for all living creatures. (This Article too was inserted within the 42nd Amendment of the Constitution).

Indian Environment Laws

A total list of important Indian laws concerned with the Environmental Protection is given in the Box B1.3.

Box 1.3

Important Indian Environment Legislations

A. Constitutional Provisions relevant to Environment protection:

- The Forty Second Amendment of the Constitution, 1976;
- Article 48A was added as Directive Principles of State Policy, which states–

  "The State shall endeavour to protect and improve the environment and to safeguard the forests and the wildlife of the country."
- Article 51A:Fundamantal Duties was added –

  " To protect and improve the natural environment including the forests, lakes, rivers and wildlife and to have compassion for living creatures." (The Supreme Court takes these aspects into consideration while dealing with the cases related to environmental protection.)

B. Various Indian Environmental Acts:

1. The Indian Forest Act, 1927;
2. The Factories Act, 1948;
3. The Insecticide Act, 1968;
4. The Wildlife Protection Act, 1972;
5. The Water Prevention and Control of Pollution Act, 1974;

Contd...

6. The Water Prevention, and Control of Pollution Cess Act, 1977;
7. The Forest Conservation Act, 1980;
8. The Air Prevention, and Control of Pollution Act, 1981;
9. The Environment Protection Act, 1986;
10. Motor Vehicles Act, 1988;
11. Hazardous Wastes Management and Handling Rules, 1989;
12. Manufacture, Storage and Import of Hazardous Chemicals Rule, 1989;
13. Scheme of labeling Environment Friendly products Ecomarks, 1991;
14. The Environment Impact Assessment Notification, 1994.

❑❑❑

2
CHAPTER

Multi-Sectoral Resources Review Towards Sustainabililty

2.1 Equitable Distribution of Resources & UN Millennium Development Goals

The concept of sustainability and the world without poverty but with equity were the core theme of the Brundtland Report 1987 under the aegis of the WCED (World Commission on Environment and Development) of the United Nations. At the advent of the 21st Century, the United Nations declared the Millennium Development Goals setting specific targets of poverty alleviations, health and sanitation, and literacy of the Third World, as presented in Box B2.1.

Box B 2.1

Millennium Development Goals

Goal 1: Eradicate Extreme Poverty and Hunger.

Target 1: Halve, between 1990 and 2015, the proportion of people whose income is less than a $1 a day.

Target 2: Halve, between 1990 and 2015, the proportion of people who suffer from hunger.

Goal 2: Achieve Universal Primary Education.

Target 3: Ensure that, by 2015, children everywhere, boys and girls alike, will be able to complete a full course of primary schooling.

Goal 3: Promote Gender Equality and Empower Women.

Target 4: Eliminate gender disparity in primary and secondary education, preferably by 2005, and in all levels of education no later than 2015.

Contd...

Goal 4: Reduce Child Mortality.

Target 5: Reduce by two-thirds, between 1990 and 2015, the under-five mortality rate.

Goal 5: Improve Maternal Health.

Target 6: Reduce by three-quarters, between 1990 and 2015, the maternal mortality ratio.

Goal 6: Combat HIV/AIDS, malaria and other diseases.

Target 7: Have halted by 2015 and began to reverse the spread of HIV/AIDS.

Target 8: Have halted by 2015 and began to reverse the incidence of malaria and other major diseases.

Goal 7: Ensure Environmental Sustainability.

Target 9: Integrate the principles of sustainable development into country policies and programs and reverse the loss of environmental resources.

Target 10: Halve, by 2015, the proportion of people without sustainable access to safe drinking water and basic education.

Target 11: Have achieved, by 2020, a significant improvement in the lives of at least 100 million slum dwellers.

Goal 8: Develop a Global partnership for Development.

Target 12: Develop further an open, rule-based, predictable non-discriminatory trading and financial system (includes a commitment to good governance, development, and poverty reduction, both nationally and internationally).

Target 13: Address the special needs of the least Developed Countries (includes tariff- and quota-free access for the Least Developed Countries' exports, enhanced program of debt relief for the heavily indebted poor countries (HIPCs) and cancellation of official bilateral debt, and more generous official development assistance for countries committed to poverty reduction).

Target 14: Address the special needs of land-locked developing countries and small Developing States (through programs of action for the sustainable development of Small Island Developing States per 22nd General Assembly Provisional).

Target 15: Deal comprehensibly with the debt problems of developing countries (through national and international measures in order to make debt sustainable in the long term).

Contd...

Target 16:	In consultation with the developing countries, develop and implement strategies for decent and productive work for youth.
Target 17:	In consultation with the pharmaceutical companies, provide access to affordable essential drugs in developing countries.
Target 18:	In consultation with the private sector, make available the benefits of new technology, especially information and communication technologies.

Sustainable development requires a strategy that meets the needs of the present without compromising the ability of the future generations to meet their needs on comparable basis. Since it may not be practicable to prophesy the specific kind of needs of the much-changed future generations, we can also use an alternative proposition of sustainability as maintenance and sustenance of the material, economic and social well being of the humans and the global ecosystems on a long-term basis of several decades or at the most a century. In any case, the broad objectives of such a global order are:

- Improve the quality of human life, abjuring the gross inequity of the kind reflected in Table T2.1.
- Conserve earth's rich diversity (see Table T2.2). Biodiversity is at the heart of bio-equilibrium. You intervene and interfere in any living sector; the eco-systems will react by malfunctioning leading to unknown and unnatural rain and weather, food chain problems and diseases.
- Conserve non-renewable resources and economize its uses. This is feasible by optimizing consumption and developing more and more renewable resources and green energy.
- Educate, empower and entrust communities to care more for their own environment. Without people's participation, long-term sustainability of the fragile eco-systems is not possible.
- Integrate the environment, economy and human health into all developmental activities and socio-political activities.
- Do not pollute or degrade the earth beyond its carrying capacity. Earth's eco-systems and environment have its own cycle of renewal. All uses of the earth's systems shall be limited to its natural renewal and recycling capacities.

In fact, the Declaration of the 1992 Rio Earth Summit expressed in the 27 Principles supported by the Agenda 21 of Implementation encompass the hallmarks of the international tasks towards sustainability. The 2000 Declaration of the Millennium Developmental Goals (MDG) has only reinforced the global resolve.

In the subsequent sections, we review the various resource scenarios aimed at the proposition of global sustainability based on the work of Mongillo (Mongillo, 2004).

Table T 2.1 Income Inequity in Selected Countries (1996/2007)

Country	Share of Income (%)		
	Poorest 20%	Richest 20%	Per Capital Income USD *
Denmark	9.6	34.5	59,000
India	8.1	46.1	3,000
United Sates	5.2	46.4	48,000
Russia	4.4	53.7	9,600
Zambia	3.3	56.6	1,600
Brazil	2.2	64.1	7,400

* world Bank 2007 (PP)

Table T 2.2 : Bio-diversity of known species

Species Name	Number of Species	Presentage of Total
Animals	1,030,000	72.9
Plants	250,000	17.7
Fungi	69,000	4.9
Protists	58,000	4.1
Moneraus	4,8000	0.3

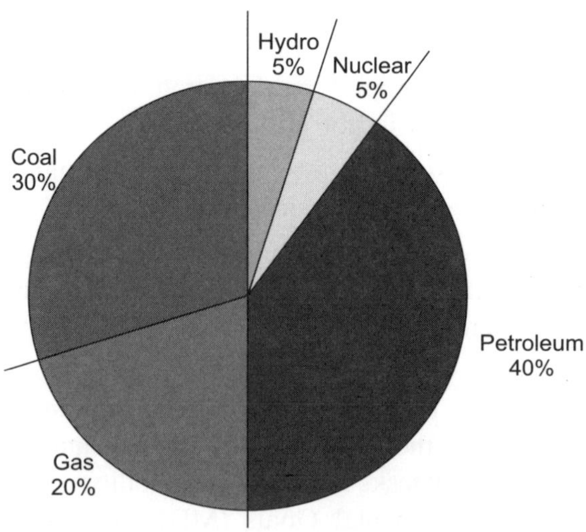

Fig. F 2.1: Global Primary Energy Distribution

Table T2.3: Sources of Fossil Carbon Dioxide Emission CO_2 (2000)

Sources	Percentage
Power Plants	24.6
Transportation	13.5
Processing Industries	26.5
Deforestation	18.3
Agriculture	13.5
Waste	3.6
Total	100.0

World Resource Institute 2000

2.2 Sustainable Energy Systems

We have explored and evaluated various energy possibilities, *e.g.* Bio-energy (Ethanol, Bio-diesel etc.), renewable wind energy, Solar energy and the upcoming Hydrogen technology and Fuel Cells in Part III. It appears that hydrogen energy may dominate the long-term future energy fuel with the infinite water resource along with the nuclear Fission and possibly also fusion energy.

Energy is the key driver of economy. Today, the total global energy consumption is estimated at about 405 Quads (405 EJ) mainly derived from fossil fuels (see Fig.F2.1). Such over-dependence on the carbon-rich fossil fuel is creating havoc emitting carbon dioxide and other Green house gases. Each major sector of the industries, without exception, is major CO_2 polluter (see TableT2.3). Therefore, for the sustainability of the world, the fossil fuel regime must change over to non-carbon or low-carbon fuel systems.

Besides, the fossil fuel is an exhaustible resource and the global stock of fossil fuel especially may not last for more than a few decades. Thus, the new energy source system should be renewable annually. As long as the consumption is controlled to within the renewability limit of the earth, there will be no shortage of fuel resources. We propose here four prominent renewable non-polluting fuels, whose uses will lead to GHG-free atmosphere:

Bio-Energy

Combustion of biomass is the oldest source of primary energy from the time immemorial, which is about as bad as the fuel fossil regarding emission of GHG gases, despite significant progress in the combustion technology. Even today, biomass provides 44 EJ per year amounting to 11 % of world's primary energy.

The landmark developments in Bio-Energy has come about by the introduction of Ethanol (by fermentation of sugarcane mostly) and Bio-diesel (by transesterification of vegetable oil and also sometimes, animal fat), which emit much less carbon oxides and other GHG gases. Brazil, among all nations, has been using bio-ethanol as motor fuel to great extent, which is dealt in more details in later Chapters of Part II and III.

Wind Energy

The fastest growing renewable power source is wind energy. Though presently wind energy produces only 2% of world's electricity, World Watch Institute estimates that wind energy could easily provide 20 to 30% of the electricity globally. In US, the American Wind Energy Association estimates by 2025, wind power will produce more than10% of all electricity generated in the US. Wind energy will be the cheapest economically, there being no raw material cost and the capital cost being also very low. More detailed treatments are given in Part III on Energy Resource.

Hydrogen Energy & Fuel Cell

Fuel Cell Technology, an electro-chemical device, promises to deliver electrical power without any combustion and without any carbon pollution. A fuel cell is analogous to a storage battery, but one that is being kept constantly charged by continuous supply of Hydrogen fuel at the cathode and oxygen (air) at the anode. The sea, with the development of appropriate technology, promises to be inexhaustible source of hydrogen and fuel cell, given the further maturization of the technology, can assuredly supply the world with the necessary energy replacing the cursed fossil fuels like coal, oil and gas.

Solar Energy

Solar energy is another sustainable energy system, which is also being developed abundantly in many countries and treated elaborately in Part III on Energy. There are three principal forms of solar energy, that is, thermal energy for direct heating application and also for generation of electricity by steam turbines and direct production of electricity power by photovoltaic route.

In 2003, Germany and Japan were the world's largest generators of solar power. In Japan, 80,000 households have installed solar roof panels to generate domestic needs of electricity. In Germany 100,000 homes have rooftops solar panels. South Africa is planning to provide 350,000 solar houses in the rural area.

2.3 Sustainable Agriculture

Human population, presently at 6 billions plus, are growing at the rate of 80 millions per year. Some population experts predict that the world population will grow to reach 8 billions by 2025. To feed the growing population, most demand will be placed on earth's land and water resources to grow more food, to transport and distribute more food.

In the circumstances, agricultural practices are to be so designed that the food productivity can be sustained, without adversely affecting the quality of the landmass, water resource and the environment. We will now briefly outline the features of sustainable agriculture.

Sustainable agriculture may compass the following three steps:

(*i*) Economize use of non-renewable resources. Enrich ecology by incorporating environ-friendly agriculture practices.

(*ii*) Extend equitable work practices among the farming community. This will enhance the quality of farming life.

(*iii*) Integrating the above two steps into an organic agriculture practices will increase food production on a sustainable basis.

Sustainable agriculture, inter alia, should also include following eco-friendly agriculture practices:

1. **Crop Protection:** Raising different crops on a land will help to increase soil fertility and reduce soil erosion. Some crops will extract nitrogen, others will add nitrogen.
2. **No-Till farming:** In no-till farming, the seeds of crop plants are planted along the remains of the previous crop. In no-till farming, soils are never bare and exposed to wind and air erosion.
3. **Contour Farming:** In contour farming, land is plowed across, perpendicular to the natural slope to create a furrow in which water that would have flowed downhill, will slow down and get accumulated. Contour farming allows top soil to grow at the rate of 1mm per year.
4. **Strip Cropping:** In strip cropping, a strong crop plant (say, sun flower) is planted alternatively with a weaker crop plant (say, wheat). Sunflower protects land from windage, besides, as a result of multiple cropping, the field is never bare. There is, thus, no soil erosion from soil erosion from wind and water.
5. **Terracing:** Sequenced Terracing of the sloping hilly terrain inhibit soil erosion.
6. **Integrated Pest Management:** One practice of IPM is biological control, which eliminates or reduces the need of synthetic pesticides. Bacteria and viruses are used to control specific pests, for example, a bacterial disease can kill caterpillar infectious virus that controls rabbits in Australia.
7. **Sustainable Waterless Farming:** This includes water conservation and storage and more use of sustainable irrigation practices, such as drip irrigation, originally developed in Israel. Harvesting of seasonal rainwater underground and Saline water irrigation where possible and applicable shall be pursued.

2.4 Sustainable Fishing & Aquaculture

Food and Agriculture Organization (FAO) of the United Nations report that out of the 15 main world Fishing Regions, 4 are depleted and 11 are declining. Several decades of overfishing have resulted in the sad condition. The collapse of Canadian cod fishery in the early 1990 left 30,000 commercial fishers unemployed. The Tables T2.4 and T2.4 A show the high growth of fishing employing high technology fishing gears and giant 'factory ships'.

Aquaculture and Fish Farming

Aquaculture is the cultivation of fish, shell fish or aquatic plants in natural or controlled marine or fresh water environments. Aquaculture is believed to produce 25 million tons of fish, crab and shellfish per year. Today fish farming is an important industry in the United States, Philippines, Japan, India, China, Israel and also Europe. The future of sustainable food regime will depend largely upon aquaculture and matching renewable farming on an organized scale.

Table T2.4: World Fishery Catch (by Aquacultute)

Regions	Catch (1000Tons)		Percentage Growth
	1970	1992	
Asia	4,802	11,012	229
Former Soviet Union	3,996	7,765	94
Europe	3,097	3,018	-3
North America	1,076	2,560	138
South America	361	816	126
Africa	244	699	187
Oceania	37	122	230
World	13,615	25,994	91

Source: FAO

Table T2.4A: Overall World Fish Production (Farm + Open Sea)

Year	1970	1980	1990	2000	2003
Production (Million tons)	60	75	100	120	130

2.5 Protecting Wildlife & Forest

Many environmentalists apprehend that by year 2100, between 20 to 50% of all wildlife species, which lived on earth since 19900 will become extinct. One environmental study reported that as many as 100 species disappear from the earth each day.

Table T2.5 shows how some significant percentages of important wildlife classes are already under threat of extinction. The main causes of extinction may be enumerated as follows:

Natural causes: Climate Change;

Natural disasters (draughts, landslides, earth quakes, volcanic eruptions, meteor impacts, etc.);

Human Activities: Releasing pollution into the air, water and soil;

Diverting water from rivers, and lakes to other uses;

Deforestation;

Destroying grazing land, farmland for development purposes;

Overhunting and poaching.

Regarding the solutions, human activities can be greatly controlled towards sustainability of wildlife and specific preservation measures can be planned to mitigate the effect of natural calamities. In particular, we may propose following measures, which will improve the situation:

Build more wildlife sanctuaries;

Protect environment and control toxic emission;

Social education about the importance of wildlife in eco-balance;

Other measures depending upon local situations.

Table T2.5: Species under Threat of Extinction

Wildlife Categories	Percentage under threat
Birds	11
Vascular Plants	12.5
Reptiles	20
Mammals	25
Amphibians	25
Fish	34

Source: Mongillo (2004)

Preserving Forest

Trees and forests are critical to the sustainability of the earth and contribute significantly in controlling the climate and rain cycling. They help the air to clean by filtering pollutants and reduce the risk of global warming by absorbing carbon dioxide.

Forests are being destroyed catastrophically with the progress of the civilization and expansion of human habitation. Natural climate change has also led to forest losses. It is believed that 20 million hectares of forestland is getting deforested every year and 10% Of all species of trees have already gone extinct. Deforestation is also affecting bio-diversity in a big way.

Managing Forest Sustainability

Sustainable forest management implies sustainable growth of forest to support environmental, economic and social needs of the community on a continuing basis.

Some of the effective measures for a growing forestation are as follows:

1. Shelter-wood Harvesting: Long and mature trees are only cut so that sheds are removed, allowing accelerated growth of other smaller trees.
2. Seed-Tree harvesting: Where trees are cut, some trees which provide the seeds for the growth of the forests are left untouched.
3. Selective Cutting: Old and mature trees are cut, leaving out the younger and growing trees allowing them to grow.
4. Strip Cutting: Trees are cut in narrow rows or strips. Uncut trees are still providing for recreation and wildlife habitation.
5. Reforestation: Reforestation is now being implemented in many countries to replace the forests that were once clear-cut mostly to produce agriculture land. Israel has undertaken reforestation and recovered 50% of nation's woodlands in an effort to slow down deforestation and desertification.
6. Tree-Farms: Tree farms are commercial reforestation programmes and are recommended for areas devasted by tree-cutting spree in the past. Tree farms can also be used for habitat of wildlife. In tree farm, controlled tree cutting is undertaken for commercial sale and manufacture for sustainability.

Table T2.6: US Share of Consumption of Key World Raw

Materials	World Production	US Consumption	(3)/(4)
	(MT)	(MT)	%
Minerals	7641	2410	31.54
Wood Products	724	170	23.48
Metals	1196	132	11.03
Synthetics	252	131	51.98
All Materials	9813	2843	28.97

Source: State of the World 1999 (WWI, New York)

2.6 Eschew Gargantuan Consumerism

Galbraith (Guha 2006) had noted the unholy interrelationship between 'private affluence' and 'public squalors', how the single-minded pursuit of wealth had driven attention and resource from nurturing true democracy, which he defined as the provision of public infrastructure, the creation of decent schools, park, and hospitals. This 'gargantuan and growing appetite' for resources are unfortunately the common features of all Western societies.

The Table T2.6 establishes the over-consumption of nature's raw stocks by USA with only less than 5% of world population and consequent over-discharge of wastes and contaminations.

German environmentalists have generally been critical of their consumer society. German Greens have always been emphatic in declaring that economic growth in Europe and North America has been possible by the economic exploitation of the Third World countries.

German philosopher Rudolf Bruno says: "the present way of life of the most industrially advanced nations stands in global antagonistic contradiction to the conditions of human existence. We are eating up what other nations and future generations need to live on."

Such critique of over consumption has also found sympathetic ears in the Dutch world of thoughts and elsewhere in Europe from time to time.

Much before such wisdom had dawned the Western minds, the great Indian savant and the supreme political leader of the country Mahatma Gandhi had prophesied:

"God forbid that India should ever take to industrialization after the manner of the West. The economic imperialism of a single tiny kingdom (England) is keeping the world in chain. If an entire nation of 300 million took to similar economic exploitation, it would strip the world bare like the locusts do."

Gandhi's critique of western civilization has of course profound implication for the way we live and relate to the environment today. He looked upon the modern society as an infinite Black hole of wants drawing in an endless stream of goods and services. His model of self-sufficient village economy is wound-up in a socio-ethical milieu of simple living and high thinking, each village unit also serving as a holistic economic entity. In support of this model, he propounded that the "world has enough for everybody's needs but not enough for everybody's greed."

2.7 The Future

What then is the prospect for the future? Eminent economist Wilfred Beckerman believes that, given the market full play, economic growth will sustain over the next 2500 years (Beckerman, 1972). Eminent Historianenvironmentalist Ramchandras Guha (Guha, 2006) believes that, given the globalisation of knowledge and free flow of information, education and heightened innovation, the world will see expanding prosperity, particularly in China and India, but he is suspicious of equitable transfer of prosperity to the less privileged and also necessary care for the environment. He also foresees the environmental menace on the scene from the incoming automobile-driven prosperity over less developed countries, China and India included. Fortunately, the longevity of pollution-emitting fossil fuel will not extend beyond few decades with growing rate of consumption and green automobile technology will emerge on the horizon.

Once higher and universal education and knowledge spread, we are hopeful of a benign environmental movement towards 'deep ecology' in this millennium itself.

❑❑❑

3
CHAPTER

Technology Solutions for Global Sustainability and Some Indian Mission Mode Initiatives

In simple language, sustainability can be looked upon as a system where life-supporting primordial resources has a surplus balance over consumption over a longer period without deterioration of the earth's environment. The system is sustainable, if there is no systemic failure. This proposition is an alternative but equivalent hypothesis to the traditional understanding of sustainability, which allocates and expends earth's resources in a manner to ensure the intergenerational equity and well-being between the present and the future.

3.1 Resource Innovation

When we talk of surplus balance of resources, we have to identify the prime resources on which the present structure of human civilization rests. In Table T3.1 are given a list of very important natural resources, whose availability is limited. Now Petroleum oil is will be exhausted, say, in 50 years. Will then the wheels of global industries and services come to a grinding halt, which is unlikely? Further, Iron will be exhausted, say, in 100 years, what will happen to global transportation, bridges, building etc. Now this is the alarming situation we face, if we get fixated on existing natural resources. Now when fossil fuel gets depleted in 50 years, new fuel will get innovated, possibly hydrogen fuel with higher efficiency will fill the vacuum!

Unlike the exhaustible resources of Table T3.1, we have fortunately another class of resources, which are apparently limitless, *e.g.,* water (sea water), Air, Earth, which after technological transformation, may give rise to superior substitute materials of construction and utility.

Now since the sustainability is a long-term consideration, the resources scenario must include existing resources and future resources. Once we are thus tele-focussed on the long term, the gloom of doomsday practitioners will vanish.

The technological transformation may be formulated by the following equations:

Unused materials = Materials (from Periodic Table)

+ Materials (from outside) ....
E3.1

Future Resources = Unused Materials × (Technological Innovation)

....E3.2

The theory of equation E3.1 stipulates:

(*i*) That we must reexamine if all known but still unused materials could be put to newer use as viable substitute, for example, to reinvent refractory as metal substitute.

(*ii*) We must explore new materials to replenish dwindling stock of existing materials. For example, identify bio-resources, e.g. bio-fuels.

(*iii*) The stock of viable unused materials shall be greater than the decrease in the stock of used materials.

The theory of equation E3.2 assures:

(*i*) That by technologically innovating unused materials, we would be able to transform them into future useable resources. For example, convert refractory to metal substitute, develop fresh water from seawater and extract aluminum from earth.

(*ii*) The stock of such new resources shall be more than deficit of existing resources arising out of current consumption.

3.2 Resource Conservation

In the short or medium term, the sustainability will greatly depend on the efficient use, maintenance and preservation of existing resources. The Maintenance Equation can be formulated as under:

Resource available – Resource Surplus

= Resource utilized + Wastage E3.3

Wastage

= Useless Byproduct + Product degradation+ Pollution (uncontrolled loss)

...E3.4

Resource Conservation is what is being widely acted upon and publicized by the conventional wisdom, by the Government, by the Environmental Activists, by the NGOs, by the Industry. We hardly need to emphasize any further.

The theory of equation E3.3 mandates following Conservation course of actions:

(*i*) We must conserve as much resources as possible *i.e.,* maximize 'Resource Surplus'.
(*ii*) We must maximize the 'Resource utilized'.
(*iii*) We must minimize the 'Wastage'.

The theory of equation E3.4 urges us to think on the following lines:

(*i*) To convert 'Useless Byproduct' to useful product by secondary processes or by catalytic conversion or otherwise.
(*ii*) To prevent degradation and toxicity and contamination of other systems by toxicity.
(*iii*) Pollution by way of uncontrolled gaseous emission or liquid effluent discharge (and in some rare case radioactive contamination) must be prevented or minimized.

3.3 Resource Recycling

The other option is recycling of 'Resource Utilized' in E3.3, after the materials have been used beyond its utility and are now to be thrown out. There are substantial amounts of throw-away of old materials, aluminum utensils and furniture, iron and steel scraps, glasses, plastics, papers, textiles, etc. These can be very economically reconverted into newer appliances of use and utility.

New resources = (discarded used resources) x (recycling technologies) ... E3.5

The theory of E3.5 reiterates and supports sustainability by economically recollecting and recycling old scraps and throwaways than by reprocessing ores and fresh materials.

3.4 Switch-over to Renewable Resources

There is a third class of resources, mostly bio-resources, which are limited in supply at any point of time but renewable on a time scale of a year, season or breeding cycle. These resources are agricultural food crops, energy crops, fisheries, animal foods, and other biomasses. As long as the quanta of consumption remain within the regenerating capacity of the earth eco-systems, this resource system is sustainable. Further, the production level at any particular level can be enhanced by adoption of appropriate technology already existing or to be innovated. In course of time, perhaps, newer engineering materials can also be developed from bio-composites.

3.5 Resource Technology Symbiosis

In E3.2, we have postulated how with the application of magic wand of Technology, new resources are created from unusable materials. Technology has converted coal and petroleum into energy resource to feed and sustain the industrial infrastructure of the modern world. Technology has converted mineral ores into iron, steel and other metals which are resource of construction to build bridges, ships, railways, which transport resources across the world.

Though green activists and sustainability practitioners have often condemned technology, having created the toxic gases, hazardous wastes, radioactive materials, these are 'wastes' under equation E3.3 theory. These are to be controlled either by catalytic conversion or by conversion by secondary processes as already enumerated in the earlier section.

Further, like everything else in this world, technology is also in flux; it is constantly evolving and transforming over the time. Table T3.2 shows several variants of dominant technologies, which have already played great transformational roles, though some of them are even today in the early evolutionary state.

Possibly, at a foreseeable future, there could be a transition from macro-mechanical process of combustion to biotechnological fermentation aided by Nanotechnological separation for most of chemical and metallurgical processes, the most notorious sector contaminating our environment. Should we call it cold reaction and cold or green metallurgy respectively.

Biotechnology is the most powerful of the modern technologies; promising revolutionary empowerment in meeting the global challenges of sustainability; raising immensely food productivity, transgenic agriculture; ridding humans and others living of fatal diseases by gene targeting; saving the environment of severe toxicity by substituting biotechnological fermentation in place of energy-guzzling and polluting combustion processes. It will help raise forestry by fast replicable tissue culture. It will generate bio-energy. In short, Biotechnology will lead to quantum technological transformation of the world in a benign environment.

Table T3.1: Some Non-Renewable (Exhaustible) Global Mineral Resources

Resources	Known Global Reserves	Projected Availability (Years)
Aluminium	1.17 x 10E9 tons	100
Chromium	7.75 x 10E8 tons	420
Coal	5 x 10E12 tons	230
Cobalt	4.8 x 10E9 tons	60
Copper	308 x 10E6 tons	36
Iron	1 x 10E11 tons	240
Lead	91 x 10E6 tons	26
Manganese	8 x 10E8 tons	97
Molybdenum	10.8 x 10E9 lbs	79
Natural Gas	1.14 x 10E15 cu.ft.	38
Nickel	147 x 10E9 lbs	150
Petroleum	155 x 10E9 bbls	31
Silver	5.5 x 10E9 troy ozs	16
Tin	4.3 x 10E6 tons	17
Tungsten	2.9 x 10E6 lbs	40
Zinc	123 x 106 tons	23

Table T3.2: Technological Evolution

Phase Variants of Technologies	Some Conspicuous Products
1 Macro-Mechanical Technologies	Prime-Movers (Steam & I.C Engine, Electrical Motor & Generator, Turbine, etc.),Ship, Aircraft, Smoke-stack Industries, etc.
2 Micro-Electronics Technologies	Transistor-Radio, TV, Telephone, Mobile Phone
3 Digital Technologies	Computer, MRI, Scanner, Digital Photography, etc.
4 Information Technologies	Software-based knowledge Products
5 Genetics & Biotechnologies	Newer Medicine, newer variant of food, recombinant products including seeds, genetic modification
6 NanoTechnologies	Semi-permeable membrane Technology

3.6 Indian Mission Mode Initiatives

The theoretical critique of the sustainability issue in the earlier sections has given us the big visionary picture and guidelines for enabling initiatives in the long run, say, over 100 years or more.

In this Section, for the present, we enunciate three specific missions-type Indian action plans, which will strengthen sustainability:

- Water Mission;
- Energy Mission;
- Environment Mission.

The Mission Management is single task focused and somewhat different from the general administration dealing with a number of development projects at the same time.

3.7 Indian Water Mission

Water is the elixir of life. Both the life and the civilization cannot survive without fresh water to drink, to clean and to run industrial processes.

Now consider 70% of earth's surface is covered with water but, of that, about 97% is sea saline water (See Table T3.3). Of the total fresh water, the useable fresh potable water, at less than 0.3%, is not plentiful.

Though the available potable water stock is very much less than 1%, the same is renewable annually by solar evaporation from the sea and evapotranspiration from the land and returning to the earth by precipitation rain.

India gets an average annual rainfall of about 1000 mm, which produces a gross of 3000-billion cum. water. Only one third of the water is saved as surface and ground, the other two-thirds flow back to the sea. Thus, the effective rate of water renewable is only 1500 cum per capita per annum. This availability of water is not bad, considering we are a very populous country (See Table T3.4).

However, our effort should be to retain a higher percentage of the total precipitation stock of 3000-billion m3 by developing newer catchments areas, interlinking rivers, greater wider water harvesting and recharging of ground water by applying technology, by better long term planning, by better general management. USA saves 40% of its total precipitation. In fact, our rainfall at 1000 mm is one of the highest in the world, though the duration of rainfall is very short and lasts only three months in a year.

The next point is to learn the economy of use. Other dry States practice dry farming, drip irrigation and make do with low consumption of water. But in India irrigation consumes the most of the fresh water (see Table T3.5). Please note that Israel does with so little water (see Part IV on Water Resources).

There is also the possibility of producing fresh water from sea water by using membrane Filtration Technology, which is slowly becoming viable with large water work plant cost.

Strategy for Solving Country's Water Problems

The following important measures are mentioned from a paper of Bhaba Atomic Research Centre (Bhattacherjee 2002):

(*i*) Large water supply scheme to meet the needs of the urban, industrial and large rural conglomerates with piped water distribution, which is economic.

(*ii*) Large scale rain water harvesting and artificial recharge of ground water sources.

(*iii*) Augmentation of water resources in coastal areas by large scale desalination of abundant sea water.

(*iv*) Proper treatment of underground water, contaminated chemically or biologically.

(*v*) Widespread recycling of domestic and industrial effluents.

We add following further strategic measures for consideration. Following the principles enunciated earlier, we shall think big and plan over a longer period, say up to 2050. The measures shall include interlinking of rivers, build large reservoirs for hydropower and /or irrigation and other large uses. Series of artificial lakes should be envisaged where technically feasible. Countrywide plan for hundreds of thousands of recharge-wells shall be made mandatory for underground water users, be it rural community or townships.

If India with annual 1000mm rainfall suffers from water scarcity, the blame must rest with the country and the Government, its economic planners, the scientists, technologists and engineers

Table T3.3: Stocks of Water on Earth

Locations	Quantity Million cubic km.	Percent of World Supply
Oceans	1350.00	97.20
Ice Caps & Glaciers	29.00	2.09
Groundwater (within 1 km)	4.20	0.30
Groundwater (beyond 1 km)	4.20	0.30
Fresh water lakes	0.125	0.009
Saline Lakes & Inland Sea	0.104	0.007
Soil Water	0.067	0.005
Atmosphere	0.013	0.0009
Water in Living Biomass	1.003	0.0002
Average in Stream Channels	0.001	0.00007

Source: Masters 2001

Table T3.4: Availability of Fresh Water per person per Year 2005

Countries	Availability per Year Cu. Metre/Person
Canada	91,419
Former USSR	31,653
USA	10,333
Japan	3,365
France	3,371
U.K.	2,474
Germany	1,866
China	2,206
India	**1,754**
South Korea	1,454
South Africa	1,106
Israel	255
Singapore	139
Kuwait	8

World Resource Institute 2005

Table T3.5: Water Consumption for various Needs 2000

Needs	Consumption km3		
	1985	2000	2010
Irrigation	470	630	770
Domestic	16.7	24.2	40
Industries	10	30	120
Power	4.3	5.8	15
Miscellaneous	39	60	105
Total	**540**	**750**	**1050**

3.8 Indian Energy Mission

Let us start with the Electrical Energy scenario. The country produces about 128,000 MW power only. Of this amount, the contributing power sectors are approximately:

Thermal	72%
Hydel	25%
Nuclear	3%

Of the thermal power, almost all are coal-based Steam-Turbine-driven, a small part of it is from Gas Turbine, which are principally captive power pants or peak power management units.

From Table T3.6, we see that India at 466 kwh per person is at the bottom of the global energy users. Therefore, in the process of development, India shall have to raise the power output to at least 1000kwh per person to reach a reasonable level of development and prosperity in 10 years *i.e.,* about 3 times the present power generation capacity to 330,000 Mwe. Presently, the target is to raise the generating capacity by 78,000 Mwe by 2012 to a total of about 200,000 Mwe.

Presently 70% of Indian consumption of Petroleum is imported. The domestic production is no doubt quite small at about 50 million tons, even then the known reserve is only 30 times the annual production (See Table T3.7, which will last less than 50 years with some new finds in the mean time.

The current annual production of natural gas is about 20 bcm and the past known reserve at 1114 bcm will again last less than 50 years but several new gas finds in the Mahanadi, Krishna and Godavari basins have high promises.

Hence, the future contributions of petroleum and natural gas are not likely to be substantial in the electrical power generation or other industrial sectors beyond a fifty year period. Thus, in growing India's electrical energy, we have to mainly look for sources from Nuclear Power;

Hydel Power and
Coal-based Power.

Nuclear Power

The nuclear source appears to have been least utilized for power generation in the past, albeit India has done real pioneering R & D work on various options of nuclear energy; One of the reasons is the international embargo on import of matching components and subsystems. The scenario now appears quite changed and India is poised to embark upon substantial expansion of her nuclear energy programmes on the following targets within 2030:

(*i*) Natural Uranium-based PHWR producing 10,000 Mwe per annum;
(*ii*) Plutonium-based Fast Breeder Reactors producing 150,000 Mwe per annum;
(*iii*) Thorium-based FBR producing 300,000 Mwe per annum.

The coal resource is quite plentiful as seen from Table T3.7 at about 7% of world reserve, so that it may last more than 100 years. There are though some problems; firstly, Indian coal has high ash content. Secondly, the mining operations are still largely under Government monopoly company Coal India; the company had suffered from lack of technology application and is not imbued with the necessary dynamism to go in for highly accelerated production and matching high productivity. Of late, all these constraints are being steadily dismantled towards possible solutions through public-private partnership in a big way.

The Hydel Power also offers good scope of producing electrical power at economic rate and with environment friendly way. Hydel power resource may include Run-of –river small type or the reservoir type larger, pumped-reservoir type (two reservoirs, one at upper level and one lower level) and many other reservoirs arising from the proposed inter-linking of rivers. The present hydel energy potential is around 148 Gwe (See Table T3.7). But the potential reserves are likely to be revised upward depending upon the implementation of the inter-river linking projects.

Combining the above three major sources; Thermal (Coal), Nuclear, and Hydel, a perspective plan of power generation is presened, generating 108 Gwe (2005), 300 Gwe (2030), 600 Gwe (2035) and 1000 Gwe (2050) as shown in Table T3.8.

The contribution of renewable type energies other than Hydel is not included in the above perspective plan. But over the years, energy sources, *e.g.,* solar, biomass, and wind should find a prominent application. Table T3.9 gives a very approximate overview of the large potentiality of renewable energy of the World and India.

Table T3.6: Per Capita Electrical Energy Consumption (2006)

Country	**Annual Elect. Consumption(kwh)**
Canada	16,279
USA	12,924
Japan	7,701
France	7,328
Germany	6,663
UK	5,773
China	2,179
Brazil	2,116
India	466

Earth Trends. Words Resource Institute 2006

Table T3.7: Recoverable Reserve of Energy Resources 2009

Type of Resources	**Annual Current Consumption**	**Total Recoverable Reserve**
Coal	500 MT	100 Bt
Crude Oil	150 MT	1500 MT
Natural Gas	42 BCM	1114 BCM
Hydropower	36Gwe	148Gwe

Mt: Million tons: Bt : Billion tons; Bem: Billion cum

USEIA, 2008/2009 Gwe : Giga watte energy

Table T3.8: Long Term Electrical Power Installation Perspectives (units 1000 MWe)

Type of Energy	2005	2020	2035	2050
Thermal	80 (72%)	210 (70%)	360 (60%)	450 (45%)
Nuclear	3 (3%)	30 (10%)	150 (25%)	450 (45%)
Hydel	25 (25%)	60 (20%)	90 (15%)	100 (10%)
Total Power	**108 (100%)**	**300 (100%)**	**600 (100%)**	**1000 (100%)**

Table T3.9: Potential Renewable Energy Resource 2004/2008

Resource Type	Global Estimate(TW)	Indian Estimate(TW)
Solar*	10,000	260
Wind*	300	3
Biomass**	600	6

* Wikipedia Encyclopedia website 2008

** Kiass 2004

3.9 Indian Environment Mission

The sustainability of human living is analogous to the maintenance of the pristine environment, with which the first life was germinated. The Environment Mission is concerned with such basic initiatives, which will preserve the environment in the coming decades and thereby improve and enhance the sustainability.

Legal Regimen

Like many countries, India has enacted a set of Statutes, which are quite adequate and comprehensive for the purpose.

The Water (Prevention and Control of Pollution) Act,1974 and the Air (Prevention and Control of Pollution) Act, 1984 and the Environment Protection Act, 1986 are three strong instruments but need to be implemented persistently for the proper maintenance of the basic composition of the water and the air environment. The Environment Protection At, 1986 and the Factories Act, 1948, amended 1987 deal comprehensively with the hazardous chemicals. The legal regimen is therefore laid out adequately, but beyond all punitive measures, what is missed, is perhaps the environmental education. Only recently, Environment as a subject has been introduced in all school curriculum and also in many courses in the college. The knowledge armour is the best safeguard to ensure that nobody, individual or the institution is entitled to tamper with the God's gift, which is the Environment.

Municipal Pollution

Politically protected municipalities are the worst polluters. Kitchen stove emissions, auto-emissions and suspended dust micro-particles pollute the town air severely. Kitchen wastes and human wastes and effluents dangerously pollute the surface and underground water and the river water as most of the cities are riverside-situated. Sound pollution is also an urban phenomenon.

The mission here is really private and public education and awareness programmes, out of which will arise policy initiatives and long-term action plans. Given the wide social sanctions, it is possible to redesign, to rectify and to remove scourge of urban pollution.

Already, supply of fresh potable water is being augmented in many cities, which will greatly improve health.

Industrial Pollution

The industrial pollution may consist of gaseous, liquid or solid pollution. The gaseous pollution could be furnace or process emissions. The gas itself could be toxic, *e.g.,* carbon monoxide or phosgene. Air or gases could be laden with particulate matter or dust creating hazard to human respiratory system. Liquid effluents from tanneries and chemical industries are equally harmful to humans and toxic to fishes and vegetables, when released untreated.

For remedy, we have the stringent Environmental laws to combat with. But for sustainability, we need extensive educative and awareness programmes as a prelude to change of mind set. We also need new technologies and may be, non-fossil fuels for the future industries or even the current industries, which do not generate pollution.

Some General Environmental Initiatives:

Some brief mentions are made of several measures, which have been identified, which help life. Also are mentioned some global phenomena, which affects life adversely.

(*a*) **Zoning of Industry:** By long-term planning, the location of industries must be segregated from the human habitation. Industrial development is much vital for sustaining modern life system as clean air. Segregation allows compliance of both the requirements.

(*b*) **Forestation:** Forestation not only facilitates rain precipitation but generates more oxygen to breathe. Forestation allows further bio-balance by offering refuge to wildlife and much needed wood as material of construction. It increases soil fertility.

Forest, Agriculture, Industry and Human Habitation, each of which are equally important for sustenance of life shall be well-zoned under long-term planning.

(*c*) **Fresh Water Management:** Scientific management and preservation of drinking water, in the urban and equally in the rural sector are called for immediately to ensure the health and sustainability of the future generation.

(*d*) **Global Warming:** Global warming has serious consequences on the undesirable change of world weather including melting of polar and glacial ice. The preventive measure is to reduce the emission of greenhouse gases like carbon dioxide. Extensive forestation reduces carbon release to the environment.

(*e*) **Ozone Layer of Environment:** Ozone layer in the atmosphere protects the life from the harmful effect of ultraviolet ray. Refrigerant CFC, which destroys ozone, has been prohibited at the Kyoto Convention.

❑❑❑

4

CHAPTER

Economics and Governance of Sustainable Development

4.1 Basis of Economic Sustainability

Sustainability is basically an economic concept. It enquires of the long-term availability of economic resources that is natural resources. Now, natural resources could be renewable resources like water, fisheries, and forest, agriculture produces, which are being continuously regenerated by the earth at certain natural rate of growth. The other broad resource type is non-renewable resource like minerals (iron, aluminum, etc), coal, petroleum, etc. The stocks of the second category resources are finite. They were formed over a million years or so, but these resources are getting fast used up over the last hundred years already being consumed. Over this time scale of one or several hundred years, these resources are technically non-renewable. There will arise crisis of economic sustainability, if there is over-consumption of even the renewable resources beyond the regeneration rate or high consumption of the non-renewable resources, depleting them fast.

There is an aspect of sustainability beyond simple economic concern. If the consumption of the natural resources leads to the pollution of the biosphere, namely, carbon dioxide pollution of the atmosphere from the combustion of the fossil fuel, such pollution is known to lead to global warming and wide spread change of weather and likely to disturb global sustainability. This is the all-pervading environmental problem, which has been greatly highlighted, in the Brundtland Report in 1987 of the UN World Commission on Environment and Development (WCED). The Report was instrumental to change the world outlook from the simple economic development to sustainable development.

There is a very important third aspect of social sustainability, which was widely deliberated and encoded in the Principles and the Agenda 21 of the Earth Summit, held at Ro de Janeiro in January1992 and the earlier Brundtland

Report. These concerns included global poverty and unequal distribution and consumption of global resources, global equity and governance. Both the WCED and the UNCED strongly resolved that unless global poverty can be alleviated, it will not be possible to attain the real sustainable world order.

4.2 Munasinghe Hypotheses of Sustainability

The sustainability model of eminent economist Munasinghe (2004) is based on three-sectoral factors, *e.g.,* environment, economy, and social concern (See Fig. F4.1). The guiding hypotheses are elaborated below:

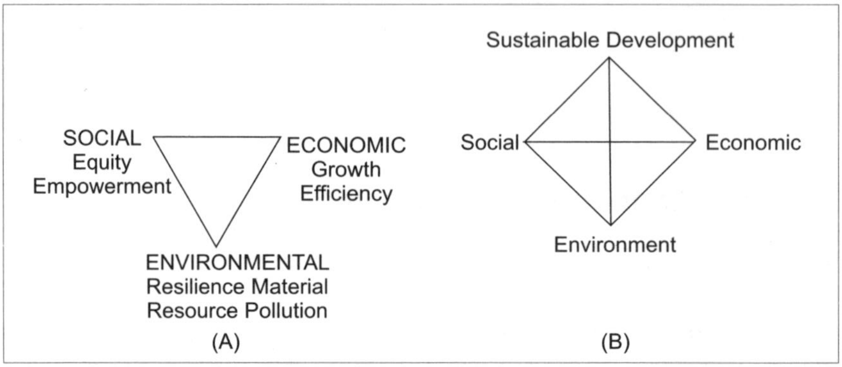

Fig. F4.1: Sustainable Development

1. Economic Hypotheses

Economic progress is often measured in terms of welfare (or utility), measured as capability to pay for goods and services. The economic policy encourages generation of higher income (consumption) and simultaneously greater and efficient production (higher supply).

Efficient economic practice on the principle of Pareto optimality implies that the welfare of at least one person improves without worsening the welfare of any other person.

The widely practiced cost-benefit criterion accepts all projects whose benefits are positive, *i.e.,* aggregate benefits exceed the aggregate costs. The implications are that the accrued benefits from the gainers can be redistributed to the losers to create a win-win situation.

2. Environmental Hypotheses

The theory of Development without adverse impact on the environment is a recent concept of sustainable development globally. Environmentalists believe that such development is viable.

It means prudent use of non-renewable resources, avoiding the degradation of environment and without creating harmful pollution. The levels of degradation and pollution shall be so limited that the nature has the necessary resilience to absorb such damages and renew and reinvigorate itself.

3. Social Hypotheses

Social development implies the improvement of both the individual and social welfare, arising out of the increase in the social capital. The quantity and quality of social interaction underpinning the human social order helps to increase the social stock. As a matter of fact social stock grows with greater use and erodes through disuse, unlike economic and environmental capitals, which are depreciated and depleted by use. Higher social interaction and social stock increase societal awareness of benefits of environmental protection and reinforces societal responsibility.

4. 3 Economics of Exhaustible Resources (Hotelling)

Some economists believe that world's stock of all finite natural resources like minerals, forestry, fisheries should be judiciously and optimally used, so that it can be conserved longer for socially and economically more purposeful and gainful public consumption. Such an approach also underpins sustainable development.

What should be the economic model of such purposeful consumption? How the consumption is to be regulated. If they were cheaply priced, it would lead to rapid consumption, depletion of stock and wastages. If they are steeply priced, it will choke its economic exploitation, denying the present generation of economic well-being. This will be a retrogressive policy, which will not meet public approbation.

In any case, what will be mechanism of price regulation or direct control of consumption of exhaustible resources? In few totalitarian type states, a governmental directive may be effective for sometime but eventually economic expediency will overtake. In capitalist type economics, free competition, initially with plentiful stocks, may push down the price to promote sales and consumption. To contain the resulting fast exhaustion of the natural resource, the Government may introduce high taxation, which will inhibit consumption. Such a situation exists in respect of tobacco products, where high governmental levy discourages high consumption but with a different purpose of public health and not to save tobacco stock. However, as the stock of the resource nears exhaustion, the price will steeply climb even under free competition, as it happens in some measure in the petroleum oil market.

Under monopolistic or cartel management of the natural resources, it will restrict the production below the market demand and extort high prices from the consumers. However, unlike the free competition, the cartel will not raise the price too heavily at near stage of exhaustion, as it will not like to delay the profit for long.

The interplays of the competitive and monopolistic market *vis-à-vis* exhaustible resources are diagrammatically represented in Fig. F4.2A.

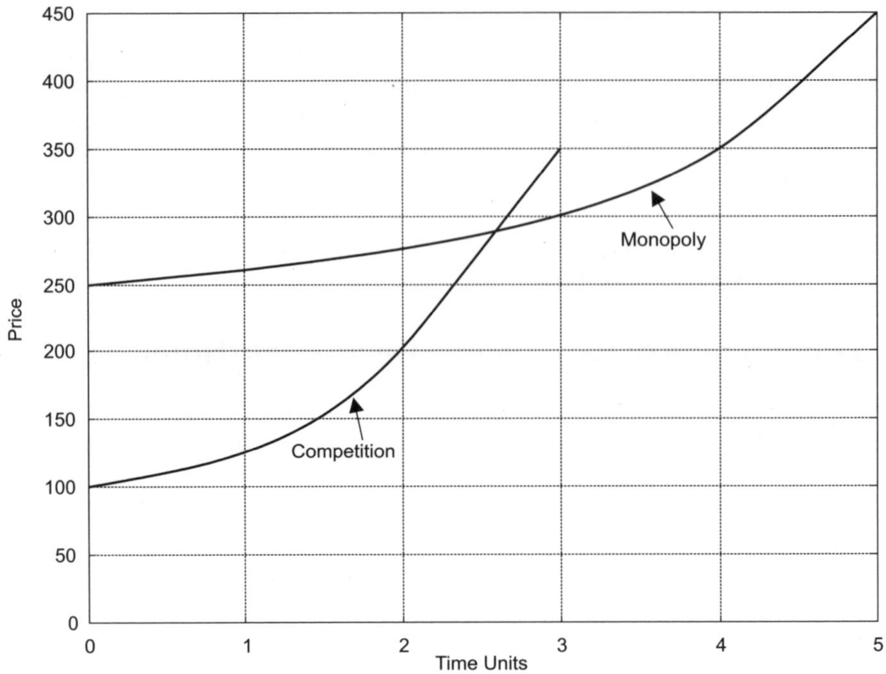

Fig. F4.2A: Price of Exhaustible Resources

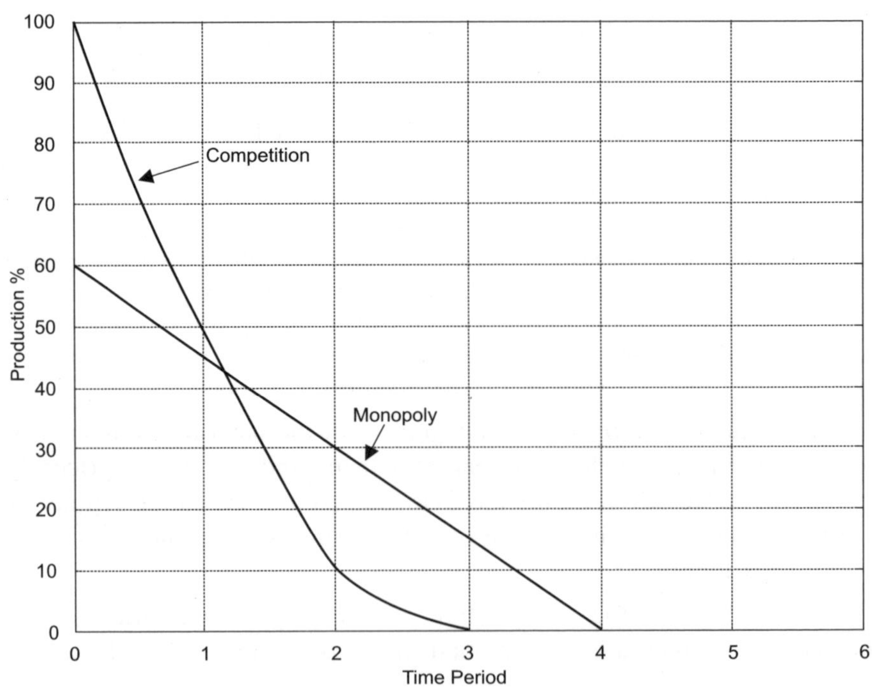

Fig.F4.2B: Production of Exhaustible Resources

In Box B 4.1, we make a brief presentation of the pricing formulation of an exhaustible resource under free competition and monopoly condition, as already indicated in Fig.F4.1B.

Box B4.1

Pricing of an Exhaustible Resource

Free Competition Pricing:

At any time, the basic price level of the exhaustible resource over the cost of extraction and distribution will depend on market demand elasticity. Assume this basic price as Po. The price P of such a commodity at any future time may be taken as

P = Po.{e**(it)},

where i is the banking rate of interest. The seller of the finite resource is thus protected from any financial depreciation.

The quantity q of the resource offered to the market will then depend on the demand function f(p,t) and

q = f (p,t)

If the total reserve in the mine is Q and T is the time when mine gets totally exhausted making the market offering q =0. Thus we get, integrating from 0 to T,

∫[q.dt] = ∫[f.(Po.{e**(it)}, t)].dt = Q, where ∫ is the integration from 0 to infinity ...E4.1

F (Po.(e)Eit, T) = 0 ... ... E4.2

Monopoly Pricing:

Under monopoly condition, the dealer will try to suppress the production to create artificial scarcity and to raise the price, so that his return R is maximized:

R = p.q, at any time t. E4.3

The monopoly price is subject, as before, to the condition

Q = ∫[q].dt, where ∫ is integration is from 0 to infinity.

The point to note here that the exhaustion time under the monopoly condition is actually extended beyond the finite lifetime T of the mine under free competition The other condition of monopoly is that the return to monopolist must be maximized:

Rmax = ∫[q.p(q).{e**(-it)}].dt E4.4

4.4 Some Further Economic Critique

The World Commission on Economic Development (WCED) originally conceived Sustainable Development as:

" Development, which meets the needs of the present without compromising the ability of future generations, to meet their own needs."

But this conceptualisation of sustainability is quite vague and too generalized. If we concern with non-renewable resource like petroleum, we can exert to optimize consumption and preserve as much as possible for the future generation. But such measures cannot be equally applied to all types of resources. That will be draconian and may result to the denial of welfare of present generation. Fortunately, present resources are generally substitutable one with the other. Sustainability should therefore not enjoin that any particular material shall be conserved by a specific percentage for the future generation, but the present economic production systems shall be less material and energy intensive.

Some economists (Solow 1991) believe that the future is not adequately represented in the market, at least not the far future. Doctrinaire market economics of supply and demand may not fully ensure long-term sustainability considerations. If we abide by the rate of interest method of discounting, the future factors in our current analytical considerations cannot but carry minor weightage.

Sustainable development is therefore multi-dimensional. It is closely related to environmental protection. If we consume too much fossil fuel, we not only deplete a precious energy resource, but we are also vitiating the atmosphere. Thus current consumption leads to the worsening of the environment and diminishing of future capacities. Therefore, some economic view is that consumption should be counterbalanced by equivalent investment, which adds to the future capability. If limestone is extracted to make cement, roads and bridges must be built connecting remote unconnected locations, which will enhance the well being of the future generations. Here again, the principle of substitutability applies. In economic sense, resources are fungible and we build up one man-made resource at the expense of another natural resource.

The most fungible resource is the development of technology. Fifty years ago, nobody could imagine that there will develop such wide air travel by Jumbo aircraft, so many wide public roads and so many automobiles, television, cell phones, refrigerators, micro ovens, and computer, information technology and global instant communications, which have made living so much rich and productive compared to that existed in the early twentieth century. Similarly, today, we are witnessing the advancement of modern science and technology in the late twentieth century leading to new path-breaking inventions in Biotechnology, nanotechnology and genetics, human Genome and modern medicine and development of new sources of energy transforming life in such dramatic ways. It will therefore be quite appropriate if we invest heavily in Research and Development. For, we repeat that technology is the most fungible economic resource.

This economic critique should therefore infer that sustainable development is viable balancing investment, conservation and consumption of natural resources. That a new regime of governance for sustainable development needs to be internationally practiced and improvised.

That poverty alleviation as a part of equity needs to be addressed by a new governance model along with the environment protection. That each nation might adopt a new Net National Product Accounting adjusted for addition and depletion of the environmental and natural resource stock.

4.5 Consumer's Cost & Benefit from Environment Change

Environmental changes may involve both improvement or deterioration of environment caused by reduction or increase in the pollution levels respectively. We will first deal with the improvement. The concept is adopted from Ulaganathan (2001).

Improvement of environment and benefit to the community

Suppose health care is a utility U enjoyed by a consumer. At the pollution level Qo, the standard of health care utility Uo at price of health care P is enjoyed by the community. The pollution level is brought down to Q1 and the community can maintain its level of health care at lower expense, the price P remaining the same. For the same utility level of Uo (original pollution level) can be written down –

E [P, S(Qo), Uo]   at the original pollution level Qo,

E [P, S(Q1), Uo]   at the improved pollution level Q1,

where S is a damage function resulting from pollution Q.

The community will be willing to pay (WTP) for the improvement of the environment proportionate to the benefit accrued to them *i.e.,*

WTP = E [P, S(Qo), Uo] –E [P, S(Q1), Uo], … … E4.5

Deterioration of environment and cost to the community

Now suppose we deal with the contra case; let the original pollution level be Q1, the corresponding health care utility be U1 and price rate P and the damage function S remains as before. Let the pollution level gets worsened to the level Qo raising the expenses to maintain the same utility standard U1. The community will likely accept such worsening of the environment if they are compensated to offset their increased cost of health care. The equivalent set of equations can be defined a under –

Expense E [P, S(Q1), U1], at the original pollution level Q1,

Expense E [P, S(Qo), U1], at the worsened pollution level Qo,

where S is a damage function resulting from pollution Q.

The minimum compensation (WTA) equal to the additional cost to the community that the community will be willing to accept,

$$WTA = E\ [P,\ S(Qo),\ U1] - E\ [P,\ S(Q1),\ U1] \quad \ldots \quad \ldots \quad E4.6$$

4.6 Some Reflection on Environmental Accounting (Bartelmus 2001)

National Accounting parameters like the Net Domestic Product mainly concern with the growth of the national economy comprising the products and services including the social services. Since the publication of the 1987 Brundtland Report "Our Common Future" under the aegis of the world Commission on Environment and Development of the United Nations, the concept of 'Sustainable Development' has got highlighted. The concept had since then been widely deliberated at the Earth Summit of the United Nations Conference on Environment and Development at Rio de Janeiro in June 1992 and reemphasized in the Declaration of twenty seven Principles and the famous Agenda 21. The sustainable development envisages economic development, which includes public welfare, alleviation of poverty afflicting a section of the society, protection of the environment, avoidance of wasteful consumption of resources and energy. The principles and treatments are developed from Bartelmus (2001).

Conventional national accounting does not cover two important aspects of sustainable development –

(*i*) Increasing scarcity of natural resources, which can hinder long-term growth of economy;

(*ii*) The degradation of the environment by way of pollution and interference from economic activities.

The United Nation Environment Programme (UNEP) and the World Bank jointly set out to develop an environmentally adjusted accounting system with the following two objectives –

(*i*) The feasibility of measuring physical and monetary accounting of the natural resources and environment;

(*ii*) The development of an alternative macro-economic model of environmentally adjusted National Accounts for Income, Consumption and Investment.

The basic outline of the proposed environmentally adjusted accounting guidelines are encoded under the following two systems –

(*i*) United nations Systems of National Accounts (SNA);

(*ii*) Integrated Environmental and Economic Accounting (SEEA) developed by the Statistical Division of the United Nations (UNSTAT).

The above two systems were also reconfirmed by the UNCED, 1992 at Rio de Janeiro. The non-produced natural resource assets are identified and indicated in Box B4.2 similar to UN practice.

The categories of environmental protection activities and expenditures thereon for their protection are shown are indicated in Box B4.3. The environmental elements include air and atmospheric pollution, surface and groundwater pollution, land and waste management.

Now comes the overall Asset Accounting under net Domestic product, which conventionally is –

NDP = Consumption + Investment + (Exports -Imports)

When adjusted for the addition, depletion, damage and renewal, the environmentally adjusted net domestic product (EDP) is given by –

EDP = Consumption + (AT-Aedd) + (X-M),

where, AT = Total assets of produced and non-produced types,

Aedd = Depletion and Degradation in environmental assets,

X and M are exports and imports respectively.

Box B4.2

Classes of Tangible Non-Produced Assets in SNA and SEEA

- Natural Assets:
  Land
  Land underlying building and Structures;
  Land under Cultivation;
  Recreational Land and Associated surface Water.
- Subsoil Assets;
  Coal, Oil, and Natural Gas;
  Metallic mineral reserves;
- Non-cultivated Biological Resources.
- Water Resources.

Box B4.3

List of Environmental Protection Activities

1. Protection of ambient air climate (prevention of air pollution, treatment of exhaust gases)
2 Protection of ambient water, excluding ground water (prevention of water pollution, industrial pretreatment for sewage, treatment of cooling water).
3. Prevention, collection, transport, treatment and disposal of wastes (collection, transport, treatment of waste, and prevention pf waste generation).
4. Recycling of wastes and other residuals.
5. Protection of soil and ground water (decontamination of soil, cleaning of ground water).
6. Noise abatement (traffic and industrial process noise).
7. Protection of nature and landscape (protection of species, habitats; erosion, fire, and avalanche protection).
8. Other environmental protection measures (education, training, administration).
9. Research and Development.

Table T4.1: Comparison of Conventional and Environmental Accounting Examples of Mexico and Papua New Guinea

	Mexico 1985			Papua New Guinea 1986-90		
Account	**Conventional**	**Integrated Green**		**Conventional**	**Integrated Green**	
Heads	**Accounting**	**Accounting**		**Accounting**	**Accounting**	
		EDP1	**EDP2**		**EDP1**	**EDP2**
NDP	42.1 P	39.7 P	36.4 P	2.760 K	2.580 K	2.580 K
EDP/NDP	-	94	87	-	92-99	90-97
C/NDP	83	88	96	89-100	93-106	95-109
+ CAP	4.6 P	2.4 P	-0.7 P	0.463 K	0.282 K	0.228 K
+CAP/NDP	11	6	-2	12 - 20	5 - 17	3 – 16
NDP/CAP	37	10	-	5	-	-

Notes: NDP = Net Domestic Product;
EDP = Environmentally adjusted Net Domestic Product;
C = Final Consumption;
+CAP = Capital Formation/Accumulation;
CAP = Capital Stock.

Examples of Environmental Accounting

In Table T4.1 is shown the EDP, which is derived after environmentally adjusting National Domestic Products for the estimated loss valuations for the depletion and degradation of the environment of Mexico and Papua New Guinea under the aegis of the World Bank. The Table is reproduced from Bartelmus (2001). After all environment costs are taken into account in EDP2, it is seen that Mexico undergoes divestment of –2%, while New Guinea has an investment range of 3 to 16 %.

4.7 Governance for Sustainable Development

Introduction

Economic development is the goal of all national Governments. But, since 1972 first UNEP Conference on Human Environment at Stockholm, the emphasis has been shifting, gradually to socially oriented projects including environment protection, poverty alleviation and equitable distribution of well being. The developmental philosophies underwent further transformation with the publication of the UN/WCED-sponsored Brundtland Report: 'Our Common Future' in 1987 and the Earth Summit at Rio de Janeiro in 1992. The developments are now to be oriented towards long-term sustainability of the global ecosystem and well-being of future generation of mankind. Such sustainable development, as the new movement has been named, is better realized with the wider community involvement and participation. As a natural corollary, it is inferred that the strategy and the implementation of sustainable development will perform better with a democratic governance framework. But, surprisingly, the pronounced social democracies of Scandinavia are faring better in global eco-projects than the free market driven democracies of US, Canada, and UK. Eventually, of course, building strong social governance Institutions is essential to carry forward the message of sustainable development and its dedicated implementation. Here again, the Scandinavian countries like Netherlands, Norway, etc. are leading the way.

Policy and Instruments

If economic development is for immediate survival, sustainable development is for long-term enrichment of life, economically, socially and culturally. Sustainable development therefore, calls for higher awareness or conceptual preparedness. It demands creating newer policy delineation or a new sets of instruments beyond the simple market forces. Apart from introducing enabling features in governance and economic instruments, the educational system should be remodeled focusing on social and ecological consciousness, rights and duties of an enlightened citizenry, conservation of global natural resources, wildlife, forestry, water, air and biodiversity in general. Social instruments should also urge prudent use of all economic resources – natural and manmade including minerals, fuel oil, energy and water. Box B1.8, adapted from Lafferty (2001), outlines some of the Policies and Instruments for promoting sustainable development.

Box B4.4

Policies and Instruments for Sustainable Development

1. Regulatory Regime: Licence /Permits; Ambient Quality Standards; Emission Standards; Process Standards; Product Standards.
2. Economic Instruments: Charges; taxes; tradable emission permits; tradable quotes; environmental subsidies; deposit-refund systems; performance bonds; non-compliance fines; resource pricing.
3. Liability & Damage: Strict Liability Rules; Compensation Funds; Compulsory Pollution Insurance; extended producers' responsibility.
4. Education & Information: New Education Policy & Public Awareness programme; Technical training programme for various environment amelioration and protection, Eco-labeling.
5. Management and Panning: Institutes for developing environment management systems; Environment zoning; Scientific & benign land management.
6. Institutional Framework: An enabling Policy for setting up a network of social and environmental transformation in the public and private domain.

Source: Lafferty (2001)

Operational Strategies

Managing sustainable development is the new challenge of the national governance. The proclamation of the 1992 Earth Summit at Rio comprising the 27 Principles, enunciating the core policies, and Agenda 21, defining the Programme of action, constitute the core strategies needed to implement the sustainable development. At the operational, strategic programme may include:

(*i*) Coordinating and integrating the social, ecological and economic programmes at the national and regional level in consonance with the international Conventions, Protocols, Agreements initiated by UN or others;

(*ii*) Participation in and partnerships with the Institutions, NGOs, Government departments in communication, information-gathering, and in promoting awareness;

(*iii*) Strengthening existing knowledge Institutions and strategising institutions for the task of sustainable development;

(*iv*) Strengthening implementation by (*a*) formulating, assessing and laying down priority issues in place; (*b*) correlating and providing adequate budget for the planned actions; (*c*) leveraging and encouraging private initiatives and investment; (*d*) a sound system of continuous monitoring and improvisation of all planned actions and strategies.

Revisiting and further Understanding Sustainable Development

Conceptually, sustainable development has social, ethical and ecological objectives than the mere scheme of economic development with promise of commercial return or otherwise. In this respect, there has been at least following three stages of developmental thinking before equity, common well being and ecological concern came at the fore at the 1992 Earth Summit or UN Millennium Declaration of 2000 or the 2002 Johannesburg Declaration:

(*i*) Free Market System is mainly concerned with the profit-driven investment and development as used to pertain in USA to an extent.

(*ii*) Social Democratic system addresses socio-economic concerns but also equally encouraging profit-motivated development in parallel.

(*iii*) Sustainable Development focuses emphasis on socio-economic development with equal concern for ecological and environmental conservation with a long term objective of preserving ecological natural resources for the future generation.

The task of sustainable development is more effectively implemented as a multi-national trans-geographic joint initiative as a part of UNICEF/UNEP programme. This is necessary as the phenomena of pollution and ecological degradation cannot be bounded within political boundaries.

Secondly, sustainable development involves a complete transformation of the mindset beyond current cold-war inflicted global milieu or electoral politics-dominated national regimes. It requires new kind of social education, awareness, and institutions.

Sustainable development is finally a holistic philosophy, a new way to manage the nations and the world, a new economic order of equitable distribution of wealth, income, ensuring public health, banishing poverty, conserving and expanding global resources.

4.8 Corporate Governance for Sustainable Development

In traditional business, efforts are generally concerned with the reduction of environmental hazards at the processing plants or from some identified products. A total approach for controlling all potential environmental damages likely to be caused during production and later from the use of the products is normally overlooked. Often the corporate governance for sustainable development is limited to efforts to control GHG emission, which is no doubt also a good contribution. For, metallurgical industries like aluminum, cement, and steel are very serious emitter of GHGs, severely polluting the environment. However, the Wuppertal Institute (Ruud2004/von Weizsacker et al 1998) has framed up a comprehensive framework of Corporate Governance norms, which go beyond specific or discrete actions:

- Reduce material intensity;
- Reduce energy intensity;
- Reduce harmful or toxic emission;
- Enhance recyclability;
- Maximize use of renewable resources;
- Extend product durability;
- Increase in service intensity;
- Enhance efficiency, and reduce wastes in all operations.

The last panacea, added by us, is an overarching aim of the corporate governance philosophy for "green business".

Corporate Strategies

From the above Wuppertal guidelines, it may be inferred that the general trend of corporate initiative towards environment benevolence should include economic use of resources, controlling all external emissions and environmental impacts and improving product life-cycle and/or recyclability.

These philosophies should be integrated into corporate business strategies. Rudd (2004) had proposed two somewhat different strategies for production and products, depicted in Fig.F4.3.

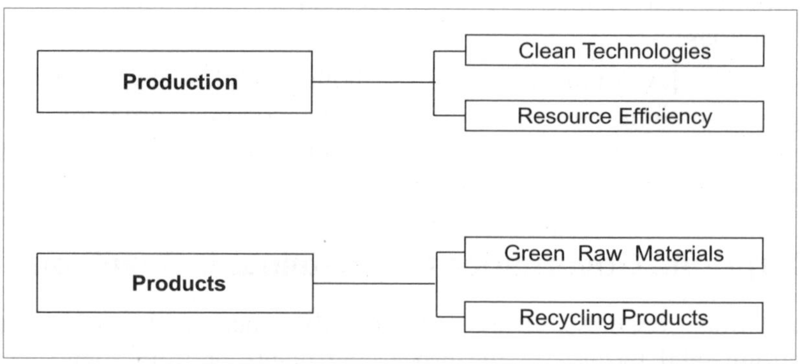

Fig. F4.3: Corporate Environmental Strategies

Clean Technology accompanied with high resource economy shall govern the Production Process. The Product System shall be sourced from 'Green Materials' as much as possible and the product lifecycle shall be augmented by adopting Recycling Technology. Both these governance outlooks highlighted in Fig. F4.3 are now integrated in Fig. F4.4. In turn, Fig. F4.4 highlights that clean technology and adoption of green materials leads to immediate reduction in the adverse environmental impacts; and the adoption of high material efficiency and recycling technology lead to long-term sustenance of the economy.

Thus, by using correct governance strategies, the corporates can add to both environmental protection and long-term sustainable development of the economy.

The 1992 Earth Summit at Rio de Janeiro recommends in its Agenda21, Chapter 30 "the strengthening the role of business and the industry towards sustainable development"

	Low Environmental Degradation	High Sustainability
Production	Clean Technology	Economise Resource Input
Products	Green Raw Materials	Recycling of Products

Fig. F4.4: Corporate Strategic Framework for Sustainable Development

In the Chapter 4 of the same agenda 21 UNCED determines changing consumption and production pattern at the heart of the sustainable development.

4.9 Visionary Agenda for Sustainable World

By 2050, the world will be burdened with a likely population of 8 millions people. Add to this the threat of global climatic change and ozone depletion and soil erosion, deforestation, water scarcity, and toxic contamination challenging the health and survivability of the planet. But we believe the word will sustain, overcome and bring solutions to all the intractable problems.

Energy source scenario

We have studied various new and pioneering sources of energy in details Part III of this treatise. We mentioned that the solar energy will contribute 10 to 20 percentage of the total future global energy needs, keeping in mind the fast development of recovery technology and that theoretically, the amount of solar energy received by the earth is of the order of 250 times the present global consumption of energy.

Similarly, it has been very reliably assessed that the wind energy recovered from mega wind-farms and innumerable small and medium turbines can contribute anther 10 to 20 percentage of global needs.

The Bio-energy in the form of Bio-diesel, Ethanol, Methanol, Butanol etc. will contribute another 10 to 20 percentage, especially for the auto-fuels.

The nuclear power is reaching a stage of high perfection in engineering, radioactivity and pollution-free waste disposal technology and safety technology and is very likely to contribute another 30 to 40 percentage, especially for generation of electrical power in ultra-large plants.

There is great possibility of Hydrogen energy from Fuel cell technology, again in dominant automotive energy sector. If this energy is realized as present R & D suggest, it can also contribute a very large percentage to that the total consumption.

The inference is that the world of post-2050 era can do without fossil fuels and without polluting the earth.

Transportation Scenario

Automotive transportation consumes more than half of all fossil fuels. We have indicated in the above that Hydrogen and bio-fuel will largely replace the fossil fuel in the automotives by the middle of this century. The electric traction is another likely solution for the railways and for the hybrid cars and trucks. Another aspect is the avoidance or reduction of wasteful private automobile transportation by highly engineering comfortable public transportation under development in several countries. The scenario is therefore technology-positive.

Zero-Waste Consumption

It is envisaged the world will gradually approach towards zero-waste economy. All used products and goods will be recycled. The recycling technology will develop so well that the cost of recycling will be far more economical than the freshly processing new raw materials, which can also generate newer pollution. Potable water is becoming a scarce resource and hence all water discharges and effluents will undergo total recycling, finally becoming a zero-discharge technology.

Food and Agriculture Scenario

How to feed likely global population of 8 million people by 2050 or earlier. The arable cropland is limited. To increase acreage, the world will have to adopt salt-tolerant and drought-resistant farming, which are being practiced even today in small measure. Then to increase productivity, special measures for increasing soil fertility by green farming and preventing soil erosion are to be adopted. Biotechnological transgenic crop can revolutionize agricultural production and double the productivity. Hence, the world will be able to feed itself well in 2050 and thereafter.

Forest Scenario

The adequate forest cover is essential for controlling the atmospheric carbon content, supplying the atmosphere with oxygen and thus purifying the atmosphere. Sufficient forest cover is also essential for evapotranspiration and stimulating rainfall. Forest prevents soil erosion, reduces flood fury. Thus the future society has to have more respect for forests and woodlands. It is estimated that annually forests are shrinking by 27 million acres, which must be stopped.

On the other hand, reforestation shall be adopted on a wider scale. Every city and village should develop a wide belt of forest and woodland around its boundary.

Apart from above forest beneficences, forest can drive high economic benefit by harvesting resins, nuts, fruits, and medicines etc.

New Economic Scenario

The employment will shift towards the new paradigm industries from mining and ore processing to recycling industries, to solar power, wind power, energy auditing and management, from metallurgical to biotechnology and technological agriculture, to services of various kinds including education, health, modern mass transportation, hospitality, socio-cultural activities and etc.

Indices of Gross Domestic Products (GDP) have to incorporate factors of environmental welfare more substantively. Materialism and consumerism will have to give way from individualism to community and national well-being. Wasteful and extravagant lifestyle will have to make way to environment-friendly culture.

❑❑❑

Part I

References

Bartelmus P *et al* 2001, *Environmental Accounting, Environmental Economics,* ed. U Shankar, Oxford University Press 2001

Brown L Flavin C *et al* 1994, *Vision of a Sustainable World, Environmental Ethics* (Ed. L P Pojman), Jones & Barlet Publisher, Boston

Dunn Seth 2004, *History of Hydrogen,* World Watch Institute, New Haven, Connecticut, *Encyclo. of Energy,* Vol 3, Ed. Cleveland, Elsvier Press

Guha Ramchandra, 2006, *"How much should a Person Consume?"* Permanent Black, New Delhi 110092

Hotelling Harold 1931, *The Economics of Exhaustible Resources, Environmental Economics* Ed. U Shankar, Oxford University Press, 2001

Lafferty W M ed. 2004, *Governance for Sustainable Development: The challenge of adapting Form to Function,* Edward Elgar, Cheltenham, U K

Mongillo John 2004, *Environmental Sciences, Vol 5, Creating a Sustainable Society,* Greenwood Press, Westport, Connecticut, USA

Munasinghe 2004, *Sustainable Development, Encyclo. of Energy,* ed. Cleveland, Elsevier Press

Solow Robert M 1991, *Sustainability – An Economist's Perspective, Environmental Economics,* Ed. U Shankar, Oxford Univ. Pres, New Delhi

Ulganathan Shankar 2001, *Introduction to the Environmental Economics,* Oxford Univ. Press, New Delhi

Part II
Challenges of Environment

5

CHAPTER

General Atmospheric Pollution

5.1 Clean Air is Vital for Life

The clean air consists mainly of 78 per cent nitrogen, 21 per cent oxygen and 0.9 per cent argon and many other trace gases as shown in the Table T5.1.

Every day a human requires to breathe 16.5 kg. of clean air. If that air is polluted, he will get less oxygen out of that quantity of air to the detriment of his health. Unfortunately, the clean air is getting today continuously and seriously polluted from industrial emissions from combustion and variously other processes. Now automotive gaseous pollution consists carbon monoxide, oxides of nitrogen NOx, oxides of sulfur and volatile hydrocarbons, which are highly injurious to human health. Besides, increasing volumes of carbon dioxides, emanating from all internal combustion engines and all other combustions and burning of fossil fuels and biomass, though non-toxic, are contributing to global warming, which again leads to serious ecological, human and social effects in the medium and long terms. All these pollutants in the air are also generating sunlight-induced secondary photolytic pollution in the lower atmosphere (Troposphere) affecting humans and biodiversity in general. With the industrialization deepening in the West and now spreading fast to the rest of the globe, the rate of pollution is greatly accelerated in the late Twentieth century. Concerned at these adverse developments, the United Nations and the member countries have come forward to frame Environmental Conventions and Protocols at the global level and to enact environmental legislations at the national levels. We briefly mention such Legislative initiatives in two representative nations USA of the developed world and India of the developing world.

Table T5.1: Compositions of Clean Dry Air

Constituents	**Concentrations** (by volume)
Nitrogen	0.7808
Oxygen	0.2095
Argon	0.0093
Carbon Dioxide	355 ppm
Neon	18 ppm
Helium	5.2 ppm
Methane	1.8 ppm
Krypton	1.1 ppm
Nitrogen oxide	0.3 ppm
Hydrogen	0.5 ppm
Ozone	0.01 ppm

Source: Masters (1991)

Clean Air Acts of USA

Even before the UN Conference on Human Environment at Stockholm in 1972, United States have taken initiatives to enact several Acts, e.g. Clean Air Act of 1963, Air Quality Act of 1967, Clean Air Amendment Act of 1970 and 1977, which together came to be known as CAA (Clean Air Act). Besides, mentions should be made of two specialist Environment Acts like Motor Vehicle Air Pollution Control Act of 1965 and the Energy supply and Environmental Coordination Act of 1974.

Table T5.2: National Ambient Air Quality Standards for Criteria Pollutants 2008

Air Pollutant	**Averaging Time**	**NAAQS (Primary)**	**Objectives**
Carbon Monoxide	8 Hours 1 Hour	10 mg/m3 (9 ppm) 40 mg/m3 (35 ppm)	control carboxyhaloglobin
Nitrogen Dioxide	Annual	100 microg/m^3 (0.053 ppm)	To improve health risk and visibility
Ozone	1 Hours.	235 microg/m^3 (0.12 ppm)	To prevent eye irritation & breathing problem
Sulfur Dioxide	Annual 24 Hours	80 microg/m^3 (0.03 ppm) 365 microg/m^3 (0.14 ppm)	To prevent respiratory disease
PM-10	24 Hours	150 mcg/m^3	To maintain visibility & prevent smog
PM-2.5 (Particulates)	Annual 24 hours	15 mcg/m^3 35 mcg/m^3	
Lead	3 months avge. 3 months Rolling Avge.	1.5 mcg/m^3 0.15 mcg/m^3	To prevent health problem

US NAAQ5, 2008

Table T5.3: Threshold Limiting Values of Gaseous Pollutants

Gaseous Pollutants	Permissible Levels (24 Hours average)
Fluoride (Fl)	0.1 ppm
Carbon Monoxide (CO)	50 ppm
Carbon Dioxide (CO_2)	5000 ppm
Sulfur Dioxide (SO_2)	5 ppm
Hydrogen Sulphide (H_2S)	10 ppm
Nitric Oxide (NO)	25 ppm
Nitrogen Dioxide (NO_2)	5 ppm
Carbon Sulphide (CS_2)	1 ppm
Cotton Flux	200 mg/m^3
Lead (Pb)	250 mg/m^3
LPG	1000 ppm
Ozone (O_3)	0.1 ppm
Ammonia (NH_3)	100 ppm

Source: Mahajan

CAA has enabled the creation of the Environmental Protection Agency (EPA), which has since done so much of the pioneering work and which is looked upon as the Global Benchmark for various National enactments around the world.

CAA has also developed two Air Quality Standards –

NAAQS: National Ambient Air Quality Standards,laying down the Acceptable concentration levels of various air pollutants. (See Table T5.2)

NSPS: New Source Performance Standards laying down the limiting emissions from different industrial sources.

Indian Environment Enactments

In pursuance of the United Nation Conference on the Human Environment held at Stockholm in June 1972, India constituted the national Committee on National Planning and Coordination, 1972.

Under the aegis of this Committee, following two Enactments came to be framed:

–The Air (Prevention and Control of Pollution) Act, 1981.

– The Environment (Protection) Act, 1986.

About the time of constitution of this committee, Government of India had already taken the initiative to enact a comprehensive water Act called " The Water (Prevention and Control of Pollution) Act, 1974.

The Water Act 1974 and the Air Act 1981 led to the formation of the Central Pollution Control Board and a number of State Pollution Control Boards at the State levels on the model of EPA of USA.

The Environment Act, 1986 widened the definition of Environment to include air, water and land and their interrelations and pollutions arising out of solid, liquid, and gaseous substances.

Permissible Standards of Principal Atmospheric Pollutants

EPA of USA has identified six main 'Criteria' pollutants of atmospheric carbon monoxide, lead, nitrogen dioxide, ozone, sulfur dioxide, and suspended particulates of aerodynamic diameter less than or equal to a nominal 10 micron (PM-10) and laid down NAAQS for them as indicated in Table T5.2.

We provide another Table T5.3, which shows threshold-limiting values of some generally toxic gaseous pollutants (other than CO2), which are followed in some institutions in India.

Allowable Standards of Gaseous Emissions from

Identified Sources

To control direct emissions from a source is another way of reducing atmospheric pollution; NSPS of USA lays down emission standards based on the input energy value. For example, for a stationary power plant, the allowable emission is expressed in pounds of pollutants per million BTU of heat input to the power plant, see Table T5.4. Similar NSPS are also laid down for cement plant, nitric acid plant, petroleum refinery, sewage treatment plant and metallurgical smelters of various kinds.

Table T5.4: USEPA Emission Standards for Steam Power Plants 2002

Pollutant Types	NSPS Limits (g/GJ)
Particulates	139
Nitrogen Oxide (NOx)	
Gas-fired	86
Oil-fired	130
Coal-fired	260
Sulfur Dioxide (Sox)	
Gas or Oil-fired	86
Coal-fired (maxm.)	500

GJ=10^9 Giga Joules of heat Input.

Table T5.5: NSPS Annual Emission Values (million tons) of Some Criteria Pollutants

Source category	Particulates	Sox	NOx	Hydrocarbons	CO
Transportation	1.4	0.9	8.5	6.5	42.6
Stationary Plants (Fuel Combustion)	1.8	17.2	10.0	2.3	7.2
Industrial Processes	2.5	3.1	0.6	7.9	4.5

Source: adapted USEPA

Table T5.6: Air Pollution in Indian Cities

Pollutants	Mumbai	Delhi	Kolkata	Hyderabad	Ahmedabad	Bangalore	Chennai	CPCB Limit
SO_2	8	8	11	5	12	9	12	30
NO_2	23	45	40	20	25	46	29	60
RSPM (PM-2.5)	79	100	119	60	22	68	42	60
SPM (PM-10)	242	268	270	154	232	138	121	140

Central Pollution Control Box, India 2008.

The Government of India, through the Central and the State Pollution Control Boards have notified limiting emission ppm of pollutant gases and particulates for various industries.

The major sources of criteria pollutants in US are given in Table T5.5. The Table shows that the transportation sector contributes the maximum carbon monoxide at 42.2 million tons and also substantial amount of NOx and hydrocarbons all of which helps to generate photolytic smog.

Air Quality in Indian Cities

The Table T5.6 shows the pollution levels in five large cities of India, Delhi the national capital, Mumbai, Kolkata, Kanpur, and Nagpur in the nineties. It will be noticed that the gaseous pollutants SO_2, NO_2, H_2S, Ozone exceed the NAAQS permissible limits of US. However, situation has since somewhat improved as LPG gas burning has replaced the coal burning in the domestic kitchens. Nagpur appears to be the most polluted city in India.

5.2 Combustion Pollution of Atmosphere

Chemically combustion is a process of oxidation of elements of carbon, Hydrogen, Sulfur, Nitrogen etc. with release of heat and other energy. Rapid combustion takes place if there is sufficient supply of oxygen. Slow combustion takes place in insufficient supply of oxygen.

We now present three cases of combustion by their chemical reaction, taking the hydrocarbon as CxHy or more generalized hydrocarbon as CxHyOzNaSb

Complete/Rapid Combustion with formation of carbon dioxide CO_2:

$$CxHy + 0.25(4x+y)\ O_2 = x.CO_2 + 0.5y.H_2O. \qquad E5.1$$

Incomplete/Slow Combustion generating more carbon monoxide CO:

$$CxHy + 0.25(2x+y)\ O2 = x.CO + 0.5\ y.H2O \quad ..... \qquad E5.2$$

More generalized combustion involving oxidation of sulfur, nitrogen besides carbon and hydrogen are given by the following formulations:

Complete combustion

$$CxHyOzNaSb + 0.25\ (4x + y + 2c + 4b - 2z).\ O_2$$
$$= x.\ CO_2 + 0.5y.H_2O + c.\ NO + 0.5.\ NO + 0.5(a\text{-}c).N_2 + b.SO_2 \qquad \ldots \quad E5.3$$

Incomplete combustion:

$$CxHyOzNaSb + 0.25\ (2x + y - 2b - 3a - 2z).\ O_2$$
$$= x.\ CO + 0.5.(y\text{-}2b\text{-}3a).H_2O + a.NH_3 + b.H_2S \qquad \ldots. \quad E5.4$$

From the above four equations, it is evident that all combustion generates carbon dioxide (CO_2), carbon monoxide (CO), and water vapour (H_2O). While carbon dioxide is non-toxic, it plays supposedly a heavy role in global warming, of which we will write more later. Similarly,

water vapour is harmless but it also does play a minor role in the global warming. Carbon monoxide (CO) is highly toxic affecting human lungs and other systems.

In hydrocarbon fuels CxHyOzNaSb, the subscripts x, y, z, a, and b represent the number of the respective elements, which are known and hence the number of required oxygen elements are determined, if the byproducts of the processes of combustion are known. Hydrogen and oxygen in the fuel combine to give 'combined water'.

Nitrogen in the fuel is generally bonded to hydrogen and in incomplete combustion, they are likely to remain combined as ammonia NH3. But in complete combustion, nitrogen converts to nitric oxide NO and N2. The extent of NO formation is unknown and hence the coefficient is indeterminate, subject to experimental evidence of the particular hydrocarbon fuel (see equation E5.3). Combustion also produces small amount of Nitrogen oxide NO_2, which are omitted in the above formulation. The whole set of various oxides of nitrogen is as NOx, which includes nitric oxide (NO), nitrogen dioxide (NO_2), nitrous oxide (N_2O), nitrogen trioxide (N_2O_3) and nitrogen tetroxide (N_2 O4). NOx formed from combustion of atmospheric nitrogen is called thermal nitrogen and that produced from combustion of fuel nitrogen is called fuel NOx.

In the presence of intensive heat of combustion, there maybe dissociation of atmospheric oxygen into oxygen radical *i.e.,* charged oxygen atom O', which in combination with the nitrogen may give rise to charged nitric oxide radical a proposed by Zildovich below:

$$O_2 + M = 2O' + M \qquad \ldots \qquad E5.5$$
$$O' + N_2 = NO' + N' \qquad \ldots \qquad E5.6$$
$$N' + O_2 = NO' + O' \qquad \ldots \qquad E5.7$$

The first reaction in E5.5 shows dissociation of oxygen into charged oxygen or free oxygen radical in presence of an independent medium m say Nitrogen to absorb additional heat released during dissociation.

Sulfur Process

During incomplete combustion, sulfur from the hydrocarbon converts into hydrogen sulphide H_2S. But during complete combustion, sulfur gets oxidized into sulfur dioxide SO_2 (See Equation E5.3). Besides sulfur dioxide, a small amount of sulfur trioxide SO_3 is also formed, which is omitted in E5.3.

When SO_2 is released to the atmosphere in presence of the high temperature of complete combustion, several charged sub-compounds *i.e.,* radicals are formed along following set of reactions:

$$SO_2 + OH' = HOSO_2' \quad \ldots$$

$$HOSO_2' + O_2 = SO_3 + HO_2' \quad \ldots \qquad E5.8$$

In the above, OH', $HOSO_2'$ and HO_2' are free radicals as they contain an unpaired electron each, which make them unstable and capable of combining other elements or radicals of opposite charge.

The sulfur trioxide in equation E5.9 in the presence of water vapour in the atmosphere give rise to the sulfuric acid H_2SO_4 and similarly sulfur dioxide into sulfurous acid H_2SO_3 as shown below.

$$SO_3 + H_2O = H_2SO_4 \quad \ldots$$

$$SO_2 + H_2O = H_2SO_3 \quad \ldots \qquad E5.9$$

The formulation of the above acids and their descent to the earth surface during rain is also known as acid rain.

5.3 Photochemical Pollution of Atmosphere (Smog)

Formulation of smog is one of the major causes of the pollution of the atmosphere. Smog is generated by the reaction of the sunlight on the atmospheric nitrogen and a large number of volatile organic and inorganic compounds present in the atmosphere. Ozone is the main constituents of smog as symbolically shown below.

Nitrogen + Hydrocarbons + other VOC + sunlight →Smog

→Ozone + NO_2 + NO + acids + other harmful organic compounds.

.. E5.10

Ozone can cause chest constriction and the irritation of the mucus membrane. But eye irritation is caused by a host of volatile organic compounds including PAN (Peroxyacetylenenitrate), PBZN (Peroxybenzoylnitrate), and acroline.

Formulation of Various NOx:

Nitricozide NO is emitted to the atmosphere from combustion. NO reacts with the atmospheric oxygen to form nitrogen dioxide. Under photolytic effect of the sun, NO2 decomposes and releases elemental oxygen radical O′, which combining with atmospheric oxygen forms O_3.

Combustion emission:	N2 + O = 2NO ...	E5.11
Thermal oxidation:	$2NO + O_2 = 2NO_2$ ...	E5.12
Photo-decomposition:	$NO_2 + hu = NO + O'$ ...	E5.13
	(hu is one photon unit of energy)	
Chemical Reaction	$O' + O_2 + M = O3 + M$	...E5.14
	(M is an external element, usually O_2 or N_2, absorbing excess energy of the reaction)	
Thermal dissociation:	$O_3 + NO = NO_2 + O_2$	...E5.15(A)

NO (like CO) combines with hemoglobin affects oxygen supply in blood, but being less than 1 ppm in the atmosphere is really not seriously harmful. But NO_2 and O_3 affect lungs from prolonged exposure.

Depending upon the diurnal position of the sun, nitric oxide is more predominant in the morning, Nitrogen dioxide is more plentiful before noon, but ozone is more plentiful just at noon. But from E5.13 to E5.15, it is seen that by virtue of photolytic reaction, the concentration levels of NO_2, NO and O_3 remain theoretically unchanged over the 24 hour cycle. But ozone accumulates through other reactions, which we will describe later.

Formulation of Hydroxyl Radical

With water vapour in the air and the photolytic heat, OH′ radical is generated.

$$O3 + hu = O' + O_2 \quad \ldots\ldots$$

$$O' + H_2O = 2OH' \quad \ldots\ldots \quad E5.15(B)$$

The highly reactive OH’ radical combines with the atmospheric pollutants carbon monoxide CO, carbonyl compounds and hydrocarbons giving rise to additional ozone and more harmful peroxyl compounds.

1. Hydroxyl Radical OH′ reactions with Carbon Monoxide CO

Incomplete combustion of fuels and hydrocarbon releases carbon monoxide in the atmosphere. Carbon monoxide reacts the hydroxyl radical OH′ present in the air producing carbon\on dioxide CO_2 and hydroperoxyl radical HO_2'. The process continues producing nitrogen dioxide NO_2, oxygen radical O′ and also OH′ radical which keeps process cycle repeating as long as carbon monoxide pollution is present in the air. The final process is the release of ozone O3, which is a net output and which ozone goes on accumulating as the cycle repeats.

The various steps of the chemical reactions are:

$$CO + OH' = CO_2 + H'$$
$$H' + O2 + M = HO_2 + M$$
$$HO_2 + NO = NO_2 + OH' \quad \ldots\ldots\ldots\ldots \quad E5.16$$
$$NO_2 + hu = NO + O'$$
$$O' + O2 + M = O_3\ M \quad \ldots\ldots\ldots\ldots\ldots \quad E5.17$$

Hydroxyl Radical OH' reacts with Hydrocarbon and other Carbonyls:

A hydrocarbon Ethane C_2H_6 can be represented by RH, where R radical C_2H_5. RH then reacts with radical OH' in the following manner:

$$RH + OH' = R' + H_2O$$
$$R' + O_2 = RO_2$$
$$RO_2' + NO = RO' + OH' \quad \ldots\ldots\ldots\ldots \quad E5.18$$

Like the reactions of the hydroxyl with CO, here nitrogen dioxide is directly formed without affecting the ozone content continuing the reactions,

$$RO' + O_2 = HO_2' + CH_3CHO$$
$$HO_2' + NO + NO_2 + OH' \quad \ldots\ldots\ldots\ldots E5.19$$

The OH′ radical is regenerated keeping the reaction cycle repeating like, CO/OH′ cycle, further continuing,

$$NO_2 + hu = NO + O'$$
$$O' + O_2 + M = O_3 + M \quad \ldots\ldots\ldots \quad E5.20$$

The last two lines of the reactions are same as E5.17.

❑❑❑

Hydroxyl Radical OH reacts with Hydrocarbon and other Gases

6

CHAPTER

Automobile Emission

6.1 Introduction

Motor vehicles emit large quantities of carbon monoxide (CO), Hydrocarbons and nitrogen oxides (NOx) and such toxic substances as fine particles and lead. Across the globe, automobile motor vehicles have been increasing at an unbelievingly high rate. For example, the data for the cars show that

in 1950, there were 53 million cars, and

in 1990, there were 430 million cars.

The above shows that there were 8 fold increase in car population globally in 40 years, growing at 9.4 millions annually. Simultaneously, the truck and bus fleet has also been growing at 3.6 millions annually. To the above must be added about 100 millions two-wheelers around the world (Two-wheeler fleet is increasing at about 4 millions per year globally). By 1996, the motor vehicle population grew to over 700 millions.

But the vehicular growth rate is slowing down in the highly industrialized countries. There is now major motor vehicle pollution in the OECD countries including European Union and Japan. There are also conspicuous increases of vehicles in the emerging economies like in China, India and Brazil, etc. It is believed that in 1986 in USA, automobile emissions contributed following national percentages (See Table T6.1):

Table T6.1: Auto-Pollutants: National Emissions (USA)

CO..	58%,
Lead	34%,
VOC	27%,
Particulates	16%.

There are three principal areas in an automobile, *e.g.,* Fuel Tank, Crankcase, and the Exhaust Pipe (See Table T6.2 and Fig. F6.1) from where are emitted the major gaseous auto-pollutants.

Table T6.2: Source Areas of Auto-Emissions

Sources	Emissions
Fuel Tank Evaporation ...	... 20%
Crankcase Vapour ...	... 20%
Engine Exhaust Pipe ...	... 60%

Fig. F6.1 Sources Of Auto Emission

6.2 Outline of the Nature of Automobile Emission

Because of the gigantic growth in the automobile population, and the large volume of emission from each vehicle, serious air pollution problems have arisen in modern life. Michael P Walsh (Walsh 2002) reviews the pollution hazards in three dimensions, locally in cities, regionally and globally in USA.

Urban Habitation

Because of very high automotive concentration in cities, there are serious health effects from all types of emissions, *e.g.,*

Carbon Monoxide: Over 90% of CO emitted in cities comes from motor vehicles; Carbon monoxide readily combines with hemoglobin to form carboxyhemoglobin (COHb), which lowers the oxygen level in the blood damaging the brain and the heart most likely.

Oxides of Nitrogen: Nearly four-fifth of air is constituted of nitrogen, which gets oxidized during combustion in the auto-engines; NOx emission produces a variety of adverse health and environmental effects. It may directly cause bronchitis in children; and combining with other environmental pollutants, it may produce smog of peroxybenzolnitrate (PBnN), peroxzyacetylnitrate (PAN) which both cause eye irritation and respiratory problems.

Lead: because lead used to be added to petrol, motor vehicles have been responsible for emission of lead in the atmosphere of most cities. Lead may affect teeth and damage brain activity.

Diesel Particulates: Diesel may emit 30 to 70 times more particulates than gasoline petrol-fueled engine equipped with catalytic converter and burning unleaded fuel. WHO has concluded that particulates less than 10 microns are probable human carcinogen.

Regional Effects of Auto-emission

Regionally, the auto-emission of hydrocarbons and nitrogen oxides affect the atmospheric environment leading to the generation of two major pollutions, Troposphere ozone and acidification of the atmosphere:

Troposphere ozone: Due to photolytic action of NOx and HC, ozone is generated in the troposphere. Ozone would cause eye irritation, chest congestion, headache, asthma and other pulmonary problems, etc.

Acidification: Due to the oxidation of NOx and SOx into such acids as Nitric acid (HNO_3) and Sulfuric acid (H_2SO_4), the atmosphere gets acidic and also results in acid rain. The sustained acidification affects and deteriorates civil structure, structural steel structure, plantation and aqua life.

Global Effects of Auto-emission

The simultaneity and cumulative adverse effects of auto-emissions from the industrialized countries, especially USA and European Union leads to worldwide deterioration of the environment in many aspects, some of which are mentioned below:

Stratospheric Ozone Depletion

While locally and regionally, hydrocarbon and NOx help to generate unwelcome and harmful ozone in the lower atmosphere of Troposphere, ozone layer in the upper atmosphere is essential to absorb and prevent the ultraviolet rays from reaching the earth. But the huge CFC released by the automobile air-conditioning and the CFC aerosol used in the auto-upholstery did destroy the stratospheric ozone layer. Fortunately, under the Montreal Protocol, 1992, all countries were required to completely phase out use of CFC by the turn of the century.

Global Warming

Global warming is considered a major problem of climatic change caused by the presence carbon dioxide, methane and nitric oxide (NO) etc, which constitute the major auto-emission.

Virtually, the entire global motor vehicle fleet runs on fossil fuel, primarily oil. For every gallon of oil consumed by a motor vehicle, about 8.6 kg of carbon dioxide (containing 2.4 kg of carbon) go directly to the atmosphere. Globally, motor vehicles account for 33 per cent of world oil consumption and 14 per cent of world's carbon dioxide emission.

Motor Vehicle Pollution: In USA, the USEPA estimates the motor vehicle causes emissions as given in Table T6.1.

In OECD countries of Europe, motor vehicles are source of following pollutions:

CO ... ... 66 %
NOx ... ... 47 %
HC ... ... 39 %

6.3 Studies of Internal Combustion Engine vis-à-vis Auto-emission

Basis of Working of an Auto-engine

An automobile engine burns fossil fuels to generate power to drive the wheels. There are basically two types of internal combustion engines:

- Petrol 4-stroke I.C. engine where the combustion of atomic fuel in air is initiated by spark ignition. The fuel is petrol or gasoline, a lighter fraction of petroleum.
- Diesel 4-stroke I.C. engine, where the injected fuel in air is ignited by the heat of compression of the air. The compression ratio in diesel engine is much higher than in the Petrol engine. This type of engine uses lower fraction of petroleum popularly known as light diesel.

All auto-engine has four operating strokes of the piston in the cylinder as detailed in Box B6.1.(See also Fig.F6.2)

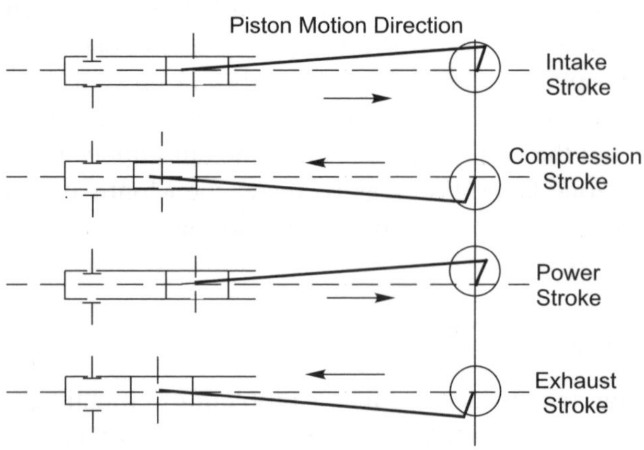

Fig. F 6.2: Piston Strokes of an IC Petrol Engine

Air-Fuel Mixture Influencing Emission

The constituents- Hydrocarbons of an auto-fuel (petrol/gasoline) are shown in Table T6.3.

For complete combustion of the fuel *i.e.,* converting carbon into carbon dioxide, the air supply has to be adequate. From the Table T6.3, we choose Heptane as the average representative hydrocarbon, so that the required air supply will be given by the following equation:

$$C_7H_{16} + 11\ O_2 = 7\ CO_2 + 8\ H_2O \quad \ldots \quad \ldots \quad E6.1$$

Table T6.3 Auto-fuel Hydrocarbons

Constituents	Boiling Points
Undecane ($C_{11}H_{24}$)	400
Decane ($C_{10}H_{22}$)	-
Nonane (C_9H_{20})	300
Octane (C_8H_{18})	-
Heptane (C_7H_{16})	200
Hexane (C_6H_{14})	-
Pentane (C_5H_{12})	100

Source: Masters1991

From the Box B6.2A, we find that that the theoretical air/fuel mixture sotichiometric ratio of 15.1. But it has been found experimentally that the air/fuel ratio is 14.5 approximately. Such stoichiometric ratio can only be obtained from an imaginary hydrocarbon C_7H_{13} without consideration of oxidation of Nitrogen (see Box B6.2B).

If the supply of air is below the stoichiometric level, the air-fuel mixture is said to be 'rich'. Because of insufficiency of oxygen in air, the combustion is likely to be incomplete, emitting CO and unburnt HC. On the other hand, if there is oversupply of air beyond stochastic level, there will be complete combustion, the mixture is so 'lean' that there is no firing *i.e.,* no initiation of combustion.

In Fig. F6.3 are shown the emission curves of CO, HC and NOx, besides curves of fuel consumption and power developed. Just below the stoichiometric level of air supply, that is when air-fuel mixture is somewhat rich, the power developed is the highest; but there are also increases of emission of CO and HC.; fuel consumption is also somewhat higher. When however, we have slightly lean mixture of fuel-air, there is fuel economy, understandable lowering of CO and HC emissions. But because of higher temperature reached, there is oxidation of Nitrogen and more formation of NOx.

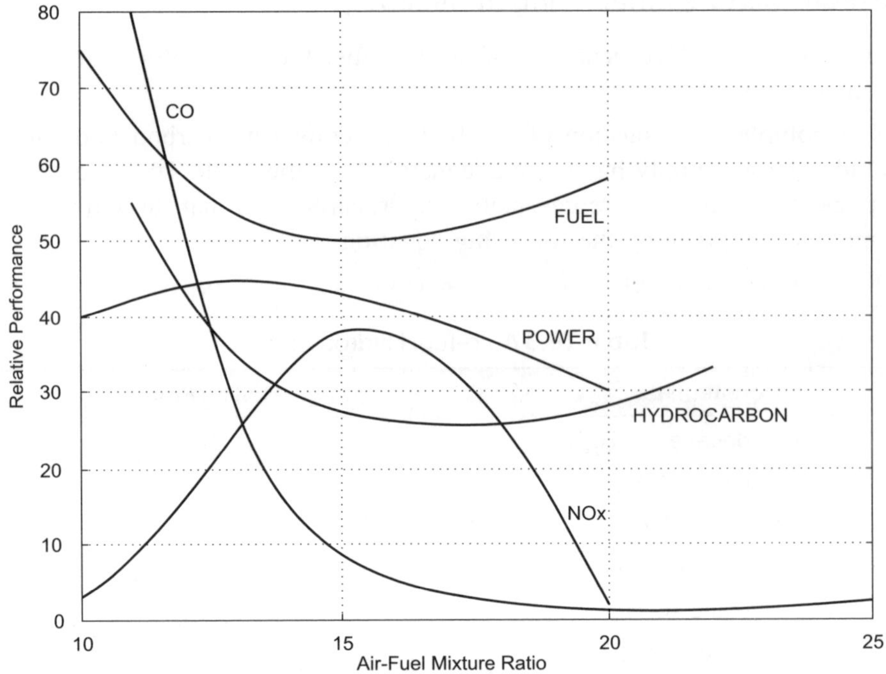

Fig. F6.3: Emission Characteristics vs. Air-Fuel Mixture

Box B6.1

Working of Internal Combustion Engine

An auto-engine has four operating strokes of the piston in the cylinder (see Fig. F6.2):

(*i*) **Intake Stroke:** Piston lowers down drawing in right mixture of fuel and air through the inlet port;

(*ii*) **Compression Stroke:** Upon completing its downward travel, the piston starts moving up compressing the fuel-air mixture as the inlet had in the mean time closed up;

(*iii*) **Power Stroke:** Towards the end of the Compression stroke, the spark plug passes a high voltage electric spark initiating the combustion of the air-fuel mixture, which expands and forces down the piston powering the flywheel to rotate. In diesel engine, there will be no sparking but the higher compression causes the mixture to heat up and auto-ignition takes place;

(*iv*) **Exhaust Stroke:** Towards the end of the power stroke, the exhaust port opens up and the energized flywheel pushes the piston up, exhausting the burnt out gases through the exhaust port till the exhaust port closes and the intake stroke starts with the intake port opening up.

Box B6.2A

Air Fuel Mixture (Fuel Heptane)

$$C_7H_{16} + 11\ O_2 = 7\ CO_2 + 8\ H_2O$$

The air consists of nitrogen: oxygen in the ratio of 79:21 by volume. Hence for every mole of oxygen burnt, air will required to be supplied with 3.76 moles of nitrogen. The stoichiometric ratio of air : fuel works out to

$$(11x32 + 11x3.76x28) / (7x12 + 16) = 15.1$$

But it has been found experimentally that the air-fuel stoichiometric ratio is 14.5 approximately, which is obtained by a hypothetical hydrocarbon C7H13 only as shown in Box B6.2B.

Box B6.2B

Air-Fuel Mixture (Fuel C7H13)

4 C7H13 + 41 O2 + 154 N2 = 28 CO2 + $2CH_2O$ + 154 N2

The stoichiometric ratio (S.R.) works out to

41×2×16 + 154×2×14) / (4×7×12 + 4×13) = 14.49

The S.R. is rounded to value 14.5.

Other Engine Design and Operation Features Influencing Emission

Other than controlling air-fuel mixture minimizing emission of CO, HC, and NOx, there are other features, which affect emission characteristics.

Among the design features, mention may be made of combustion Chamber geometry, compression ratio, initiation time of combustion (firing of spark for petrol engine) etc.

Among the operational features, the important aspects are idling of engine, accelerating and decelerating frequencies and steady cruising time.

1. Ventilation of Engine to Control Hydrocarbon Emission

During the compression and the power strokes, the pressure above the piston is high and some of the mixture, called 'Blow by' escapes through the space between the cylinder and the piston rings to the crankcase. Generally, the crankcase ventilates to the atmosphere, vitiating it with unburnt hydrocarbons. As a control measure, the crankcase vapour is pumped back into the intake manifold for admission back to the cylinder through a PCV (positive crankcase ventilation) valve.

Similarly, the fuel tank continuously evaporates hydrocarbons, which may escape to the atmosphere causing pollution. Here again, the fuel tank may also be ventilated to the engine intake filter. In the US, when the fuel tank is filled with gasoline, fuel station has a system of drawing out the HC vapour back into the Station Fuel Storage tank.

Carburetion Chamber is another source of hydrocarbon evaporation and pollution. Emission from the float chamber is particularly high jut after the engine turns off. Emission during this period is called hot soak emission. This emission is prevented from escaping the atmosphere by provision of venting to carbon canister. As the engine turns off, the activated carbon absorbs the hot soak vapour. On starting the engine, the intake air is filtered through the canister, regenerating the activated carbon.

Another way to prevent carburetion pollution is to replace the carburetor with a fuel injection system, used mostly in Diesel engine.

2. Catalytic Control of Exhaust Gas Pollution

In the Internal Combustion engine, the combustion takes place very fast and some unburnt HC is generally carried with the exhaust gas to the exhaust tailpipe. This is controlled in two ways –by use of Thermal Reactor or more commonly by Catalytic Converter.

The thermal reactor is basically an afterburner. Upon the exhaust manifold, a pump vents the gas to the afterburner unit, which burns out the residual hydrocarbons to CO_2 and water at about 1100 Celsius.

The three-way catalytic converter, the second method, is manufactured of platinum and rhodium, and is a throwaway converter oxidizing CO, NOx and HC per chemical reaction below:

$$HC + CO + NOx = H_2O + CO_2 + N_2 \ldots\ldots \qquad \ldots\ldots\ldots E6.2$$

The three-way catalytic converter is also the favoured method used by the American Automobile Manufacturers to achieve the emission standards of the Clean Air Act. The other technological advantage is that the three-way converter works best at about the stochiometric ratio of air-fuel mixture. Thus, effective emission control and fuel economy are achieved simultaneously with the three-converter system.

Fig. F6.3 shows the optimum operating region of the three-way converter, is in the region between the schiometric ratio of 14.8 and 14.9.

Fig. F6.4 shows the schematic working principle of the three-way converter.

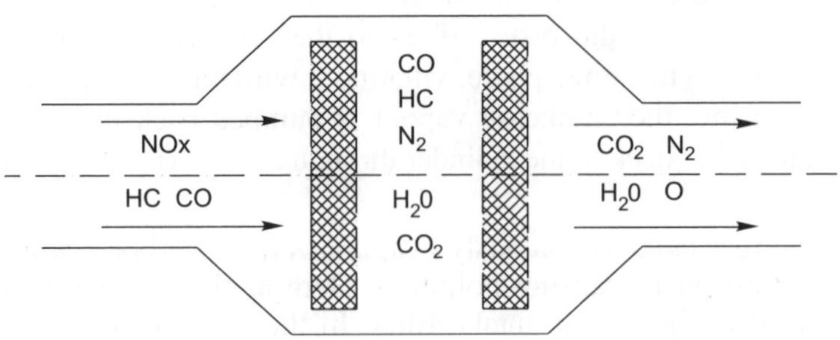

Fig. F6.4: Three Way Catalyst

3. Control of NOx – further consideration

The nitrogen in the air gets oxidized into NOx, when the temperature of combustion during the power stroke is high. The combustion temperature can be lowered by recirculating exhaust gas as part of the intake air. The exhaust gas will not burn like the rest of the intake air-fuel mixture. Thus the average temperature rise during combustion will come down.

6.4 Control of Automobile Emission in USA

In 1959, California adopted legislation requiring installation of pollution control devices in automobile as soon as three independent workable control devices were developed. In 1964, California was able to certify that three independent workable add-on devices had been developed by three manufacturers. This was followed by the legal requirement that beginning from 1966, all new car models are required to be fitted with emission control devices. Shortly, the large car manufacturers announced that they have developed their own technologies to reduce emission, which they will use in their own new models.

In view of California's pioneering work and the national nature of the auto-pollution, the US Congress enacted in 1965 the Clean Air Act Amendments, which required that from 1968 onwards the Californian Auto-pollution Standards of 1966 would become nationally applicable.

In December 1979, the US Congress further amended the Clean Air Act "to protect and enhance the quality of the national air resources", taking in particular view the automobile pollution. The new Amendment required 90% reduction in the emission standards of carbon monoxide from 1971 onwards.

In the light of oil embargo of 1973 and the apprehension that the auto-emission standards may impact fuel economy adversely, a new act 'Federal Energy Supply and Coordination Act 1974' postponed the emission standards till 1978-79. The Amendment Act 1977 further postponed compliance to 1981.

US Auto-emission as they existed in 1989 are briefly reported in Table T6.4

Table T6.4 : US Auto-Emission Standards (gm/mile)

Emission Type	Automobile	
	Federal	California
Hydrocarbon (Tail-Pipe)	0.41	0.41
Carbon Monoxide	3.40	3.40
Nitrogen Oxide	1.00	0.40
Evaporative Hydrocarbon	26 g/test	
Particulates	0.2 g/test	

Present US Auto-Emission Standards

Two Sets of standards have been adopted for the light-duty vehicles in the clean Air Amendmts Act (CAAA) of 1990.

- Tierl standards, which were published as a final rule in June 1991 and phased progressively between 1994 are 1997.
- Tier 2 standards. which were adopted in December 1999 with a phased in implementation schedule for 2004-2009.

Tier I Standards

These applied to all light-duty vehicles (LVD) of less then 8500 all (3864 kg) GVWR (Gross vehicle weight rating) including.

- Passenger cars, Suv (spots utilily vehicles), minibus our pickup;
- Light light-duty trucks (LLDT), below 6000lb GVWR;
- Heavy Light-duty trucks (HLDT), abouve 6000lb GVWR;

The emission standards under FTP (Federal Test Procdure are given in Table T6.5A.

Tiev 2 Standards

The Tier 2 standards apply to all vehicle types covered by Tierl and additionally to MDPV (Medium Duty Passenger vehicles) between 8500 and 10,000 lbs. GVWR and are used for personal transportation.

The Tier2 regulations introduced more stringent numerical emission limits relative to previous Tierl.

The same emission limits, apply to all vehicles irrespective of fuelss used, gasoline, diesel or alternative fuels.

The Tier2 regulations brought new requirements for fuel quality, also needed for advanced emission devices. (e.g. catalysts and particulate filters).

- Sulphur level in gasoline should meet corporate average of 120 ppm with a cap of 300 ppm beginning in 2004 and an average of 80 ppm with a cap of 80 ppm in 2006.
- Sulphur in diesel is to reduce to 15 ppm in 2007-2010 as a part of emission regulation for heavy-duty engines.

Table T6.5A : EPA Tier 1 Emission standards under FTP 75
Limits: gm/miles (Implementation by 1997)

Category	50,000 miles or 5 years						10,000 miles or 10 years					
	HC	NMHC	CO	NOx diesl.	Nox gasol.	PM	THC	NMHC	CO	NOx	NOx diesl.	PM gasol.
Passenger car	0.41	0.25	3.4	1.0	0.4	0.08	–	0.31	4.2	1.25	0.6	0.10
LLDT (<3750lb)	–	0.25	3.4	1.0	0.4	0.08	0.8	0.31	4.2	1.25	0.6	0.10
LLDT (>3750lb)	–	0.32	4.4	–	0.7	0.08	0.8	0.40	5.5	0.97	0.97	0.10
HLDT (<5750lb)	0.32	–	4.4	–	0.7	–	0.8	0.46	6.4	0.98	0.98	0.10
HLDT (>57750lb)	0.39	–	5.0	–	1.1	–	0.8	0.56	7.3	1.53	1.53	0.12

HC: Hydro-carbon; NMHC : Non-Methone Hydrocarbon
THC: Total Hydrocarbon;

Table T6.5B : EPA Tier 2 Emission standards (FTP 75)
(Implementation 2004-2009)
Limits: gm/miles

Bin No.	Intermediate life (5 years or 50 miles)					Full useful life				
	NMOG	CO	NOx	PM	HCHO	NMOG	CO	NOx	PM	HCHO
Temporary Bin	–	–	–	–	–	0.280	7.3	0.9	0.12	0.032
11										
10	0.125	3.4	0.4	–	0.015	0.156	4.2	0.6	0.08	0.018
	(0.160)	(4.4)			(0.018)	(0.230)	(5.4)			(0.027)
9	0.075	3.4	0.2	–	0.015	0.090	4.2	0.3	0.060	0.018
	(0.140)					(0.180)				
Permanement	0.100	3.4	0.14	–	0.015	0.125	4.2	0.2	0.02	0.018
bin										
8	(0.125)					0.156)				
7	0.075	3.4	0.08	–	0.015	0.090	4.2	0.15	0.02	0.018
6	0.075	3.4	0.08	–	0.015	0.090	4.2	0.10	0.01	0.018
5	0.075	3.4	0.05	–	0.015	0.090	4.2	0.07	0.01	0.018
4	–	–	–	–	–	0.070	2.1	0.04	0.01	0.011
3	–	–	–	–	–	0.055	2.1	0.03	0.01	0.011
1	–	–	–	–	–	0	0	0.0	0	0

NMOG : Non-Methane Organic Gas

6.5 Control of Automobile Emission in Western Europe

Light Duty Vehicles

In December 1997, the Environment Ministers of Europe adopted the Particulate emission standard for light commercial vehicles (LDV). At the instance of Germany, which have a larger number of Diesel Vehicles, this standard was upgraded for diesel vehicles by the end of 1989.

There followed two other upgradation of particulate emission standards in February 1990 and a second upgradation in October 1990 at the Council of Ministers' meeting following critical comments in the European parliament.

But at the European Council of Environment Ministers in December 1990, there were finalized automobile emission standards of HC, NOx, CO, and Particulates.

Currently, emissions of Nitrogen Oxide (Nox), Total Hydrocarbon (THC), Non-Methane Hydrocarbons (NMHC), Carbon Monoxide (CO), and Particulate matter (PM) are prescribed for light commerical vehicles up to Euro 6 standards for implementation over period till 2015 and Tables T6.6A to T6.6C:

Table T6.6A : European Emission Standards for Passenger Cars (GVW < 2500 Kg)
Units : gm/km

Standards	Year	Diesel						Gasoline (Petrol)					
		CO	THC	NMHC	NOx	HC+NOx	PM	CO	THC	NMHC	NOx	HC+NOx	PM
Euro 3	2000	0.64	–	–	0.50	0.56	0.05	23	0.20	–	0.15	–	–
Euro 4	2005	0.50	–	–	0.25	0.30	0.025	1.0	0.10	–	0.08	–	–
Euro 5	2009	0.50	–	–	0.18	0.23	0.025	1.0	0.10	0.06	0.06	–	0.005
Euro 6	2014	0.50	–	–	0.08	0.17	0.005	1.0	0.10	0.068	0.06	–	0.005

Table T6.6B : European Emission Standards for Light Commercial Vehicles (GVW 1309-1760 kg) & Passenger Cars (GVW >72500 kg)
Units : g/km

Standards	Year	Diesel						Gasoline (Petrol)					
		CO	THC	NMHC	NOx	HC+NOx	PM	CO	THC	NMHC	NOx	HC+NOx	PM
Euro 3	2001	1.25	–	–	–	1.0	0.12	4.17	0.25	–	0.18	–	–
Euro 4	2006	0.80	–	–	0.65	0.72	0.07	1.81	0.13	–	0.10	–	–
Euro 5	2010	0.63	–	–	0.33	0.39	0.04	1.81	0.13	0.09	0.075	–	0.005
Euro 6	2015	0.63	–	–	0.105	0.195	0.005	1.81	0.13	0.09	0.075	–	0.005

Table T6.6C : European Emission Standards for Light Commercial Vehicles (GVW 1760–3500 kg)
Units : g/km

Standards	Year	Diesel						Gasoline (Petrol)					
		CO	THC	NMHC	NOx	HC+NOx	PM	CO	THC	NMHC	NOx	HC+NOx	PM
Euro 3	2001	0.95	–	–	0.78	0.86	0.10	5.22	0.29	–	0.21	1.7	–
Euro 4	2006	0.74	–	–	0.39	0.46	0.06	2.27	0.16	–	0.11	0.7	–
Euro 5	2010	0.74	–	–	0.28	0.35	0.05	2.27	0.16	0.108	0.082	–	0.005
Euro 6	2015	0.74	–	–	0.125	0.215	0.005	2.27	0.16	0.108	0.08	–	0.005

Heavy Duty Vehicles

After considerable deliberation in light of the consolidated emission standards of Light Duty Vehicles, the European Commission agreed for the following EC norms for the Heavy Duty Diesel Trucks, shown briefly in Table T6.7:

Table T6.7: European Emission Standards for Large Diesel Vehicle (Heavy Duty) Units: g/kwh

Standards	Year	CO	NOx	HC	PM
Euro 2	1995-99	4.0	7.0	1.1	0.15
Euro 3	1999-2005	2.1	5.0	0.66	0.10
Euro 4	2005-2008	1.5	3.5	0.46	0.02
Euro 5	2008-2012	1.5	2.0	0.46	0.02

6.6 Indian Auto-Emission Standards

The Government of India constituted a Committee in August 2001 to study the Auto-Fuel Policy for the country. The Committee made a broad review of the suitable auto-fuels, their availability and security of supply, vehicle technology and emission reduction in a cost-effective manner, fiscal measures and institutional framework for bringing about progressive improvements by reducing vehicular emission on ambient air and laying down a road map for implementation.

The Committee recommendations have been made with two broad objectives being given due importance.

(*i*) They should make a decisive impact on air quality;

(*ii*) They should not place an undue burden on the consumer.

The Committee recommendations included, inter alia, following important issues:

1. Issues regarding emission standards, vehicular technology, and fuel quality to be considered and implemented by various agencies of the Central and State Governments.
2. The creation of a National Automobile Pollution and Fuel Authority (NAPFA), which should be duly empowered by an Act of Parliament for research, policy formulation and for implementation guidelines.
3. Periodic revision of Auto-fuel policy, preferably every five years.
4. The main fuel for which the regulations will be prescribed in the first instance are Petrol, Diesel, natural Gas, and LPG.
5. Developing models for Impact Assessment.
6. The following grades of Auto-Emission Standards with their graduated dates of implementations were recommended.

On 6 October, 2003, the National Emission Standards were announced. which envisaged a phased implementation of four standards I to IV (equivalent to Euro 1 to Euro 4) over a decads from retrospective effect from 2000 to 2010 as set out in Tables T6.8 to 6.11.

Table T6.8: Implementation Schedule of New Indian Auto Emission Standards (4 wheel vehicles)

Standards	Reference	Date	Region
Bharat I	Euro1	2000	Nation worde
Bharat II	Euro 2	2001	Delhi, Mumbai, Kolkata, Chennai
		2003	Delhi capital Region & Ten other ...
		2005	Nationworde
Bharat III	Euro 3	2005	Delhi & Ten other cities
		2010	Nationworde
Bharat IV	Euro 4	2010	Delhi & Ten Other cities

Table T6.9: Indian Emission Standards for Heavy Duty Diseal Truck & Bus GVW ≥ 3500 kg. Units : g/kwh

Year	Standards	Test	CO	HC	NOx	PM
2000	Bhart I	ECE	4.5	1.1	8.0	0.36
2005	Bhart II	ECE	4.0	1.1	7.0	0.15
2010	Bhart III	ESC	2.1	0.66	5.0	0.10
		ECT	5.45	0.78	5.0	0.16
2010	Bhart IV	ESC	1.50	0.46	3.5	0.02
		ETC	4.0	0.55	3.5	0.03

Table T6.10: Indian Emission Standards for Light-Duty Diesel Vehicle (GVW < 3500 kg) Units : g/kwh

Year	Standards	CO	HC	HC+NOx	NOx	PM
2000	Bharat I	2.72-6.90	–	0.97-1.70	0.14-0.25	–
2005	Bharat II	1.0-1.5	–	0.7-1.20	0.08-0.17	–
2005/10	Bharat III	0.64	–	0.56	0.50	
		0.80	–	0.72	0.65	0.07
		0.95	–	0.86	0.78	0.10
2010	Bharat IV	0.50	–	0.30	0.15	0.025
		0.63	–	0.39	0.33	0.040
		0.74	–	0.46	0.39	0.060

Note: Lowest Values apply to Passenger cars (GVW≤ 2500 kg/6 Passengers)

Table T6.11A: Indian Emission Standards for Low-powered Diesel Vehicle (speed <90 Km/h). Units : g/kwh

Year	Standards	CO	HC	NOx	PM
2000	Bharat I	4.5	1.1	8.0	0.36
2005	Bharat II	4.0	1.1	7.0	0.15

Table T6.11B: Indian Emission Standards for Gasline vehicles (GVW ≤ 3500kg) Units : g/kwh

Year	Standards	CO	HC	HC+NOx	NOx
2000	Bharat I	2.72-6.90	–	0.97-1.70	–
2005	Bharat II	2.2-5.0	–	0.50-0.70	–
2010	Bharat III	2.3	0.20	–	0.15
		4.17	0.25	–	0.18
		5.22	0.29	–	0.26
2010	Bharat IV	1.0	0.10	–	0.08
		1.81	0.13	–	0.10
		2.27	0.16	–	0.11

Note : Lowest values apply to Passenger values (GVW < 2500 kg; 6 Passenger)

Table T6.11C: Indian Emission Standards for 3-GVW ≥ 3500 kg. Units : g/kwh

Year	Standards	CO	HC	HC+NOx
2000	Bharat I	4.0	–	2.0
2005	Bharat II	2.25	–	2.0
2010	Bharat III	1.25	–	1.25

6.7 Alternative Auto Fuels

Globally, there are several sources of primary energy as shown by the Pie Diagram in Fig F6.5, of which Petroleum oil constitutes the largest share at 35.3%. Petrol and Diesel derived from Petroleum constitutes the primary sources of auto-fuels today. Petroleum is a fossil fuel, which are geographically located mainly in few countries in the Middle East Asia. In view of fast dwindling reserve of petro-oil and the concern of energy security, countries are working seriously various alternative auto-fuels including the developmental efforts of automobile manufacturers to produce flexi-fuel cars and hybrid cars.

Fig. F6.6 shows a matrix of potential auto-fuels, their feedstock sources and conversion processes to auto-fuels. As on today, promising auto-fuels can be listed as under:

Natural Gas (CNG), Propane (LPG);

Ethanol, Methanol;

Bio-diesel;

Hydrocarbons;

Electricity

Table T6.12 gives an outline of the main Emission pollutants of LPG and CNG. We will discuss now the properties and technical and commercial feasibility of these potential auto-fuels.

1. Natural Gas

Natural Gas is a mixture of various hydrocarbons, mainly methane (CH_4), but also contains ethane (C_2H_6), propane (C_3H_8), and other gases such as nitrogen, helium, carbon dioxide, hydrogen sulphide and water vapour. It is produced either from gas oil or in conjunction with crude oil production. Natural gas is consumed in the residential, commercial, industrial and utility applications.

The interest in natural gas as an alternative fuel comes from its clean burning properties, with extremely low sulphur and particulate components. In 2002 in the US market, it was priced at USD 1.5/mmBtu with the spot price at USD 2.1-3.5/mmBtu. Another reason for its preference is that natural gas is more evenly distributed worldwide than Petroleum.

2. Propane/LPG

Liquified petroleum gas consists mainly of propane gas with other hydrocarbons propylene, butane and butylene in various mixtures. LPG is a byproduct from two sources: from natural gas processing and crude oil refining. The propane market is spread worldwide and approximately 1.3 billion barrels of propane is produced worldwide.

Table T6.12: Comparative Emission rates of Alternative Fuels.(gm/kwh)

Emissions	Gasoline	LPG	CNG
Carbon	1.12	0.91	0,45
Hydrocarbon	0.15	0.12	0.36
Nitrogen Oxide	0.15	0.21	0.13
Particulates	0.015	0.005	0.025

3. Ethanol (C_2H_5OH)

Ethanol is produced by fermentation from agricultural products such as sugarcane, corn, and molasses. Ethanol is produced mostly from sugarcane in Brazil and from corn in US.

Ethanol is used as a high octane fuel in vehicles. More than 4 million cars run on hydrated ethanol in Brazil. All gasoline is blended with anhydrous ethanol (20-25 % ethanol) as a Government programme since 1970. In US, similar programme is being started under influence of corn-growing States.

Brazilian sugarcane ethanol is cheaper than US corn ethanol. Ethanol makes an excellent motor fuel. It has high octane number than petrol. Its vapour pressure is lower, which results in lower evaporative pressure. Ethanol's flammability is lower, hence it has cold starting problem. This problem is overcome in E85 mixture containing 85 % ethanol and 15 % petrol.

4. Methanol (CH_3OH):

Methanol (Methyl Alcohol) was being used as an alternative fuel in vehicle as M85 (a blend of 85 % methanol+15 % gasoline). It has since been discontinued as it causes health hazard.

Methanol is produced by steam reformation of natural gas to crate a synthesis gas of CO and hydrogen, which is fed into a reactor vessel in the presence of a catalyst to produce methanol and water.

World demand of methanol is 32 million tons per year, one quarter of which is produced in USA. MTBE (Methyl-Tart-Butyl-Ether) is derived from Methanol in USA.

In 2002, methanol cost 64 cents/gallon in US, 61.7 cents/gallon in Europe and 55.9 cents in Asia. There is overproduction of methanol in the world.

Table T6.13 summarizes the basic operational properties of Ethanol and Methanol vis-à-vis Gasoline.

Table T6.13: Comparative Properties of Ethanol, methanol and Gasoline.

Properties	Methanol	Ethanol	Gasoline
Chemical composition	CH_3OH	C_2H_5OH	C_4-C_{12} Mixture
Carbon: Hydrogen Ratio (by weight)	3.0	4.0	5.33 (for C_8H_{18})
Molecular weight	32	46	114-224
Carbon % (by weight)	37.5	52	84
Oxygen % (by weight)	49.9	34.7	0
Density at 20 (Celsius)	792	789.5	702
Boiling Pointing (Celsius)	65	78.3	127.3
Freezing Point (Celsius)	-94.9	-117.6	-57.3
Calorific Value (MJ/kg)		21.2 - 23.4	30.1 – 34.9
Latent heat of Vaporization (KJ/kg)	1101.1	841.5	297.3
Latent heat of fusion (KJ/kg)	100.5	104.7	180
Critical temperature (C)	240.2	243.3	296.2
Critical Pressure (bar)	97.1	61.8	24.2
Critical density (kg/m3)	358	280	274
Hydrogen % (by weight)	12.6	13.2	16

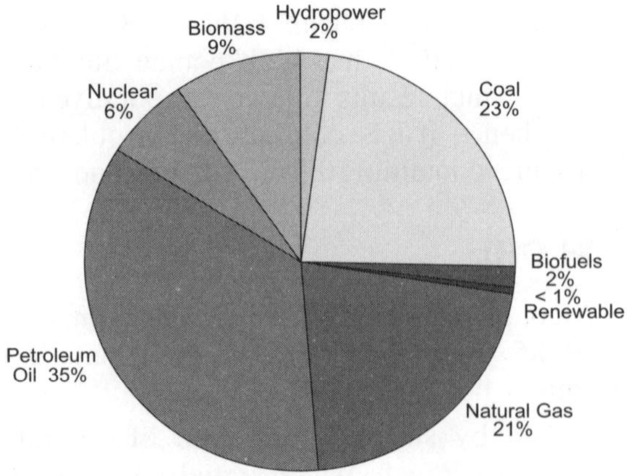

Fig. F. 6.5 : Present Global Primary Energy Consumption

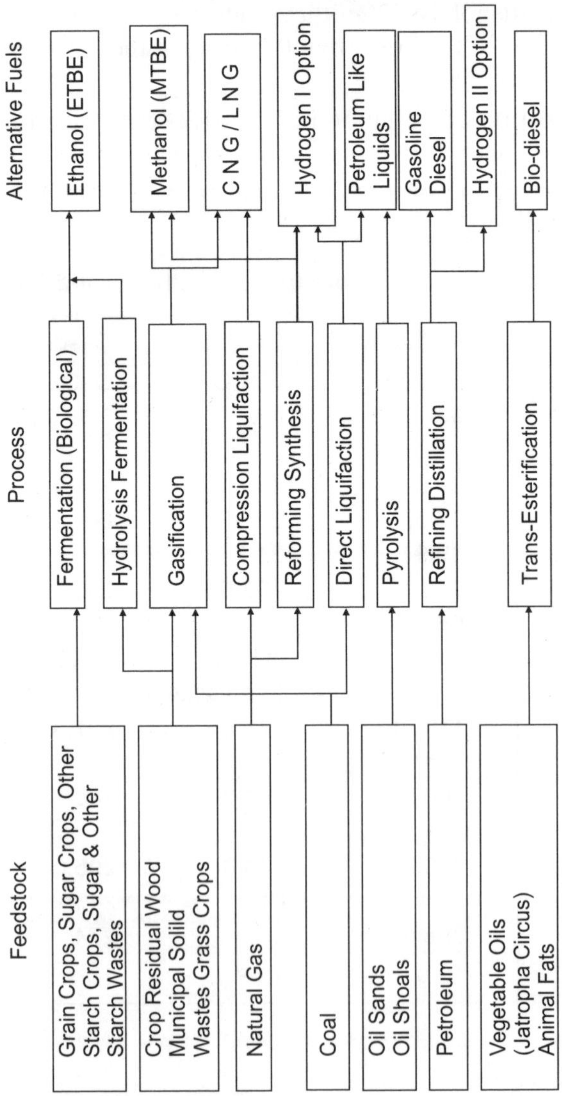

Fig. F6.6: Chart Of Alternative Auto-fuels, Their Source Feedstocks & Conversion Processes.

5. Bio-diesel

Bio-diesel is a mono-alkyl ester of vegetable oil or fat CH3-O-CO-R, where R represents a fatty acid chain.

In many developed countries, edible oil is used to produce bio-diesel, like corn oil in USA, soybean oil in Germany and even in Malaysia. But in India, edible oil is in shortage because of very high domestic consumption. It has been experimentally verified in India that bio-diesel can also be processed from non-edible oils from the following plants:

Madhuca Indica (Mahua) Pongamia Pinata (Karanja)

Azadirachta Indica (Neem) Jatropha Curcus (Ratanjot/Veranda)

Of the above, Jatropha has the highest potentiality as it grows in abundance in dry fallow land and does not require much care. This oil source shall be cheap and hence the bio-diesel cost comes quite economic.

The vegetable oil is actually a glyceride of the respective fatty acid. The glyceride is alcoholized at about 60 to 70 C in presence of catalysts KOH or NaOH to convert it into bio-diesel i.e. an monoalkyl ester.

The comparative properties of jatropha oil, its bio-diesel and the petro-diesel are presented in Table T6.14. Bio-diesel has somewhat higher viscosity compared to petro-diesel, giving rise to engine starting problem in cold weather. Similarly, its calorific value is somewhat lower than that of petro-diesel. These marginal shortcomings can be made good by mixing 20 % bio-diesel with fossil diesel to B20. No engine modification is required for B20 fuel.

But the advantages of bio-diesel are manifold and quite conspicuous. Firstly, it is a renewable fuel unlike fossil fuel. Secondly, it will allow higher energy security to many countries, who have no national sources of fossil fuel. Thirdly, its pollution level is lower than fossil fuel. Table T6.15 shows that it generates much less emission of HC, CO, particulates. Though it generates 10 % more NOx, which can be compensated by mixing with diesel as in B20.

6. Hydrogen

The most important use of hydrogen as an alternative fuel is in the electric-drive vehicles in the future. Fuel cells produce electricity similarly to battery. Like a battery, it converts chemical energy produced from a chemical reaction directly into useable electric power. However, unlike a battery, a fuel cell needs an external fuel supply, hydrogen gas and generates electricity as long as fuel is supplied and unlike a battery it does not need recharging. Since, there is no combustion, there is no pollution. Moreover, it is silent, fuel efficient and dependable.

The problem is with the production and storage of hydrogen gas. The common method of making hydrogen gas is by reform reaction of primarily natural gas but also of ethanol, methanol .The reform reaction involves chemical reaction that generates hydrogen after separation from CO. Generally, hydrogen is provided in a vehicle from a high-pressure hydrogen tank. Hydrogen can also be produced from gasoline or diesel by reform reaction but in that case, it will not remain an alternative fuel.

Today the capital cost of fuel cells is higher than other power sources, costing USD 1000-5000 per kilowatt. Hydrogen may cost 17-55 US cent per 100 cubic feet.

Table T6.14: Comparative Properties of Jatropha-sourced Bio-diesel & Petroleum Diesel

Fuel Properties	Jatropha Oil	Bio-diesel (Jatropha)	Euro-Standards (Bio-diesel)	Petroleum Diesel
Density at 15 C (kg/m3)	920	880	860-900	820
Pour Point (C)	5	4	-	-23
Viscosity at 20 C (m-Pa-S)	75.6	4.8	3.5-5	2.6-4.1
Calorific value (MJ/kg)	37	39	-	42
Flash Point (C)	340	191	>120	74
Acid Value (mg KOH/gm)	8	0.44	<0.5	0.35
Carbon Residue %	4.2	0.025	<0.3	0.17
Ash Content %	1.1	0.014	<0.02	0.17

Table T2.20: Emission Improvement (Percentages) with Bio-diesel

Engine Test	HC	CO	PM	NOx	CO_2
Cummins N-14	-95.6	45.3	-28.3	+13.1	-78
DDC S-50	-83.3	-38.3	-49.0	-11.3	
Cummins B5.9	-74.2	-38.0	-36.7	+4.3	

❑❑❑

7

CHAPTER

Depletion Of Ozone Layer

7.1 Preamble

Ozone, as such, is toxic and does cause chest congestion and irritation in the mucous membrane. Since it exists in traces in the nearest atmosphere (Troposphere), it is generally not harmful. EPA of USA and Indian authorities prescribe upper limit of ozone concentration at about 0.1 ppm. The general concentration of ozone in the Troposphere and Stratosphere is given in Fig. F7.1 based in USEPA data (1986).

Ozone is continuously generated in the stratosphere in the presence of heat of absorption of U-V radiation and again gets dissociated due to photochemical reactions with other pollutants we will describe.

7.2 Stratospheric Ozone Shield and Protection against UV Radiation

The greatest beneficial property of ozone is that it absorbs shortwave length UV radiation up to 320 nm, emanating from the sun. Fig. F7.2 shows how the ozone shield prevents solar radiation of microwave ultraviolet B range from reaching the earth's surface. This radiation is carcinogenic affecting animal skin and can even damage the DNA molecule.

The ozone shield is really a very diluted layer of ozone, in the concentration of 0.01 to 0.001 ppb, extending some 10 to 25 kilometres above the earth. Any destruction of this shield can cause skin cancer but can also endanger human food supplies by retarding the growth of the basic unicellular form of life called phytoplankton. Phytoplankton, smallest form of life in the sea is created from the photosynthesis of CO_2, water and solar energy. Ultraviolet radiation can stall or retard the growth of phytoplankton on which depends the sustenance of the entire marine livestock and food chain.

7.3 Process of Destruction of Ozone layer

Ozone gets destroyed and converted into molecular oxygen by catalytic reaction with Chlorine, Bromine, and Nitrogen oxide. The following reactions may be envisaged:

$$O2 + hv \ = 2O' ; \qquad hv \text{ is the solar energy.}$$

$$X + O_3 \ = XO + O_2;$$

$$XO + O' \ = X + O_2;$$

Net reactions: $X + O' \ + O_3 = O_2 + O_2 + X,$ E7.1

where X is any catalytic gaseous atom Cl, Br, N or OH radical.

Since X acts as a catalytic agent, it gets free at the end of each cycle of conversion of O3 molecule to O2 molecule. One molecule of X can go on destroying thousands of O3 leading to the depletion of ozone later in the stratosphere. This is a very dangerous proposition for the human race, rather of the whole animal kingdom.

The main source of chlorine is from the CFC (Chlorofluorocarbon) class of gas that escapes to the atmosphere; the CFC gas then releases chlorine atom in presence of solar heat energy in the stratosphere. A typical reaction of CFC-12 (CCl2F2) is

CCl2F2 + hv = Cl + CClF2 ... ... E7.2

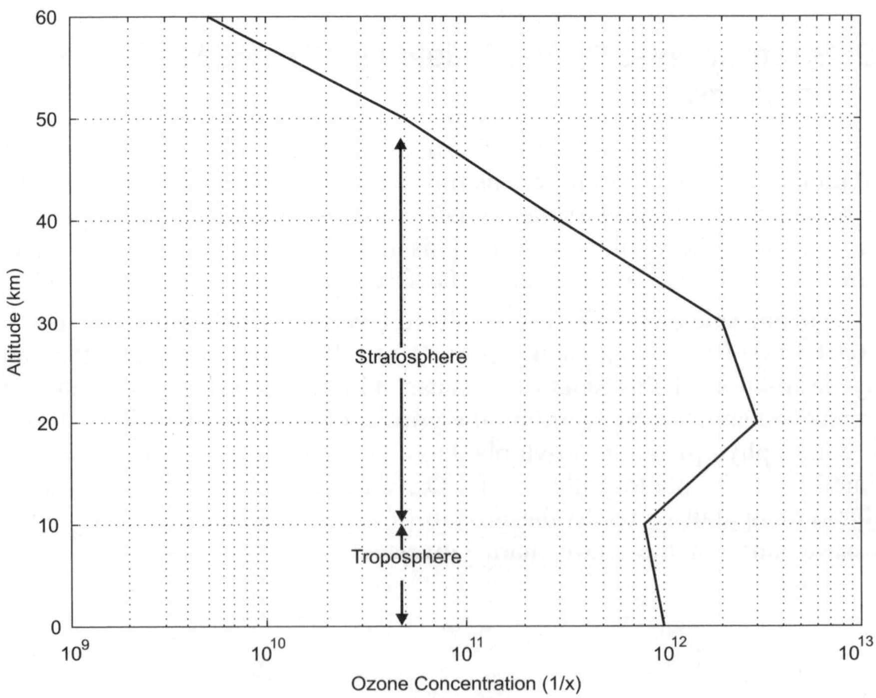

FIG. F7.1: Concentration of Ozone in Atmosphere

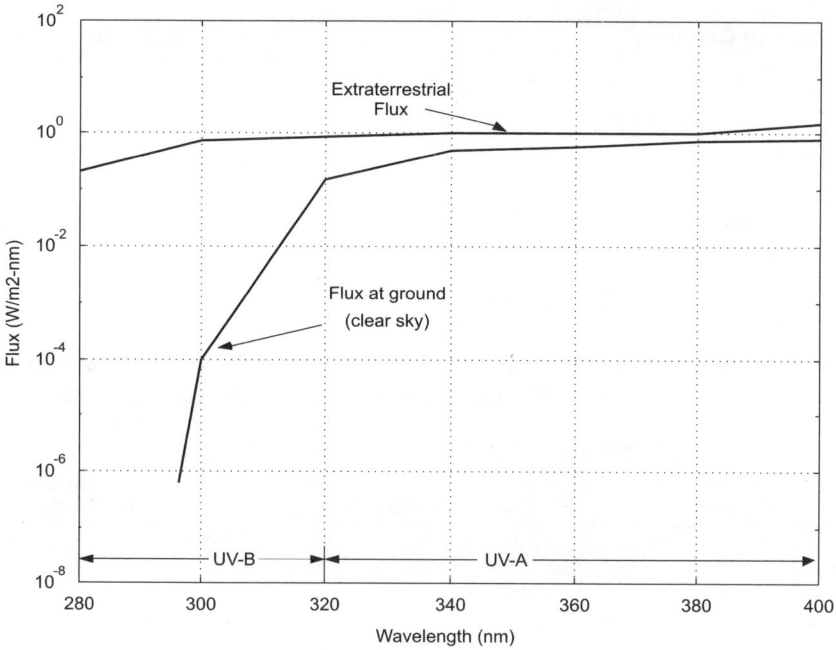

Fig.F7.2: Solar UV Flux At Extraterrestrial And Ground Levels

7.4 Chlorofluorocarbon Gas, its Properties and Uses

Chlorofluorocarbon (popularly known as CFC) are man-made gases, constituted of carbon, chlorine and fluorine. The most common CFCs are CC13F (CFC-11), CC12F2 (CFC-12) and C2C13F3 (CFC-113). They are also known as Green House gases, having much higher capacity of absorbing infrared heat radiation from the earth than carbon dioxide, in fact 15,000 times more than carbon dioxide.

They are inert and insoluble in water. They, therefore, do not get removed by precipitation from the troposphere. They travel upward to the Stratosphere and can reside there for many years. In the stratosphere, they are eventually photolyzed upon by radiation and release free chlorine atom.

Free chlorine then acts as a catalytic agent destroying thousands of molecules of ozone in the stratospheric layer.

CCs were mainly used as refrigerants (CFC-12 or Freon12), foaming agents for polymer insulation, aerosols as deodorants, hairspray and paint spray etc. Table T7.1 shows the percentage wise use of ozone in 1985. The major use is in aerosols and in foaming, which are though banned in USA since 1987.

Table T7.1: Various uses of CFC

Use of Category	Percentage Use
Aerosol	25
Foaming Polymer	26
Refrigerant	12
Air conditioning	8
Solvents	19
Miscellaneous	10

Source: Shea1989/Masters1992

The production and consumption of three most commonly used CFCs are given in Table T7.2 for 1985, but their uses are sine banned since 1987.

Table T7.2: Global Consumption of CFCs (1985)

CFC Types	World Consumption ($X10^3$)	USA Consumption ($X10^3$)
CFC-11	341	75
CFC-12	443	135
CFC-113	163	85

Source: Hamit et al 1986

7.5: Ozone Hole and Further Scientific Work

It was for the first time in 1974 that Molina and Rowland (Molina et al 1974) first hypothesized that anthropogenic CFC in the atmosphere travels up to the stratosphere, where chlorine gets dissociated from CFC by solar photolysis. This chlorine then gets oxidized from ozone, converting it to molecular oxygen and gradually destroying the ozone layer. However, this finding did not get, at that time adequate attention of the scientific community.

But the most startling revelation came in 1985, when the British Antarctic Survey came up with the evidence of a 40 per cent drop in the springtime ozone concentration from the 1964 level over the Halley Bay, Antarctica; the size of which was as large as the land area of the United States. This so-called Ozone hole then travels northwards during summer and gets further thinned mixing with fresh air masses. In effect, there is then a global thinning of the ozone shield, endangering the risk of global bombardment of carcinogenic UV-B radiation. During 1990, further scientific evidence (Sandler 1997) of occurrence of springtime ozone hole over the North Pole was reported.

There is now new scientific explanation to the formation of Ozone Hole over the Polar region. CFC is a stable compound, which remains unaffected in the troposphere layer and travels up to the stratosphere, where it gets photolyzed to release chlorine by UV radiation.

For CFC-12,

$$CCl_2F_2 + hu = Cl + CClF_2, \qquad \text{hu is solar energy}$$
$$Cl + O_3 = ClO + O_2$$

The Chlorine Monoxide then reacts with Nitrogen oxide forming Chloride Nitrate,

$$ClO + NO_2 = ClONO_2$$

Chlorine Nitrate $ClONO_2$ is an inert compound, which does not react with ozone any more and chlorine remains entrapped as $ClONO_2$.

Over the Antarctica, a polar vortex forms, which entraps cold air and prevents warmer air from lower latitude mixing with and warming polar air. Consequently, the polar air in the vortex cools down further to about –90 C and the ice crystals form (Kugt 1983). The ice crystal creates additional surfaces for chlorine nitrate to react with water to form hypochlorous acid HOCl and Nitric acid,

$$ClONO_2 + H_2O = HOC_1 + HNO_3$$

As long as polar vortex persists in the winter, hypochlorous acid goes on accumulating. But as the spring sets in August/September, hypochlorous acid photolyzes, breaking free chlorine initiating a continuing cycle of destruction of ozone:

$$HOCl + hu = Cl + OH'$$
$$Cl + O_3 = ClO + O_2$$
$$OH' + O_3 = O2 + HO_2'$$
$$ClO + HO_2' = HOCl + O_2 \qquad \text{...E7.2}$$

The destructive cycle processes through equations E7.2 and repeats itself generating chlorine all the time. In the process, thousands of molecules of ozone get depleted to molecular oxygen.

7.6 International Conventions and Protocols

The dissemination of scientific knowledge globally about the ozone depletion and its serious ill-effects awakened multinational concern about the need of some urgent international initiatives to control CFC emissions This concern was heightened by the Report of USEPA revealing that USA could save USD 6.4 trillion by 2075 in reduced costs by 50% reductio in CFC emission.

Vienna Convention, 1985

Coinciding with the discovery of the ozone hole over the Antarctic, the Convention for the Protection of the Ozone layer came into being on March 22, 1985 at Vienna. The Vienna Convention enjoined on the signatory members to monitor CFC emission and its harmful effects on depletion of the ozone layer, to exchange relevant scientific information and to take measure to control domestic production of ozone depleting chemicals. Twenty seven countries ratified the Convention by 1989.

Montreal Protocols, 1987-89

Extending the above Vienna Convention, the above Montreal protocol was also signed in September 1987 and came into force on January 1,1989. This Protocol set specific limits to emission of ozone destroying gases, particularly CFC-11 and CFC-12. It also envisaged that later protocols would set limits to several other harmful substances like methyl Chloroform, carbon tetrachloride, etc. The principal requirements of Montreal protocols are enumerated in Box B7.1.

The signatories to the Montreal Protocols were deemed to be also the signatories to the Vienna Convention. Developing countries enjoyed an inducement to ratify the Montreal protocols, as they were entitled to an exemption from compliance with the provisions of the Protocols for a period of 10 years from the date of ratification of the same; they will still escape Trade Sanctions under the Protocols. As on 31 October 1995, 149 countries ratified the Protocols. India ratified the Montreal Protocols in 1992.

The 1990 London Amendments strengthened the provisions of the Montreal Protocols with new stipulations.

The Copenhagen Amendments, Nov. 1992: These amendments set limits of emissions and phasing out of Hydrochlorofluorocarbons (HClFCs) and Hydrobromofluorocarbons (HBrFCs) and carbontetrachloride and methyl chloroform by 1996

Box B7.1: Principal Requirements of Montreal Protocols

1. From the date of ratification to 1 July 1993: All signatory countries shall reduce their CFC emissions to 1986 levels.
2. From 1 July 1993 to 1 July 1998: All Protocols Ratifying countries shall continue to limit their annual emission to 80% of 1986 levels.
3. From 1 July 1998 onwards: The emission of CFC from the ratifying countries shall be limited to 50 % of the 1986 levels.
4. From 1990 onwards: The ratifier countries cannot import controlled substances from non-ratifier countries.
5. From 1January 1993 onwards: Ratifier countries cannot export ozone depleting substances to non-ratifier countries.
6. For Developing Countries: Developing countries whose annual consumption is less than 0.3 kg. per capita on their ratifying date, a ten year's delay in compliance of the provisions of the protocols is permitted.
7. Emission Transfers: To induce ratification globally, countries ratifying the Protocols can exchange reduction in emission benefit between them.

Contd...

London Amendments, 1990

The Provisions of the Montreal Protocols were further strengthened by the following stipulations:

1. From 1995: The CFC emissions are to be reduced to 50% of 1986 levels.
2. From 1997 to 1999: The CFC emissions are to be further reduced to15% of 1986 levels.
3. From 2000: the CFC production and emissions are to be reduced to zero levels except for recycling of previously produced CFCs.

Further amendments and enhancements included the followings:

- Provisions regarding Trade restrictions were further clarified.
- Multilateral Fund was established to help developing countries.
- Twelve more ozone-depleting substances were included for control like CFC.
- Additional 34 chemicals were listed for observations and future control.

The London Amendments of protocols came into force in August 1992.

Table T7.3: National CFC Emission

Country (Ktons)	Emission1989 Percentage	1989 (Ktons)	Emission 1991 Percentage	1991
United States	130	22.4	90	22.5
Japan	95	16.4	64	16
USSR	67	11.6	44	11
Germany	34	5.7	23	5.8
U K	25	4.3	17	4.3
Italy	25	4.3	17	4.3
France	24	4.1	17	4.3
Spain	17	2.4	11	2.8
China	12	2.1	8	2.0
Canada	11	2.5	8	2.0
Australia	8	1.4	5	1.3
S. Africa	7	1.2	5	1.3
Total	455	78.4	309	77.3
World Total	580	100	400	100

7.7 Future Scenario

The success of the Montreal Protocols under the Vienna Convention is partly due to the fact that there are only few CFC-emitting nations. In 1985, only three countries USA, Japan, and then Soviet Union accounted for 46 % of total global CFC emission. Table T7.3 shows that in 1989, 12 countries amounted to 78 % of total global emission.

The second reason is the effective leadership by USA. The 1987 EPA Report found that USA could save USD 6.4 Trillion by 2075 from reduced costs of Skin cancer by merely reducing CFC emission by 50 %. This fact encouraged positive US initiatives.

The other important fact for the success of the Montreal Agreement is good substitutes of CFC could be developed such as HCFC and HFC. HCFC presents 90% less risk of ozone depletion compared to CFC. But HCFC did contain one atom of Chlorine. In this respect, HFC contained no carbon and hence 100 percent safe, posing no threat at all to ozone. Both these gases though are GHGs and could produce global warming. Finally, however, under the Copenhagen Agreement, HCFC will be required to be eliminated by 2030.

❑❑❑

8

CHAPTER

Problem Of Acid Rain

Unlike global dimension of global warming and the ozone depletion, acid rain is basically a regional problem, being studied here to know the various remedial actions in Europe and America. In actual practice more actions have been taken to control sulfur dioxide emission, which give rise to sulfuric acid rain than nitrogen oxide emission, which causes Nitric acid rain.

8.1 Occurrence of Acid Rain

The acid rain contains harmful proportions of sulfuric and nitric acids. As it is, rainfall is generally slightly acidic. Water vapour in equilibrium with atmospheric carbon dioxide forms a weak solution of carbonic acid (H_2CO_3) with a pH value of about 5.6. Consequently, acid rain is usually defined to be a precipitation with pH less than 5.6 (completely neutral water has a pH of 7 at 25° Celsius).

Oil and coal generally contain appreciable quantities of sulphur and also nitrogen. Thus, during combustion, sulfur and nitrogen oxides are discharged and pollutants from power plants, automobiles, and other industrial plants, etc. The percentages of such emissions are estimated as shown in Table T8.1.

Table T8.1: Sources of Sulfur and Nitrogen Oxides. (%)

Emission	Power-Plant	Automobiles	Industry	Residential	Miscellaneous	Totals
Sulfur (SOx)	47.8	3.7	37.4	10	1.1	100
Nitrogen (NOx)	23.5	53.5	15.4	6.1	1.31	100

Source: Sandler 1997

Sulfur is released mostly as sulfur dioxide (SO_2, but also small amounts as sulfur trioxide (SO_3). Sulfur dioxide, once released, can convert to SO_3 reacting with free radical OH¢

$$SO_2 + OH' = HOSO_2$$

$$HOSO_2' + O_2 = SO_3 + HO_2'$$

$$HO_2' + NO = NO_2 + OH' \quad \ldots \qquad \ldots \qquad \text{E8.1}$$

In equation E8.1, the radical OH may again be reproduced by reaction with hydrogen peroxide.

The Nitrogen Oxide formed as pollutant gets oxidized with the atmospheric oxygen and finally reacts with radical OH¢ to form Nitric acid HNO_3:

$$2NO + O_2 = 2NO_2$$

$$NO_2 + OH' = HNO_3 \ldots \qquad \ldots \qquad \text{E8.2}$$

The transformation of SO_2 to SO_4 Particulate also takes place gradually. Along with those particulates, polycyclic aromatic hydrocarbons (PAH) are also health hazards.

8.2 Effects of Acid Rain

The acid rain adversely affects the ecosystems very widely. Firstly, acid degrades buildings, construction materials (limestone, marble, cements etc.). This is manifested in accelerated weathering of building surfaces.

Secondly, the forests are considerably stressed leading to stunted growth, untimely weathering and decay. Agricultural productivity is likewise adversely affected due to lack of fertility from the acidification of soil.

Thirdly, aquatic organism and fish life are very sensitive to pH value. Aquatic systems are stressed seriously in water medium with pH value below 5.5 and fishes will die out with pH under 5.

Finally, acid rain in association with the troposphere ozone, sulfur dioxide, nitrogen oxide and volatile organic compounds present considerable threat to human health.

8.3 Areas of Acid Rain

Acidic rain generally turns waterways acidic affecting and stressing aquatic lives. But there are different types of lakes – Bicarbonate lakes have high pH value, Acid lakes which have water highly acidic with low pH and Transitional lakes having water that gets acidic by acid rain (Fig. F8.1). Bicarbonate lakes have reserve alkalinity and do not easily get acidic, till very sustained acid downpour destroys all alkalinity.

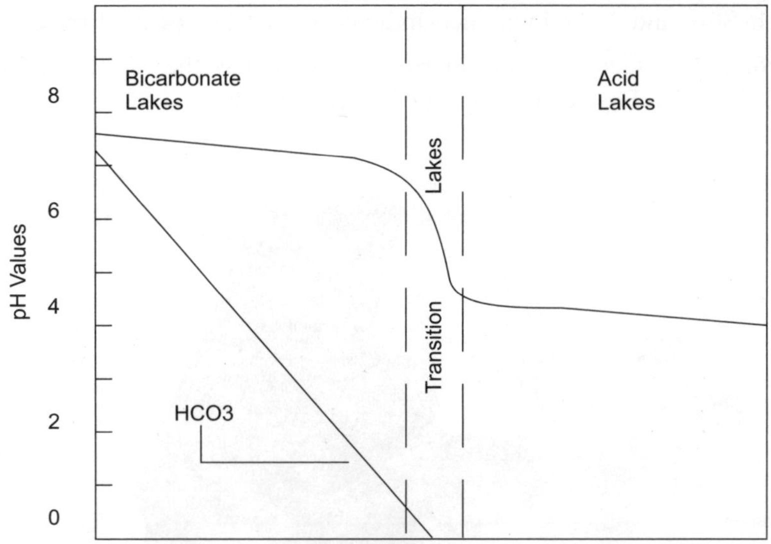

Fig.F8.1: Bicarbonate Buffering

Fig. F8.2 shows acidic lake areas in the western and northern USA and also eastern Canada. Here the soil does not have lime ($CaCO_3$) base. The areas have been determined by bedrock geology.

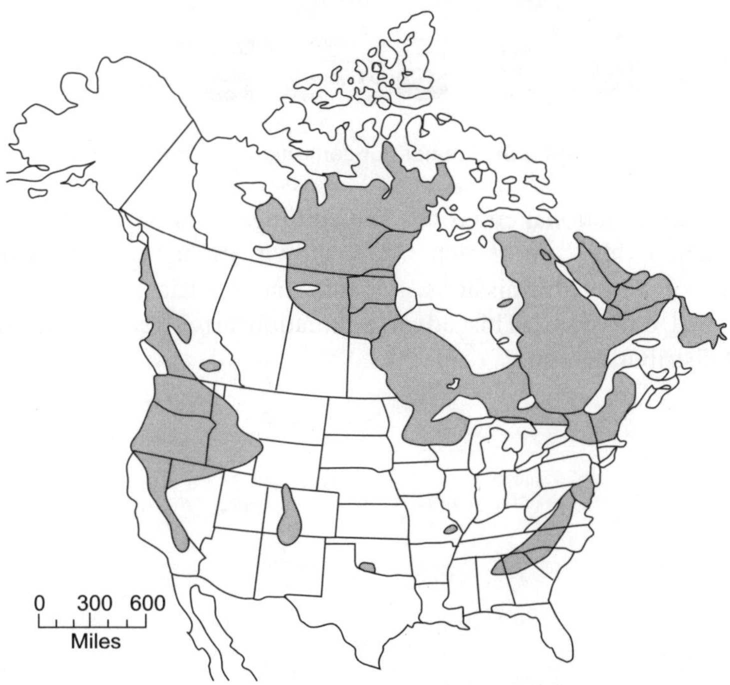

Fig. F 8.2 : Acidic Lakes in N. America

Now we present SOx and NOx comparative distribution (Table T8.2) for Europe. High SOx and NOx incidence indicate possibility of acid rain.

In Fig. F8.3, it is seen that former Soviet Union has the highest sulfur emission followed by East Germany and United Kingdom.

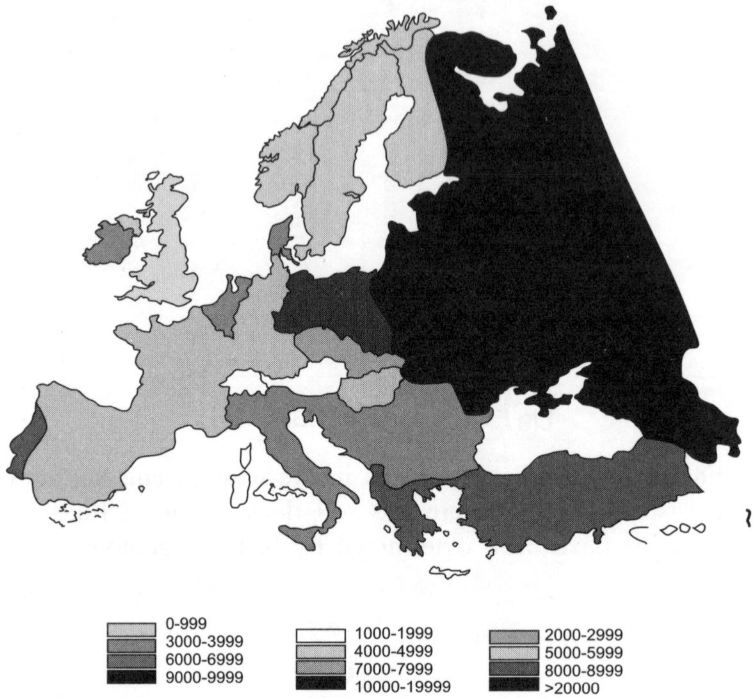

Fig.F8.3: Sulphur concentration in Europe

In Table T8.2, national emissions and spillings from other nations of sulfur and nitrogen oxides are shown. Also shown are national depositions as percentage of national emissions. Albania and Austria have more spillings than national emissions. This adverse situation arise because the nature of the wind distribution

Table T8.2: Sources of Emissions of Sulfur and Nitrogen Oxides (SOx & NOx) (1990)

Countries	Sulfur Dioxide			Nitrogen Oxide		
	Own Emission 100T	other Spillings 100T	Own Depositions %	Own Emission 100T	Other spilling 100T	Own Deposition %
Albania	76	172	53.9	4	70	25
Austria	256	1311	44.9	275	677	16.7
Belgium	1465	404	28.4	615	216	4.4
Czech	7861	2573	39.6	1632	834	13.9
Denmark	419	341	24.1	440	222	4.5
Finland	745	830	526	442	443	26.7
France	3956	1941	52.1	2803	1065	308
Germany	19369	4434	40.5	5742	2164	14.2
Italy	4325	2385	64.4	1593	708	38.7
Netherlands	574	61.5	26.8	951	280	5.7
Norway	104	900	57.4	206	718	28.8
Poland	10529	5555	49.5	2264	1795	22.8
Sweden	414	1413	55.6	568	971	27.6
Switzerland	167	480	50.3	185	304	18.4
USSR	22309	10724	92.1	5043	4888	88.6
UK	8189	320	57.4	2984	232	19.9

Source: Sandler 1997

8.4 Control of Acid Rain -Various Regional Protocols

1. **Long-range Transboundary Air Pollution (LRTAP) 1983**
   Under the aegis of the UN Economic Commission for Europe, draft LRTAP Convention was prepared and agreed for the protection of the environment. LRTAP agreement mandated scientific investigation by a set of monitoring stations throughout Europe. LRTAP Convention was duly ratified on 16 March 1983 with 12 member nations signing it.
2. **Helsinki Protocols, 1985**
   Helsinki Protocols to LRTAP Convention was adopted on 8 July 1985 requiring signatory nations to reduce sulfur emission by 30% below the 1980 emission levels by 1993. The Protocols came into force on 2 September 1987.
3. **Sofia Protocol, 1988**
   On 31 October 1988, the Sofia Protocol to LRTAP Convention was signed requiring member nations to reduce NOx emissions to 1987 levels by 31 December 1991.

4 **Oslo Protocols 1994**
   Oslo Protocols adopted on 14 June 1994 required further cutback in sulfur omissions beyond Helsinki Protocol based on minimization of Control coat. This assigned varying percentage cuts beyond 1980 levels for different nations.

5 **Geneva Protocols, 1995**
   The Geneva Protocol, adopted on 11 November 1995, sets emission reductions for the first time to certain volatile organic compounds (VOCs) to improve surface ozone pollution.

There was more pan-Europe support for reduction and control of sulfur emission than NOx. This was partly due to the fact that sulfur emission caused more localized depositions and pollution than NOx. Secondly, sulfur emissions were more concentrated to large power plants and process industries, while NOx emissions were from numerous private automobiles and were difficult to contain. Since then, of course, automobile emissions norms have become very stringent all over the world to prevent smog, and transportation pollution.

❑❑❑

9

CHAPTER

Global Warming And Climatic Changes

9.1 A short Overview of Global Warming

Basic Principle of Global Warming

The earth is receiving the heat energy from all the solar radiation all the time. At the same time the earth is radiating away heat to the open universe. So assuming heat balance, the earth should remain at the same temperature.

But the earth is covered with a deep blanket of atmosphere, which is interfering with the transfer of heat radiation. The basic atmosphere consists principally of Nitrogen (79%) and oxygen (21%) as shown in Table T5.1 approximately, which does not affect heat transfer. But there are some polluting gases, which are mainly generated as a result of human activities and which pass the high frequencies solar radiation to reach the earth surface but absorbs and obstructs the low-frequency heat radiation from the earth to the atmosphere. These gases – carbon dioxide (CO_2), nitrous oxide (N_2O), methane (CH_4), and CFC (chlorofluorocarbons) are called Greenhouse gases (GHG). Because, they work in the similar way to the plant greenhouse, *i.e.,* they allow heat in but stops heat out.

Greenhouse Gases

Carbon dioxide is generated with the burning of any fossil fuels in industrial furnaces and domestic kitchens; Deforestation also causes decrease in capturing of carbon from the atmosphere and hence increases of carbon dioxide in the atmosphere.

Methane is produced from anaerobic decomposition of solid waste, coal mining, oil and gas production, live stock excreta, agriculture paddy fields, marshlands and also substantially from auto-emission.

Nitrous oxide is produced primarily from use of fertilizer, automotive combustion and stationary fossil combustion at the powerhouses and process industries.

CFCs are generated from the refrigeration industries and plastic-process industries. CFC generation has been controlled very effectively under the international agreements. Of the GHGs, CFCs trap 15000 times as much heat as an equal volume of carbon dioxide, while methane traps about 25 times and nitrous oxide traps 200 times more than that CO_2 does.

But because of higher concentration of carbon dioxide, it contributes about 70% to the global warming compared to 7% by CFCs, and 14% by N_2O and 8% by methane. These are tentative and average estimates. Table T9.1 shows the global concentrations of the four main GHGs in the years 1982,1986 1990, and 1994. Please note gradual increases in their concentrations.

Table T9.2 shows annual emissions of carbon dioxide from various countries in year 1991. United States have the ignominy of highest contribution at 4931 million tones annually amounting to 21.8 % of total pollution of carbon dioxide.

On the other hand, it is seen from Table T9.3 that China and India contribute 16% and 14% respectively of total global emissions of methane, though USA and USSR follow closely at 11.6 and 11.2 percentages respectively.

Table T9.1: Atmospheric concentration of Principal Green House Gases

Year	CO_2(ppm)	CFC(ppt)	NOx(ppb)	CH_4(ppb)
1982	341	500	301	NA
1986	347	603	305	1600
1990	354	718	307	1645
1994	359	760	309	1666

ppm=parts per million; ppb=parts per billion; ppt=parts per trillion.

Source: Sandler 1997.

Table T9.2: GHG* Emissions from various Countries, 2000

Countries	CO_2 Equivalement Emissions (million Tons)	Per Cent Global Emissions
United States	6928	20.6
China	4938	14.7
EU-25	4725	14.0
Russia	1915	5.7
India	1884	5.6
Japan	1317	3.9
Oermany	1009	3.0
Brazil	851	2.5
Canada	680	2.0
United Kingdom	654	1.9
Global	33,665	100

World Resource Institute 2005

* GHG includes CO_2, CH_4, N_2O, HFC, PFC_5* SF_6

Table T9.3: Methane Emissions from Various Countries

Countries	Methane Emissions (Million Tons)	Per Cents Global Emissions
China	40	16
India	35	14
United States	29	11.6
U S S R	28	11.2
Brazil	10	4
Bulgaria	9	3.6
Indonesia	7	2.8
Bangladesh	6.2	2.5
Thailand	6	2.4
Australia	4.5	1.8
Global	250	100

Source: Sandler 1997

Problem of Ozone Depletion

Besides, the problem of Green House Gases causing global warming, there is the problem of depletion of ozone layer in the stratosphere. The ozone layer absorbs the dangerous ultraviolet radiation, thus preventing it from reaching the earth's surface. We have already dealt with this problem in some details in Chapter 7 of this Part II.

Atmospheric Temperature Effect

It has been estimated that because of green house effect, the earth surface has today an average temperature of 288 K (15 C), instead of 255 K (-18 C), which is welcome making life more comfortable.

But the earth is covered with a varied atmospheric gases consisting of mainly nitrogen and oxygen but also many other polluting gases in small quantities including the GHGs (see Table T5.1). Above the earth surface, the temperature varies. Following upon the temperature profile, the atmosphere seems to be divided into four horizontal layers:

Troposphere: temperature range 288 K to 210 K at top;
Stratosphere: temperature range 210 K to 260 K at top;
Mesosphere: temperature range 260 K to160 K;
Thermosphere: temperature range 160K to very high value.

In Fig. F9.1 is shown the temperature map of the atmosphere, which also shows the above four layers. In Table T9.4 are depicted the properties of above atmospheric layers.

In the Troposphere, most of the pollution accumulates. But here again precipitation from the clouds takes place and rains washes away much of the pollution back to the earth.

Stratosphere is the residual of the atmospheric ozone, which absorbs the carcinogenic ultra-violet rays, thus saving the earth and the humans.

Thermosphere (Ionosphere) has the unique property of reflecting the radio waves from the earth back to the earth making possible global wireless communication till the advent of satellite communication.

National Acts, and International Convention & Protocols on Global Warming & Ozone Depletion

As there are no common laws among countries at the international level, it has always been difficult to arrive at agreements on environmental matters. Because of the different levels of development, nations have different priorities on environmental problems.

The first comprehensive United Nations initiative was the UN Conference on Human Environment held at Stockholm in 1972. But the UN Conventions and protocols directly relating to the problems to global warming are the followings:

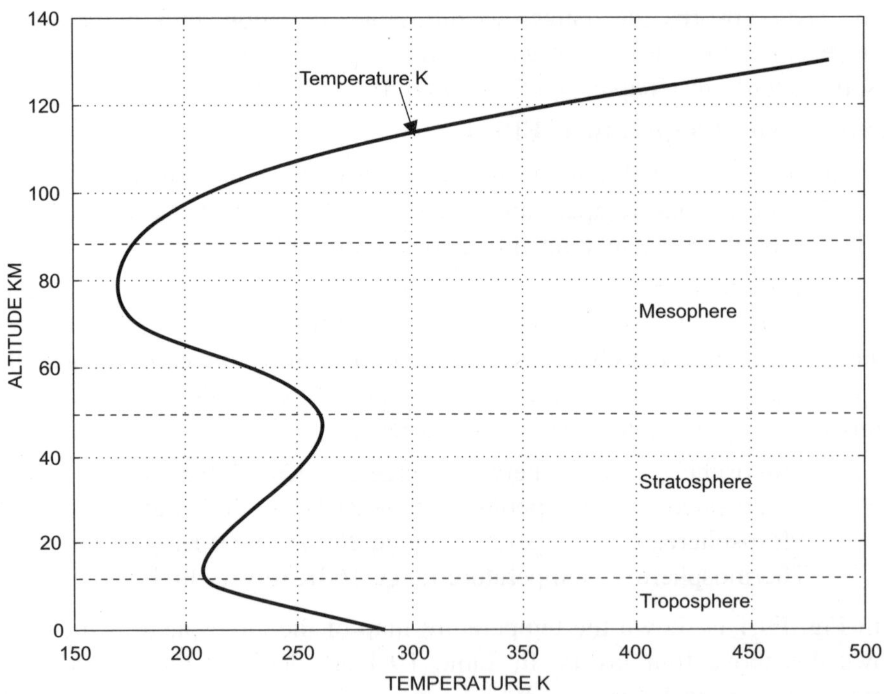

Fig. F9.1: Temperature Profile of Atmosphere

1985 Vienna Convention under the Stockholm Declaration dealing with protection of the Stratospheric ozone layer;

1987 Montreal Protocol on substances that deplete ozone layer;

1992 U N Convention on Climatic Change laying down the guiding principles, *i.e.,* Framework Convention on Climate Change (FCCC)

1997 Kyoto Protocols under FCCC laying down quantative limits of GHGs for Annex I developed countries (OECD, Eastern Europe, Russia, USA, & Japan)

Under the imperatives of the 1972 UN Convention, Government of India constituted the Indian National Committee on Environmental Planning and Coordination in 1972. India acceded to the Montreal Protocols in 1992 and also carried out the 73rd and 74th Amendments to the Indian Constitution following Bhopal tragedies in 1984 and 1992 Rio Conference. The Air (Prevention and Control of Pollution) Act, 1981 and the Environment (Protection) Act 1986 were enacted with the same purpose off national effort towards the protection of the Air Environment, which helps the cause of Climate Change.

Table T9.4: Properties of Atmospheric Layers.

Atmospheric Layers	Avg. Distance from Earth Surface (km)	Temperature from Earth side (K)	Main Contents	% Total	Characteristics
Troposphere	0 – 12	+15 to -60	Water vapour, clouds, Precipitati-on, (air)	80	Turbulence facilitating air mixing
Stratosphere	12 – 50	-60 to –13	mainly Ozone	19.9	Stable region, practically no air movement; ozone absorbs UV rays and gets heated causing temperature inversion & stability.
Mesosphere	50 – 90	-13 to –113	Negligible	Negligible	Some Turbu-lence because of higher temperature at bottom
Thermosphere (Ionosphere)	90 - 140	-113 to +180	Negligible	Negligible	Contains ionized oxygen helping wireless commu-nication reflecting radio-waves back to earth

Source: Gilbert 1991

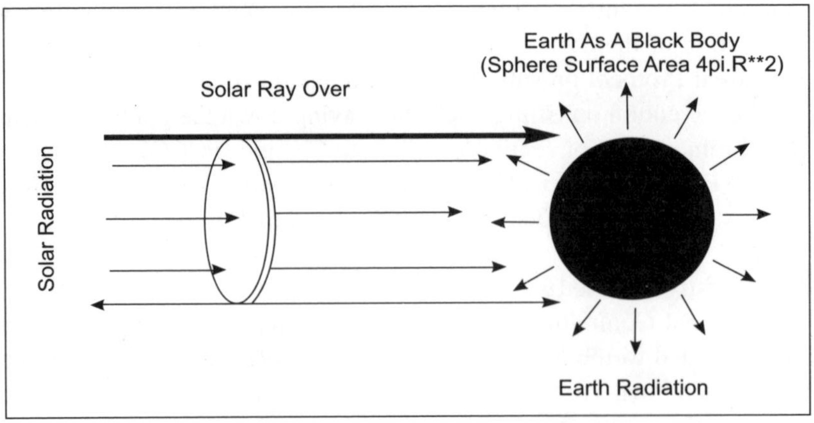

Fig. F9.2: Earth as a Black Body Without Atmosphere

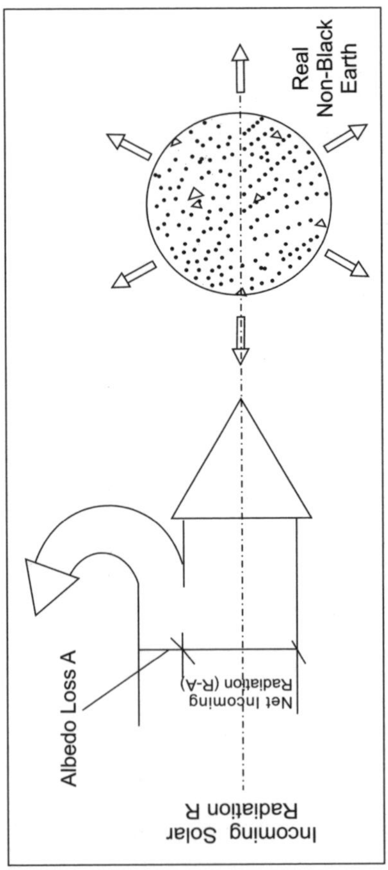

Fig. F9.3: Solar Radiation on Real Earth with Albedo Loss A

Anthropogenic Character of Global Warming

Global Warming has now come to be known as the anthropogenic in character. Nevertheless, it has become intractable to control the problem through international agreements among nations because of varying interest. The cooperation among nations remains mainly elusive. The global warming arises, because earth's atmosphere polluted with certain gases, allow the sunlight to pass through; but absorb the returning infrared radiation from the earth. The captured heat, between the atmosphere and the earth, warms up the atmosphere and the earth. The warming effect is similar to the Green House effect and hence, these gases are called Green House Gas (GHG). The principal GHG gases are carbon dioxide, methane, CFC (chlorofluorocarbon) and nitrous oxide (N_2O).

The carbon dioxide is formed and emitted by the combustion of fossil fuel in the automobiles, power plants, in residences and etc. Carbon dioxide also escapes to the atmosphere because of extensive deforestation. Methane is generated from solid waste, coal mining, oil and gas production, livestock and rice agriculture. Nitrous oxide is generated from use of fertilizer and fuel consumption at the power plant. CFCs are released from the refrigeration and certain chemical reactions.

Thus it is seen from the above that the greenhouse gases are generated mainly as a result of human activities, economic, industrial, and agricultural.

Adverse Effect of Global Warming

The world community has now come to realize that unabated and unregulated accumulation of GHGs can raise average global temperature by about 1.5 to 3 C by the turn of the century. In Section S2.5.2, we have tried to develop an empirical relationship between the GHG concentration and global temperature, but the exact relationship is yet to be fully evidenced and proven. Further, the contributions of other atmospheric pollutants need also to be fully assessed scientifically. Thus there is considerable doubt of about the estimate of global warming.

Given a certain level of temperature rise, there is no global model of climate change beyond a computer weather modeling. Even such computer meteorological sometimes does not hold good at all. However, generally speaking, global temperature rise should affect quality of rainfall, duration of rainfall and timing of rainfall, affecting agricultural production schedule and even productivity, glacial melting and rise of sea level, occurrence of storm and possibly other geological and meteorological phenomena. But the effect of the global warming will not be uniform, some countries will have more rainfall and more agriculture and other countries will suffer from intense aridity and long-term fear of desertification.

It is then likely that the potential gainer countries might prefer global warming, while the potential loser nations will wish to prevent global warming, which can be achieved only by reduction of emission of GHG. The reduction of GHG will be our next topic of consideration.

Some Reduction Measures of GHG Emission:

We will deal more comprehensively with the technology and economics of the methods of reduction of carbon dioxide and other GHG emissions in a subsequent chapter. Here, we propose to give an introductory outlines of these problems.

Carbon dioxide contributes about 70 per cent or more to global warming among all principal GHGs. The main source of CO_2 emission is due to combustion of fossil fuels (petroleum oils, Petroleum gases and coal) in automobiles, Transportation in general (aviation and railways), power plants and process industries. It is very difficult to change all these plants and automobiles to suit newer carbon-free energy sources, if any. Thus, what can be tried is undertake optimization, efficiency- enhancement and preventing system wastage through energy audit. These measures might help to reduce GHG emissions marginally, say by an order of 10% or 20%. The cost of these marginal reductions could be about, say less than 1% of global GDP and may amount to figure of many billion dollars as is apprehended in some quarters.

Other measures, which are feasible, are to switch over more and more to nuclear power and hydroelectric power, increasing their proportion in the national grid. **Firstly,** though both these forms of energy are totally GHG-free, but there is still strong public opposition to nuclear energy. **Secondly,** the disposal of radioactive nuclear waste is a big problem technically, because of long half-life of radioactive waste. **Thirdly,** both these energies are highly capital-intensive; and large-scale change over may cost a trillion dollars or so annually. **Fourthly,** the availability of hydropower is comparatively limited.

Fortunately, there is a recent development of Bio-diesel, which has about 50% less carbon content than the conventional fossil petroleum diesel. The bio-diesel is produced from vegetable oils, which can be extracted from the seeds of the non-edible shrubs like Jatropha circus grown profusely on arid cultivable land and fallow hilly terrain. The bio-diesel is presently useable up to 20% with petroleum diesel without any change of the plant or engine. Hence the change over can be effected economically. Germany, USA and Brazil are introducing this energy form quite extensively. India has big plan for developing bio-diesel technology in a big way.

Now, all these measures call for large capital investment today, but the results in the form of carbon dioxide reduction will take place after some years. It will be seen from the Table T9.2 that three developed countries United States, old Soviet Union and Japan together account for 48.8 % of global CO_2 emission; and naturally they were to control the emission first, by deploying necessary large investments, which they were not till recently very willing to incur. Under the 1992 UN Convention on climate Change, Kyoto Protocols, 1997 placed the developed countries, producing heavy emission, under Annex I and LDC (the Less Developed Countries) under the Non-Annex Countries. Under the Kyoto Protocols, Annex I countries are required to lower their CO_2 emission by an average of 5.2 per cent below the 1990 levels by the period 2008-21, while the non-Annex countries, having quite low per capita carbon emission, are not required to take any action. USA, the largest producer of CO_2 emission, has not signed the Kyoto Protocols, insisting that the large LDC countries like China and India must also give commitment of controlling CO2 emission, as they are likely to become large economies in the next 25 to 50 years.

Again fortunately, the situation in respect of CFC emission and consequent problem of ozone hole is much better controlled. Most countries including India have acceded to the 1985 Convention and the 1985/19987 Montreal Protocol on substances that deplete the ozone layer. The production and uses of CFC have now almost gone out of practice since the Montreal Protocols. One reason of this success is due to the fact that only few countries were producing CFCs to start with. The number of production units was also limited.

9.2 Basic Physics of Global Warming

The earth gets heated up receiving solar radiation, the sun being the primal source of heat energy. We determine the temperature of the earth's surface by applying Stephan-Boltzman equation to two simplified energy transference models. The first model considers direct heating of the earth as a black body without the atmosphere and the second model includes the existence of the earth's atmosphere and also treats the earth as a gray body with 'albedo' effect

Earth without Atmosphere

The earth is assumed a black body. It therefore absorbs all the incident radiations from the sun and emits an equal amount of heat energy. Referring to Fig. F9.2, the solar radiation hits the earth within a radius R equal to earth's radius, over an area (Pi)R^2 .But the intensity of solar radiation at earth's surface is known as I=1372 W/m2. The heat energy absorbed Ea is therefore:

$$Ea = I.\pi R^2, \quad \dots \quad \dots \qquad E9.1A$$

But the earth emits energy over all its global surface of area of area $4.(p).R^2$. If we now assume that the earth's surface is isothermal, having temperature T, the energy radiated Er is given by Stephan-Boltzman equation:

$$Er = S.4.(\pi)R^2.T^4, \dots \quad \dots \qquad E9.1B$$

Where S= the Stephan-Boltzman constant = 5.67×10^{-8} $w/m^2.K^4$ for the black body.

Since the earth's temperature is assumed constant, there is no change of internal energy and hence equating energy absorbed (E9.1A) with the energy emitted (E9.1B), we get:

$$I.(\pi)R^2 = S.4.(\pi)R^2.T^4$$

$$\text{Or, } T = 279 \text{ K}, \quad \ldots \quad \ldots \quad \ldots \quad \text{E9.1C}$$

This temperature of 279 K is quite close to the actual average temperature of 288 K (15 C) of the earth. But the above proximity of result is accidental. For the above model is not real, being simplistic without atmosphere and the earth being assumed a black body.

Earth as the Modified Blackbody with Modified Atmosphere

Here the assumptions are –

(*i*) The earth and the atmosphere being not perfect blackbody, part of the incident radiations are reflected back from the earth and the atmosphere.

(*ii*) Though the atmosphere takes part in reflecting the incident radiation, there is no greenhouse gas effect from the atmosphere *i.e.,* secondary warming of the earth by the atmosphere.

The reflected solar energy is called 'Albedo', denoted by a and amounts to 30 % approximately on annual average basis. The albedo is shown in Fig. F9.3, which diagrammatically represents the second earth model.With the above assumptions, the equality equation E9.2C of the incoming and emitting energy stands modified as;

$$I.(1\text{-}a).(\pi).R^2 = S.4.(\pi).R^2.Tm^4 \quad \ldots \quad \ldots\text{E2.11}$$

Where 'a' is the albedo and the Tm is the modified temperature of the earth and from E9.2.

$$Tm = 255 \text{ K}$$

The above revised temperature of 255 K (-18 C) is still further away from the actual temperature of 288 K (+15 C) of the earth's surface. But this difference of 33 C is actually a correct evaluation, for even the second model has included the albedo loss of radiation but did not include the secondary warming of the earth by the polluted atmosphere of the earth, otherwise known as the Greenhouse Effect. We will deal with the Greenhouse effect next.

Greenhouse Effect

Now one of the important reasons of the Greenhouse effect is that the solar radiation is of higher frequency (*i.e.,* lower wave length) energy wave, while the thermal energy wave emitted by the earth is of lower frequency (*i.e.,* comparatively of higher wave length).

Stephan-Boltzman Law gives the rate of energy radiated from a blackbody. But the radiated energy is spread over a range of Wavelengths or frequencies forming a spectrum similar to the two spectra shown in Fig. F9.4, one for the sun radiation and the other for the earth radiation. But using the temperature of the radiating body, the Wien's Law (E9.3) defines the wavelength at the mode of the spectrum, *i.e.,* at the maximum of the spectrum intensity,

Lemax = 2898/T microm E9.3

Where L = wavelength in micrometer; emax = maximum energy intensity; T = temperature of the radiating body in degree K.

Assuming the sun as a blackbody and having surface temperature of 6000 K, we get from E9.3.

[Lemax]sun = 0.48 micrometer... E9.3A

Similarly, taking the earth's surface temperature at 288 K, we get from E9.3,

[Lemax]earth = 10.1 micrometer ... E9.3B

From the Fig. F9.4, it will be seen that the solar radiation spectrum is distributed over the wavelength from 0.15 microm to 4.5 microm, having its modal maximum at 0.5 microm. Similarly, the earth's radiation spectrum is distributed from 3 to 90 microm wavelengths, having its modal maximum at 10.1 microm wavelength.

Now the earth is covered with atmospheric air of oxygen, nitrogen etc., which freely allows solar radiation to reach the earth and the radiation from the heated earth to pass out to the space. But the earth's atmosphere is polluted with anthropogenic gases and other gases. In particular, the immediate atmospheric layer Troposphere is polluted with water vapour, clouds, carbon dioxide, Nitrous oxide, CFC. These pollutant gases have the property that they allow the low wavelength solar radiation below 4 microm to pass through to the earth but absorb the earth's radiation of relatively higher wavelength above 4 microm (see Fig. F9.5). This phenomenon leads to the gradual warming up of the earth. This is the reason why the earth is warmer by 33 C than what was the derived temperature arrived at from the second global warming up model above.

The secondary effect is known as Greenhouse Effect, similar to the greenhouse effect, which saves the plants and saplings from the severe winter of the western countries.

Generally, all gases, which absorb energy radiation of wavelength above 4 mcrom, are called Greenhouse gases. The followings are the principal greenhouse gases (often abbreviated as GHGs) and they are shown with energy spectral ranges absorbed by them:

Table 9.5A: Energy Absorption by GHGs

GHGs	Wavelength Absorption Range (Micrometer)
Water Vapour	above 18
Carbon Dioxide	from 13 to 18
Ozone	from 9.5 to 10.5
Methane	from 3.5 to 7.5
Nitrous Oxide	from 8 to 13

From the above Table T9.5A, it may be seen that there are two windows 7.5 to 9.5 micrometer and 10.5 to 13 micrometer, over which there is no GHG effect and through which energy radiation can transmit through in either directions.

Table T9.5B: Contribution to Global Warming by Principal GHGs

Name of Gases	Atmospheric Concentration (ppm)	Annual Increase of Concentration %	Relative Greenhouse Effect	Current GHG Effect Contribution		Principal Gas Sources
				Weightage (2) × (4)	Percentage (5) × 100/498	
(1)	(2)	(3)	(4)	(5)	(6)	(7)
Carbon Dioxide	351	0.4	1	351	70.5	Fossil fuels, other combustions, Deforestation
CFCs	0.00225	5	15,000	34	0.8	Foams, Aero sols, Refrigerants, Solvents.
Methane	1.675	1	25	42	8.4	Wetlands, Rice, Livestock.
Nitrous Oxide	0.331	0.2	230	71	14.3	Fuels, Fertilizer, Deforestation
Totals	-	-	-	**498**	-	

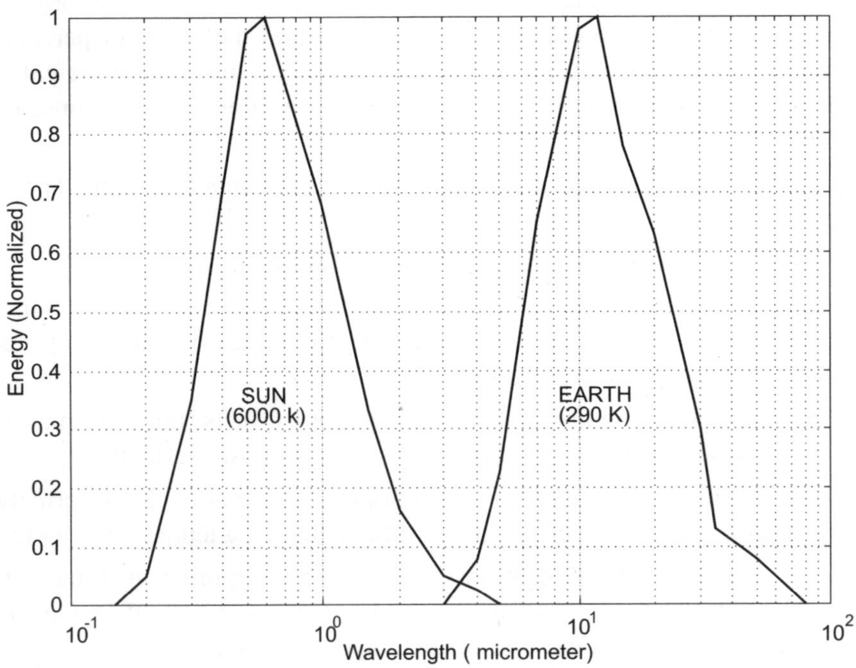

Fig. F9.5: Energy Spectrum of Blackbody Radiaon of Earth & Sun

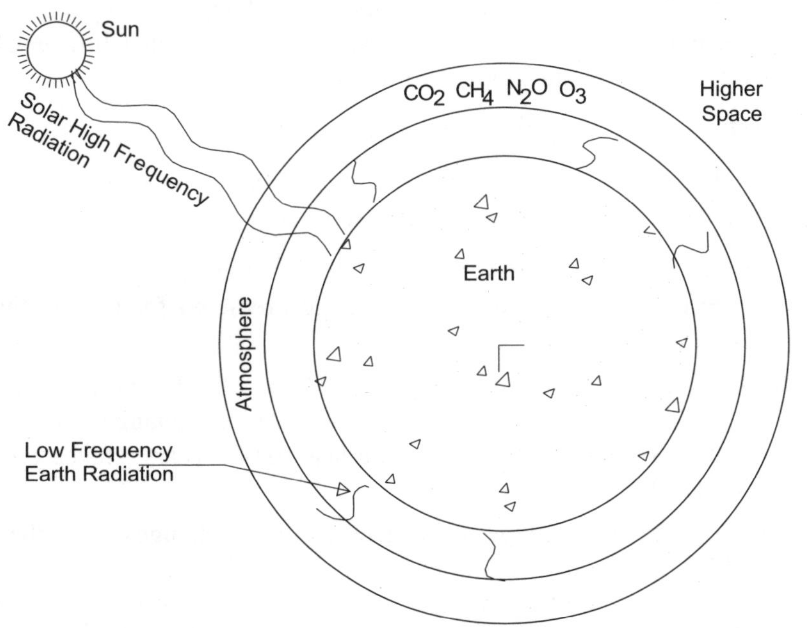

Fig. F9.5 Greenhouse Phenomenon

Table T9.5B shows the four principal GHGs with their relative absorbability of power *vis-à-vis* carbon dioxide in column 4 and their overall percentage thermal absorption in column 5. Carbon Dioxide is the first principal gas, which is responsible for 70% of the global warming and its consequent problems.

Water vapour is not included as the GHGs as it cannot accumulate in the atmosphere, as it precipitates and comes down as rain.

Relationship of Atmospheric GHG and Temperature

There is considerable uncertainty in modeling the atmosphere admixed with GHGs *vis-à-vis* temperature change. The complexities arise, inter alia, from some of the following factors:

GHGs raise global temperature by greenhouse effects, but they may also increase the rate of albedo, which may create opposite effect;

Global warming may intensify evaporation and cause cloud formation, which has also greenhouse effect, thus further enhancing temperature increase;

As the ocean is a large heat sink, the long-run equilibrium temperature takes a long time, perhaps many years, to stabilize much after the GHG concentration leads to some short-run temperature rise.

There are also many other effects of atmospheric dynamics – adiabatic lapse rate effect, which change temperature profile continuity.

Temperature Change due to CO_2 Concentration

The temperature model estimates temperature change over every doubling of CO_2 concentration. The National Research Council on Climate Change recommends the following model formulation:

$$DT = [DTd] / [\ln 2] \times \ln[(CO_2)t / (CO_2)o] \quad \ldots \quad \text{E9.4}$$

Where, DT = average global equilibrium temperature increase at time t;

$(CO_2)t$ = concentration of CO_2 at time t;

$(CO_2)o$= concentration of CO_2 at initial time;

DTd = equilibrium temperature change predicted for the doubling of CO_2 concentration.

Typically, DTd is considered between 1.5 C and 4.5 C. Even if the lower limit of 1.5 C is taken as the quantum of temperature change, such change is believed to have not taken place in the last 10,000 years. Hence our view is to accept DTd = 1.5-2.0 C.

Fig. F9.6 gives the average surface temperature changes over the years as recorded by the Goddard Institute for Space sciences. From the regression analysis curve, it is seen that between 1885 to 1985, the surface temperature increased by about 0.5, that corresponds approximately with the calculated result using the equation E9.4 as shown in Example X2.1 for about hundred years 1850 to 1984 (actually 135 years). Firstly, the actual temperature rise in Goddard Institute data of Fig.F9.6 might not have been done due to increase in CO_2 concentration level only, for other factors like situational location of the Institute, etc.

Example X9.1 (From 1850 - 1985)

Predicted Temperature Change in 1985.

Using equation E9.4, predicted temperature rise in 1985 is given as under:

CO_2 concentration level in 1850	280 ppm
CO_2 concentration level in 1985	345 ppm

Assume DTd =1.5 C

DT = [DTd / ln2] × ln[345/280] = 0.602 C

Temperature Change due combined effect of Greenhouse Gases

The earth-atmospheric temperature changes are considerably influenced by Greenhouse gases other than CO_2. Extending the temperature change model in equation E9.4 of National Research Council to other GHGs using work of Ramnathan (1985), Muitzer (1987), Masters (1991), we present below the extended temperature change model:

$$DTt = [DTd/\ln 2]\times\ln[(CO_2)t/(CO_2)o] + 0.057\times[(N_2O)t^{**}0.5 - (N_2O)o^{**}0.5]$$
$$+ 0.14\times[(CFC11)t - (CFC11)o] + 0.16\times[(CFC12)t - (CFC_{12})o]$$
$$+ 0.019\times[(CH4)t^{**}0.5 - (CH4)o^{**}0.5] \ldots \qquad E9.5$$

where concentrations for N_2O, CFC and CH_4 are expressed in ppb and that for CO_2 in ppm.

The application of Equation E9.5 and E9.6 (a proposed simplified version of E9.5) have been worked out details in Examples X9.2 Part A and X9.2 Part B respectively .The two sets of temperature rises due to effects of various GHGs are shown in Fig. F9.7A and Fig. F9.7B respectively. The equation E9.6 appears to give more realistic flatter temperature rises.

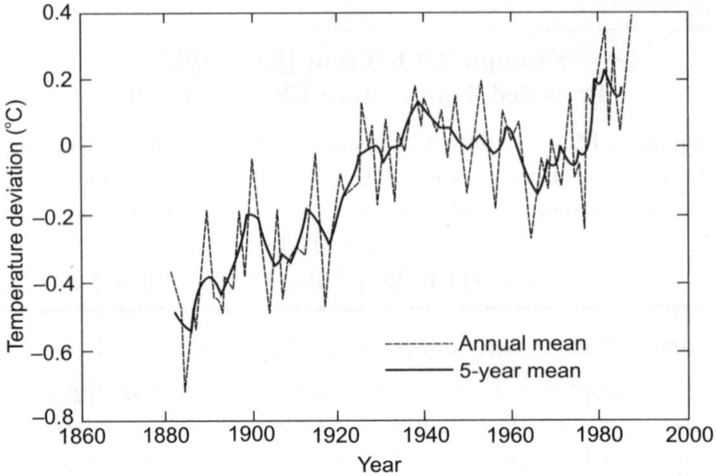

Fig. F9.6 : Global Surface Temperature Change (Munna Lao Observatory)

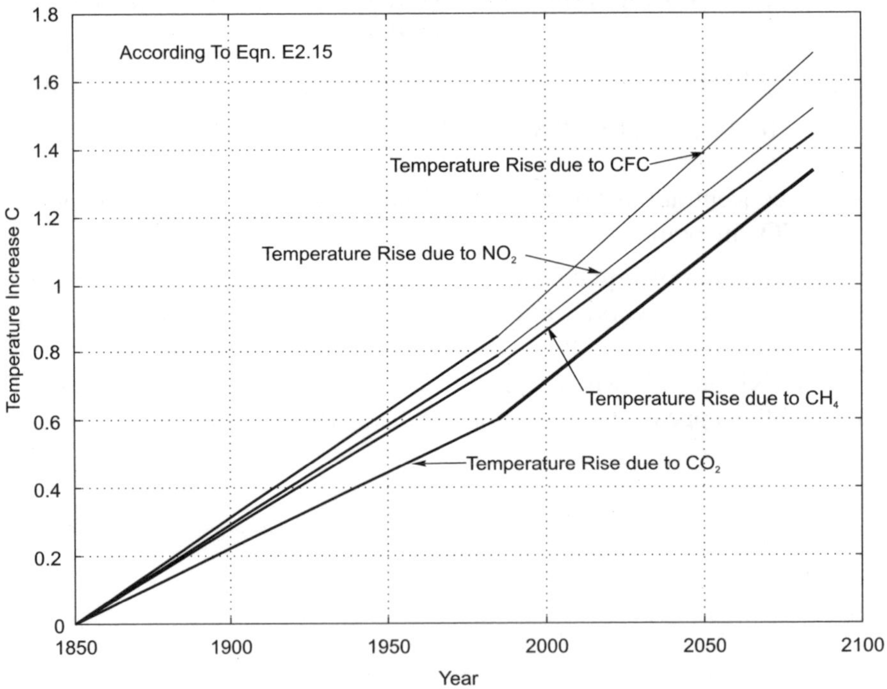

Fig. F9.7A: Temperature Rise due to various GHG Effects

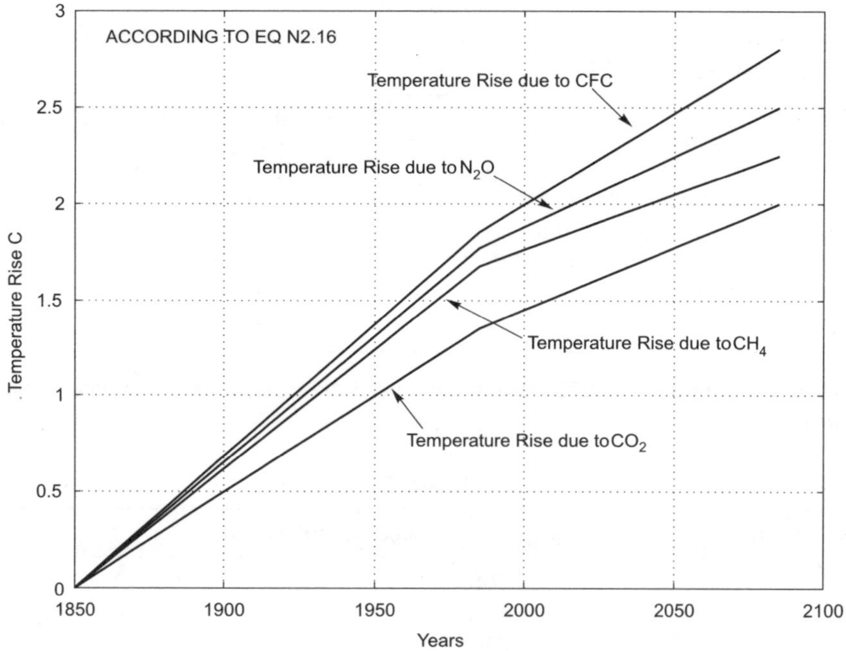

Fig. F9.7B: Temperature Rise due to various GHG Effects

Example X9.2 (Part A)
Temperature Changes due to GHG Warming

Task: Draw Earth-Atmosphere Temperature Change graphs for one hundred years from 1985 to 2085, given the following data in Table A.

Table A

GHG Name	GHG Concentration		Annual Incremental Concentration (Period 1985 –2085)
	1850	1985 (Increase /year)	
Carbon Dioxide (CO_2)	280 ppm	345 ppm (0.5 ppm/Y)	1 ppm/annum
Methane (CH_4)	1150 ppb	1790 ppb (4.7 ppb/Y)	5 ppb/annum
Nitrous Oxide (N_2O)	285 ppb	305 ppb (0.5 ppb/Y)	0.5 ppb/annum
CFC-11 & CFC-12	0 ppb	0.3 ppb (0.002 ppb/Y)	0.01 ppb/annum

Table B

As discussed in Section 9.2, DTd is taken as 2°C and earth-atmosphere temperature is derived from equation E9.5:

Formulation	Temperature Change DT C Period 1850- 1950
DT(CO2) = [DTd/ln2] x ln[345/280]	0.602
DT (CH4) = 0.019 x [1790**0.5 – 1150**0.5]	0.159
DT (N2O) = 0.057 x [305**0.5 –285**0.5]	0.033
DT (CFC) = 0.16 x [0.3- 0]	0.048
Total DT	0.842

The GHG concentration in 2085 are derived from the guidelines in the data Table A and are shown in Table C.

Table C

GHJG Name	GHG Concentration	
	1985	2085
CO_2	345 ppm	445 ppm
CH_4	1790 ppb	2290 ppb
N_2O	305 ppb	355 ppb
CFC	0.3 ppb	1.3 ppb

The temperature change ranges in 2085 are derived from equation E9.5 in Table D:

Table D

GHG Names	Temperature Changes DT Period 1985 - 2085
$DT(CO_2)$ = [2.0/ln2]x[ln(445/345)]	0.735 C
$DT(CH_4)$ = 0.019[(2290)**0.5 – (1790)**0.5]	0.105
$DT(N_2O)$ = 0.057[(355)**0.5 – (305)**0.5]	0.078 C
DT(CFC) = 0.160[1.3 – 0.3]	0.160 C
Total DT	1.078 C

The temperature change DT graphs between 1985 and 2085 per Eq. E9.5 are shown in Fig. F9.7A

Example X 9.2 (Part B)

From Table T9.5B, we know that the greenhouse effect multiplier with respect to CO_2 (taken as 1) for CFC, Methane and Nitrous Oxide are respectively 15,000, 25, and 230 respectively. Using these known properties and using the benchmark value of temperature change DT=2 C for doubling of CO_2 concentration, we develop a set of alternative temperature equations E9.6 similar to E9.5 but simpler.
Using the data of Table A of Example X9.2, we get alternative results of Temperature changes for the period between 1885 and 1985 in Table CC and for the period between 1985 and 2085 in Table DD. These results seem to be more realistic than the earlier results by equations E9.5.

$[DT]co_2$	= 0.02.[d.CO_2],	d.CO_2 change in ppm	(a);
$[DT]ch_4$	= 0.0005.[d.CH4],	d.CH_4 change in ppb	(b);
[DT]nox	= 0.005.[d.N2O],	d.N_2O change in ppb	(c);
[DT]cfc	= 0.30.[d.CFC],	d.CFC change in ppb	(d);

...E9.6

Table CC

Temperature Change between 1885 and 1985 (Celsius)

$DT(CO_2)$ = 0.02(345 - 280)	= 1.35
$DT(CH_4)$ = 0.0005 (1790 -1150)	= 0.32
$DT(N_2O)$ = 0.005 (305 - 285)	= 0.10
DT(CFC) = 0.30 (0.3 - 0)	= 0.09

Table DD

Temperature Change between 1985 and 2085 (Celsius)

$DT(CO_2)$	= 0.02 (445 - 345)	= 2.00
$DT(CH_4)$	= 0.005 (2290 - 1790)	= 0.25
$DT(N_2O)$	= 0.005 (355 - 395)	= 0.25
DT(CFC)	= 0.30 (1.30 – 0.3)	= 0.30
Total DT (1985-2085)		=2.80 C

The temperature changes DT (1985-2085) per Eq. E9.6 are shown graphically in Fig. F9.7B

❑❑❑

10

CHAPTER

Carbon Emission Scenarios

10.1 Introduction

The release of carbon in the atmosphere in the form of carbon dioxide (CO_2), methane (CH_4), carbon monoxide (CO) and indirectly by deforestation is a cause of much concern, generating health hazards as smog, global warming etc. Here in this section, we will study atmospheric carbon problems, keeping in mind the following salient aspects:

- Present carbon accumulation in the atmosphere;
- Current annual carbon build-up in the atmosphere;
- Future scenario of energy mix and carbon pollution;
- How to reduce the atmospheric concentration by natural effort and under International Agreements

10.2 Estimation of Present Carbon Accumulation in Air

We determine carbon content from the CO_2 source, which is the major source. From the Mauna Loa Observatory data, we know that the 1990 level of CO_2 is about 355 ppm, say increasing 1 ppm per year to 365 ppm.

Box B2.4: Amount of Carbon in the Atmosphere (2000)

Now 1 gm-mole of CO_2 has a mass of (12+2x16)=44 gm.
Again, 1 gm-mole of CO_2 has a volume of 22.4 litres at NTP under Avogadro Hypothesis.
Total mass of atmospheric air is known to have a mass of about 5.1x10**18 kg;
And the density of at NTP is 1.29 kg/m**3.
Therefore total volume of air is (5.1x10**18/1.29) = 3.9535x10**18 m**3.

Contd...

Now, 1 million cu. Meter of air contains 365 cu. meter CO_2 equal to (365x44)/(22.4)=716.96 kg CO_2.
Then, total atmospheric air volume 3.9535x10**18 cu.m. contains –
716.96 x 3.9535x10**12 kg CO_2
= 2834.5 Gigatonne CO_2
= 2834.5x12/44 Gigatonne Carbon = **773 Gigatonne Carbon**

From the above Box B10.1, we see the total mass of carbon in the atmosphere amounts to about 773 GT. Further, the total carbon content of 773 GT at 365 ppm level of concentration is equivalent to

1 ppm Carbon concentration in air = 773/365 =2.12 GT Carbon

i.e., 2.12 Gt C / ppm.

Thus, in year 2000, the atmosphere already contained 773 GT carbon, which is equivalent to only 1.5 kg carbon per sq. metre of gross earth surface (Earth radius is 6378 km). It is theoretically possible to absorb the complete carbon from the atmosphere by forestation as the above represents only 15 tonnes biomass per hectre (equivalent to 10**4 sq. metre).

10.3 Annual Build-up of Carbon in Atmospheric Air

Carbon is continuously emitted to the atmosphere by plant photosynthesis and by fuel combustion. During daytime, the plants absorbs solar energy and inhales CO_2 and water vapour and by photosynthesis converts them to cellulosic glucose $(C_6H_{12}O_6)$n and releases oxygen.

$$6CO_2 + 6H_2O + \text{solar energy} = (C_6H_{12}O_6) + 6O_2 \qquad \text{...E10.1}$$

The complex glucose is the cellulose building material of the plant body. Thus there is reduction in the CO_2 level of the atmosphere.

During the night the process reversed. That is the plants derive energy by inhaling oxygen and by the break-up of the glucose back to CO_2.

$$C_6H_{12}O_6 + 6O_2 = 6CO_2 + 6H_2O + \text{released energy...} \qquad \text{E10.2}$$

However, CO2 released is always less than the CO_2 absorbed, the difference is represented by the biomass of the plant body. Thus, a plant is a carbon sink and any form of forestation leads to net carbon reduction.

Thus, as a contra hypothesis, any form of deforestation leads to building-up carbon in the forms carbon monoxide(CO), carbon dioxide(CO_2), Methane(CH_4) etc. in the atmosphere. Siegenthaler et el (1987) and Bolin (1986 estimate such biotic contribution to atmosphere carbon to a statistical mean of 1.2 gigatons per annum somewhat roughly. We have indicated in the earlier section that the total annual carbon build-up is of the order of 2.12 gigatonnes.

Annual Contribution of Fossil Fuels to Atmospheric Carbon

The world consumption of primary energy other than biomass in one year (1985) is given at 302×10**15 BTU (*i.e.,* 302 Quadrillion BTU) or 302×10**18 Joules (302 ExaJoule) as given by EIA, USA and shown in the Pie diagram F10.1. Of the energy consumption, 88% (226 Quads) are supplied by fossil fuels (86 quads coal, 60 quads natural gas, & 120 quads petroleum).

Table T10.1 shows the conversion factors of generation of carbon in million tonnes per Quad energy (equivalent to tonne per Terajoule). It will be seen that the conversion of gas emits the least amount of carbon per unit energy at 14.5 tonnes per million megajoule, which coal produces 25.2 tonne of carbon per million megajoules.

The annual release of carbon from the fossil fuel amounts to 5.6 gigatonne from Table T10.2 and that from the biomass energy amounts to 1.2 gigatonnes from the earlier section. The total carbon released is 6.8 GT, but about 50% is absorbed by the sea and biomass and 50% goes into the atmosphere as seen from Table T10.3.

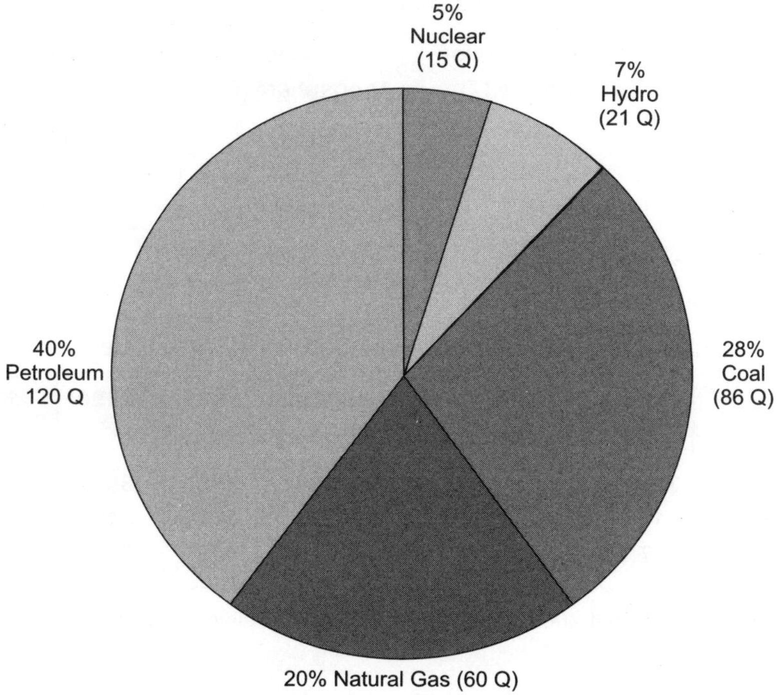

Fig. F10.1: World Primary Energy Production (Without Biomass)

*Source:*US EIO 1985

Table T10.1: Carbon Emission Factors for various Fuels

Energy Source	Carbon Emission Rates (Million tons per Quad)
Natural Gas	15.3
Petroleum	20.0
Coal	26.8

WRI/BP Report 2005

Table T10.2: Release of Carbon from Energy Consumption (Non-Biomass)

Energy Sources	Energy Consumption Quads (ExaJ)*	Carbon Content 10^6 tones/Quad	Release of Carbon Gigatons
Natural Gas	108	15.3	1.65
Petroleum Oil	169	20.0	3.38
Coal	124	26.8	3.32
Total	401	-	8.35

Note: 1 Quad = 1×10^{18} BTU; 1 ExaJ = 1×10^{18}J 1Q = 1 ExaJ.

* Derived from WRI/BP Rept 2005, Fig. 8.7, p.43.

Table T10.3: Annual Release of Carbon to Atmosphere

Energy Sources	Annual Generation of Carbon (GT)	Part released to Atmosphere (GT)	Percentage%
Biomass*	1.20	0.6	126
Fossil Fuel* *	8.35	4.18	87.4
Total	**9.55**	**4.78**	**100**

*Flavin(1990) / Siegelthaler (1997)

* * W.R.I./ B.P. 2005

Table T10.4: Measured Annual Carbon Release Rate

Year	1960	1965	1970	1975	1980	1985	1990	2000
Carbon ppm								
Upper Limit	321	322.5	327	333	339	348	357	–
Lower Limit	314	317	323	327.5	334	341.5	349.5	–
Median ppm	317.5	320	325	330	336.5	345	353	380

Source: Mauna Loa WRI 2005 Obsevatory, Hawaii / Keeling et al.(1990) /Masters (1991)

Annual Carbon Emission Rate:

Fig.10.2 shows the highly fluctuating CO_2 concentration data from the work of C D keeling et al.(1989) at Mauna Loa Observatory for years from 1960 onwards to 1990. On the same Figure we draw by dotted line upper tangential curve and the lower tangential curve touching the crests and the troughs of the fluctuating waves. From the data of the upper and lower limit lines, we also derive the median values.

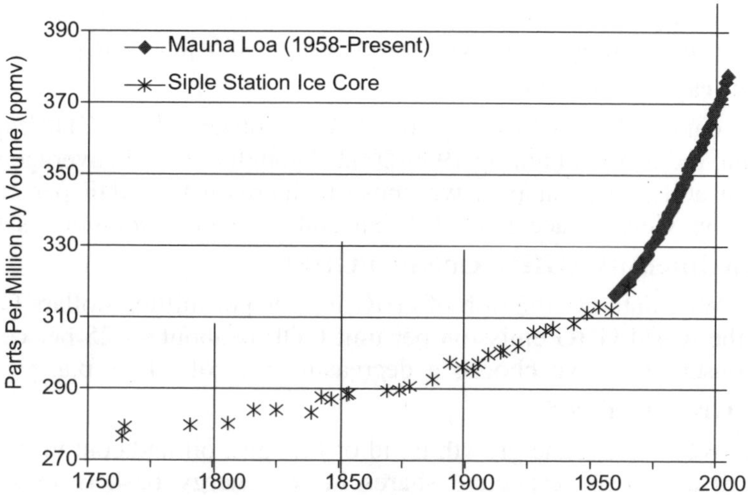

Fig. F10.2: Concentration of Atmospheric CO_2 (Mauna Loa Observatory 2002)

The median and the two limit values are now compiled in Table T10.4 with the hope that they may be used to derive a relationship equation ppm-year.

Building Carbon Emission Scenario

To envisage carbon emissions in the coming decades, we need to know the type of global social, economic and technological states in short and long terms. In general, the underdeveloped countries will work for more education, health, and better housing and more economic well being through industrialization, modern agriculture, water resource and infrastructure development and finally enhanced awareness for preservation of the natural environment. All these measures will demand higher energy consumption. This means carbon emission will grow faster, as indicated in the Fig.F10.2 earlier. This steep upward trend in CO_2 emission is already exemplified by the growth of CO_2 by 60 ppm in 1950-2000 compared to 15 ppm in 1900-1950.

Fortuitously, because of dwindling stocks and spiraling costs of fossil fuel, newer non-carbon energies, both renewable and non-renewable, are progressively emerging. To mention, Nuclear Energy, Solar Energy, Wind Energy and still developing Hydrogen-based Fuel Cell energy. The development of non-fossil fuels will push down the carbon emission more and more in the longer perspectives.

Keeping the above overall perspectives, we develop the growth trends of several parameters, which directly affect the futuristic carbon emission scenarios, *e.g.*, Energy, Population, GDP, Emission intensity and Energy intensity as under:

Population Growth Trend

From the data we observed the growth trend of India, China, US, Japan and EU, which vary from 0.5 to 2 percents per annum. We concluded that the average in the coming years will hover around 0.5 per cent per annum.

GDP per capita Growth Trend

From the data of World Resource Institute we observed the GDP/capita for US, China, India, World during 1980-2002. Though the world average in 2002 was given at 1.3 percent p.a., we chose to increase the GDP per capita to 1.5% p.a. in view of accelerated Asian and African developments today.

Emission Intensity (GHG emission/GDP)

From the WRI data of emission of GHG in tons per million dollars GDP, we derived the world GHG emission per unit GDP is about –1.25 per cents p.a. To be conservative, we choose a decreasing rate of –1 % p.a.

Energy Growth Trend

From the WRI data of the growth trend of fuel gas, oil and coal in the period 2002-2030 and their respective shares in the energy baskets in 2004 we derived the energy growth trend at 1.8% p.a.. To be conservative, we choose the Energy Growth rate at 1.5% p.a.

Energy Intensity (Energy/unit GDP)

Having,	Population growth	@ 0.5 % p.a.
	GDP per capita growth	@ 1.5 % p.a.
	Energy growth	@1.5 % p.a.

we derive

Rate of Energy Intensity Growth = Energy/GDP,
= (Energy Rate)/[(Population Rate).GDP/pop.rate)],
= (e**1.5)/[(e**0.5).(e**1.5)]
= [e**(-0.5)]

Hence we observe that energy intensity, *i.e.*, energy consumed per unit GDP is decreasing @ -0.5 % per annum.

Having developed the normative growth trends of population, Energy, GDP per capita, Emission intensity, Energy Intensity, we now present two Scenario studies of Energy and carbon emission.

Scenario I: Carbon emission in 2050

The details of the working of the Scenario are given in Example X10.1. We present the highlights here –

(*i*) Energy grows at 2 % p.a. till 2050.

(*ii*) Fossil fuels Mix in 2007 and 2050 are assumed as below.

Energy Mix	Year 2007	Year 2050
Coal	24 %	20 %
Petroleum oil	35 %	15 %
Natural Gas	21 %	15 %
Totals	80 %	50 %

The results of the scenario studies are as follow:

Parameters	Year 2007	Year 2050
Global Energy Demand	450 EJ	1053 EJ
Gross Carbon Emission	5.5 GT	11.5 GT
Net Carbon Emission (@50%)	2.8 GT	5.6 GT
CO2 Emission Increase	1.3 ppm	2.6 ppm

Observations

Even after assuming that the Fossil fuel mix coming down from 80% in 2007 to 50 % in 2050, the yearly growth in carbon dioxide emission will increase by 2.6 ppm in 2050. This is an unacceptable proposition. Use of Fossil fuel mix in 2050 must come down to 25% of the energy basket, if we are to allow energy growth at 2 %.

Scenario II: Energy & Carbon emission in 2100:

The details of the working of the Scenario are given in Example ×10.2. We enumerates the highlights and observations here.

Principal Parameters	Percentages	Units
Population Growth Rate	0.5 % p. a.	-
GDP per capita	1.5 % p. a.	($ ppp)
Carbon emission per unit GDP	-1 % p. a.	(Tons /million dollars)
Energy Growth Rate	1.5 % p.a.	(EJ)
Energy per unit GDP	-0.5 % p.a.	(EJ/million dollars)

Results:

	2007	2050	2100	Total Growth Ratio(times)
Energy Demand (EJ)	450	858	1815	4.0
Gross Carbon Release(Gt)	5.5	6.8	8.7	1.6
CO_2-release (ppm)	365	428	520	1.4

Observations

(*i*) The energy increases by over 4 times, CO_2 (ppm) increases 1.4 times over almost 100 years till 2100. The CO_2 increase being less than 2 times, the global temperature increase will be below 3 C. The Scenario II predicts that the world is not veering to any major disaster.

(*ii*) The Scenario II being parametrically built, it is supposedly more reliable than the scenario I.

❑ Example X10.1: SCENARIO I

Carbon Emission in Year 2050: With given Fossil Fuel Mix of the Scenario Given the following data:

(*i*) Considering that highly accelerated development is going on in all developing countries, energy demand in next 43 years increasing at 2% per annum upto 2050 will increase *i.e.,*

Total Energy increase in 43 years

[1.02 **43] × 33 234 %

(*ii*) Again considering that concerted efforts are going on globally to diversify into non-fossils like Bio-diesel in Transportation, fuel cells in automobiles, nuclear fuels in major power plants, renewable solar power in domestic and minor power plants, the fossil fuel consumption scenario in 2050 could be envisaged as under:

	Year 2007	Year 2050
Coal ...	24 %	20%
Petroleum Oil ...	35%	15%
Natural Gas ...	21%	15%
	80%	**50%**

(*iii*) Assume present energy and emission scenario in 2007 as:

Energy Consumption (in 2007) ... 450 ExaJoules

Carbon Emission (in 2007) ... 5.5 Gigatonnes.

❑ Solution

To build up the energy & Emission Scenario for 2050:

Total Energy Consumption in 2050, 450 x 2.34 =1053 EJ

Using Carbon Emission Factors from Table T10.1.

Carbon Emission from coal =25.0x0.20x1053 ... 5625 million tones C/Y

Carbon Emission from Oil =20.5x0.15x1053... 3238 -do-

Carbon Emission from Gas =14.5x0.15x1053... 2290 ... -do-

Total Carbon Emission in 2050 ... 11,153 million tones C/Y

Assuming exponential rate of increase of carbon emission per year,

C = Co (E**43r)

11.5 = 5.5 (E**43r) or r = 0.0172

Therefore, rate of increase of carbon emission of Carbon per year

1.7 % per annum

The quantitative release of carbon per year,

[11.5-5.5] /43 Giga tones 0.14 Giga tonnes per annum

Note: Please note that 1 ppm concentration of carbon in the atmosphere is equivalent to 2.12 giga tonnes at present.

❑ Example X10.2: SCENARIO II

Carbon Emission and Growth in Energy Supply in Next 93 Years to Year 2100

A. Scenario II: Notes:

1. Population growth is slowing down and stabilizing.
2. Global GDP growth is accelerating because of emerging economies.
3. There is increasing energy efficiency and hence less proportionate carbon emission

B. Scenario II: Data:

Population grows @ 0.5% per annum

GNP per capita grows @1.5 % per annum

Carbon Emission per unit energy decreases @-1.0 % per annum.

Energy per unit GNP grows @ -0.5 per annum

C. Current Data:

CO_2 Emission airborne ..	58 %
2007 Carbon Emission ...	5.5 Gt per annum
2007 CO_2 Concentration in atmosphere	365 ppm
2007 Primary Energy Supply	450 ExaJ

❑ Solutions

SA: Energy Growth: (See Fig. F10.3A) 1.5% per year

Rate of energy supply is derived,

Energy Supply = Population x (GNP/Population) x (Energy/GNP)

Assuming continuous exponential growth rate of all factors,

E**renergy = (E**rp) x (E**r.gnp/cap) x (E**r.eng/gnp);

r(energy) = r(pop) + r(gnp/pop) + r(eng/gnp)

= 0.5 + 1.5 – 0.5 = 1.5 % pr year

SB: Carbon Growth: (See Fig. F10.3B) 0.5 % per year

Similarly, Rate of Growth of Carbon Emission,

Carbon emission rate = (Pop.) x (GNP/Pop.) x (Energy/GNP) x (Cemi./Energy);

Again, assuming continuous exponential growth of all factors,

r(carbon) = r(pop.) + r(gnp/pop.) + r(eng/gnp) + r(carb./eng.)

= 0.5 + 1.5 – 0.5 – 1.0 = 0.5 % per year.

SC: CO_2 Emission: (See Fig.F10.3C)

Now carbon emission in any year t,

= 5.5 x (E**0.005t) Gt per year;

Further carbon content per 1 ppm CO_2 in the atmosphere = 2.1 Gt C;

Therefore, addition of carbon from base year 2007 to year number t in Gt,

= 5.5.[Io-t (E**0.005t) , where Io-t is the integral from 0 to t ;

= (5.5/0.005) x [(E**0.005t) - 1]

Addition of carbon from base year 2007 to year t in ppm

= (1100/2.1) x [(E**0.005) - 1]

Add total concentration of CO_2 in ppm in 2007 @365,

Concentration of CO_2 in year t = 365 + 524.[(E**0.005t) - 1]

❑ SD: Equations:

The equations that follow from the exercises in SA, SB, and SC above are:

Et = 450 [exp (0.015 t)] exaJoules in year t;

Ct = 5.5 [exp (0.05t)] gigatonnes in year t;

CO_2 (ppm) = 365 + 524 [exp (0.005) - 1] ppm in year t.

The numerical values are tabulated below

Scenarios: Growth of Energy Consumption, Carbon Emission and CO_2 Emission

Year	2007	2010	2020	2050	2100
Energy Growth (EJ)	450	471	547	858	1815
Carbon Growth (GT)	5.5	5.6	5.9	6.8	8.7
Carbon Dioxide (ppm)	365	369	383	428	520

The above data are shown graphically in Figs. F10.3A, F10.3B and F10.3C.

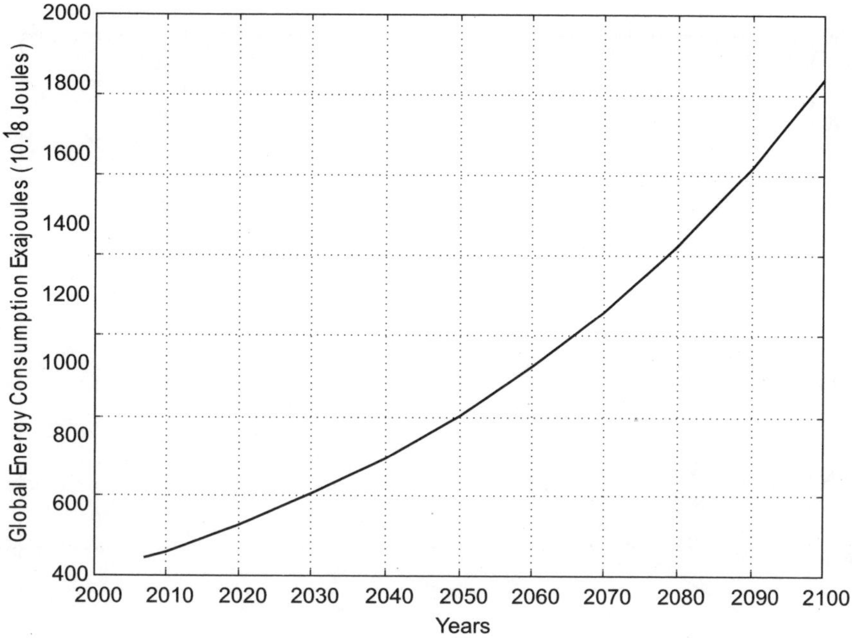

Fig. F10.3A: Trend of Energy Growth 2000-2100

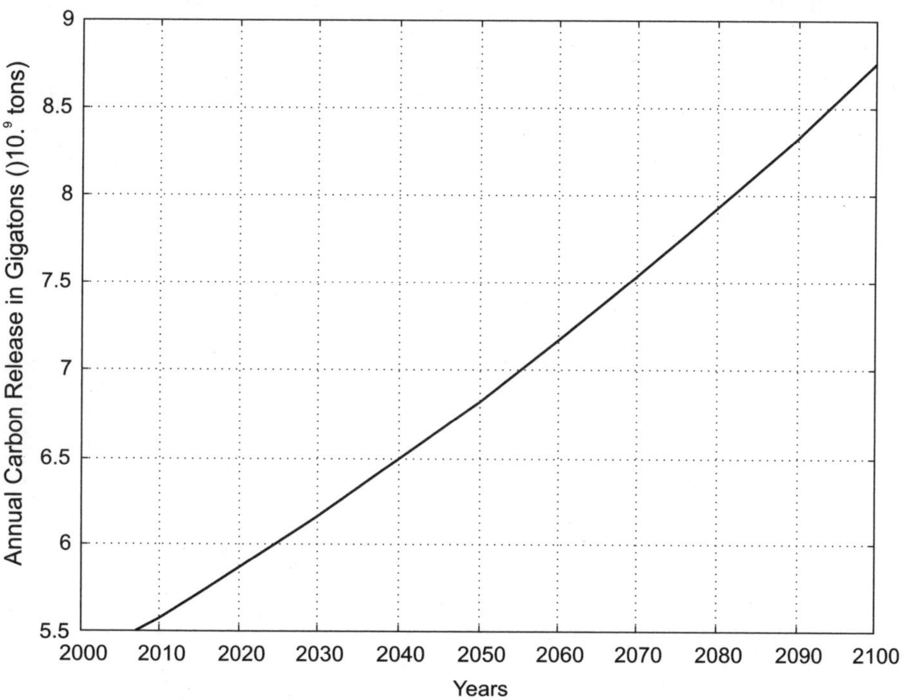

Fig. F10.3B: Trend of Carbon Release in Atmosphere in 2000-2100

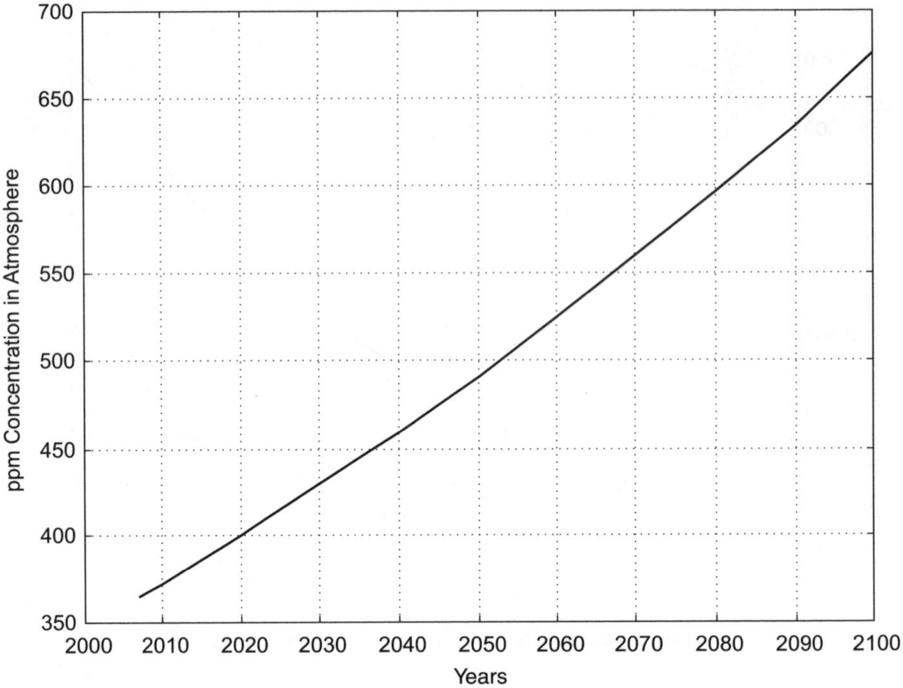

Fig.F10.3C: Trend of ppm Concentration of CO_2 in Atmosphere 2000-2100

❑❑❑

11

CHAPTER

Climate Change And Carbon Trading

11.1 International Conventions on Climate Change

The emission of greenhouse gases is widely contributed from all countries of the world but the developed countries release more than half of the global total with much less than half of the world population. Further, the developed countries also posses the economic and the technological capabilities to control the GHG emission. Because of these uneven responsibilities and differentiated capabilities, it was found rather difficult for a long time to find an equitable consensus among nations for effective international measures for control of GHG emission.

To overcome the above difficulties, the United Nations Environment Programme (UNEP) and the World Meteorological Office established the International Panel on Climate Change (IPCC) in 1988. Under the aegis of the United Nation Conference on Environment and Development (UNCED), the IPCC, in 1992, was instrumental to organize the Framework Convention on Climate Change (FCCC) following the Earth Summit at Rio de Janeiro. The task of FCCC was to monitor and investigate the various equitable and effective measures to stop the increase of carbon dioxide concentration in the atmosphere, which is principally caused by anthropogenic activities. The guiding principles of FCCC may be broadly stated as under:

(*i*) the need to protect the climate system on the basis of equity and in accordance with the States' common but differentiated responsibilities and capabilities; accordingly, the developed country parties should take the lead;

(*ii*) the specific needs and special circumstances of developing country parties, especially those who are particularly vulnerable;

(*iii*) the need for the precautionary measures in the absence of full scientific certainty, qualified by the need to be 'cost effective' and comprehensive; that is, address all sources and sinks, adaptations and all economic sectors;

(*iv*) parties have a right to and should promote sustainable development, integrated with the national development programmes;

(*v*) parties should cooperate to promote a supportive and open international system; measures taken should not constitute a means of arbitrary or unjustified discrimination or a disguised restriction on international trade.

In determining the share of responsibilities and capabilities of various countries, following considerations were kept in view:

Responsibility: These should be reflected by the percentage of total carbon emission.

Capability: Basically, economic well-being of a country, which indirectly also reflects technological ability. Economic well-being is measured by the per capita income for that country.

Based on the above two criteria, the Convention on Climate Change divided the countries in two categories –

Annex I countries:OECD countries, Eastern Europe and former Soviet Union.

Non- Annex Countries: Other than Annex I countries, generally all developing countries.

Table T11.1 gives the names of Annex I countries and some of the non-annex countries with their population, per capita income (PPP US Dollars), per capita carbon emission. Some of the findings are very revealing. For example, Annex I countries emit 63.9 % carbon, while the large number of other countries emit only 36.1. The per capita income of Annex I countries were 15,986.00, whereas the others had per capita income of only USD2904.00 (PPP).

Table T11.1

	Population in 1996 (millions)	Per Capita Income (1994) (PPP $)	Per capita emissions (1996) (Tons)	Share in carbon emissions (%)
Annex I Countries	**1299**	**15,986**	**3.1**	**63.9**
USA	262	26,397	5.6	24.4
European	369	18,570	2.5	15.1
Japan	125	21,581	2.5	4.9
EE/FSU*	411	403	2.0	14.1
Non-Annex I Countries	**4438**	**2904**	**0.5**	**36.1**
China	1232	2606	0.6	13.4
India	951	1348	0.8	12.6
Other Asia	892	3763	0.8	12.6
Africa	708	1134	0.3	3.3
Latin America	478	5873	0.4	3.3
WORLD	**5666**	**5798**	**1.1**	**100.0**

Source: U Shankar et al (2001) * European Union/ Former Soviet Union

11.2 Kyoto Protocol, 1997

As a sequel to the 1992 Convention on Climate Change, the parties to the Convention drew up a Protocol of obligations in quantitative terms for control of carbon emissions in December 1997 at Kyoto in Japan, known as Kyoto Protocols 1997. The principal terms of the Protocol were laid down as under:

(*i*) Firstly, quantitative limitation or commitment of emission reduction by Annex I countries as to reduce the aggregate GHG emission by an average of 5.2 per cent below the 1990 levels by period 2008-21.

(*ii*) Commitment by Annex I countries to show demonstrable progress in this regards by 2005.

(*iii*) Thirdly, commitment by Annex countries to prepare an inventory of GHG emissions and reporting of national plans of emissions to IPCC/FCCC.

(*iv*) Introduction of following four flexible mechanisms, which will allow compensatory emission reductions beyond geographical borders of an Annex country making commitment to the Protocol -

Joint Implementation (JI);

Bubble;

Internal Emission Trading (IET);

Clean Development Mechanism (CDM).

(v) Lastly, no requirements of any commitments by any non-Annex I countries.

The main reason of creating the above four project-based mechanisms is to provide Annex countries opportunities to meet their Protocol commitment of emission reduction in a cost-effective manner to the extent possible. At the same time, these mechanisms enable non-Annex I countries to undertake emission reduction measures with the economic and technical assistance of Annex I countries.

JI enables an Annex I country, with GHG emission limits, to contribute financially and technically with a project of emission reduction or creation of additional sink in another country with a national commitment and earn credit commensurate with its contribution to the project.

CDM is similar to JI, but it is more flexible in that the second country even need not have a national plan.

IET is tradable quota regime, where the number of quota defines its committed emission limits. Naturally, the quota can be achieved in a second non-Annex I country.

There were a number of issues and procedures, which require to be clarified or correctly defined like baseline emission, monitoring and verifying requirements, and registration requirements of projects.

At the end of the decade today the Kyoto Protocol still remains to be ratified due mainly to the USA opposition. The USA opposes the exemptions from all commitments given to large emerging economies like China and India, which countries are already contributing to large carbon emissions and which emissions are likely to substantially increase in the near future.

11.3 Carbon Trading

Carbon Trading under the Kyoto Protocol enables a company or an institution in one country to buy credit of carbon emission reduction achieved by another company or institution in another country to the benefit of its own commitment of reduction of carbon emission. This mechanism allows economic optimization of cost of controlling carbon emission without compromising the global objective of emission reduction.

The trading media unit is CER. Like the Dollar, a CER is the basic unit in carbon trading markets. Each time a company, through its actions like using lesser or cleaner fuel reduces the emission of one tonne of carbon, it accumulates a CER. It can then sell this CER for a price – 12 Euros per CER at the moment – to another company in another part of the world that might want to reduce the emission, but unable to do so due to some economic or technical reasons.

In India, companies like Tamil Nadu Newsprint and Paper Ltd. and Gujarat Flurochemicals have taken a lead in developing projects to reduce emissions and accumulate CERs. A project – a Clean Development Mechanism (CDM) project per Kyoto Protocols – can be something as simple as planting more trees to something more complex such as adopting a new manufacturing process or an environment-friendly carbon-reducing waste-disposal system.

With the CDM benefits becoming more defined and achievable, a large number of Indian companies are undertaking investments in emission reduction projects. Companies in China and Brazil are still more active.

One Indian Public Sector Undertaking, ONGC are getting 14 new CDM projects approved by IPCC administration, which include doing away with gas flaring, increased energy efficiency and waste heat recovery. But let it be appreciated that such projects ae not easy to set up. The process of getting approval, registration and implementation is quite rigorous.

11.4 Bali Summit, 2007

Over 100 countries met at Bali, Java Island in Indonesia in December 2007 to carry out the task of halting and eliminating carbon emission in the atmosphere beyond the Kyoto Protocol. Fortunately, this time, USA and Australia, the two countries that did not sign the Kyoto Protocol joined the Bali Summit. The Summit agreed to carry out mutual consultation over two-year time period to work out global consensus to break a new path. A new Summit has been planned under aegis of UNCED/IPCC at Copenhagen in December 2009.

11.5 Compenhagen Accord 2009

In terms of the Bali Action Plan. One hundred and ninety two countries assembled in the United Nations Climate Change Conference at copenhagen, known popularly as the Copenhagen Summit, hold between 7 and 18 December, 2009.

As a comprehensive legally binding unanimous agreement could not be arrived at, United States in conjunction with China, India, Brazil and South Africa authored a "Copenhagen Accord", which was hailed by the UN Secretary General as a new beginning towards adaptation and intigation of the climate change programme.

The Copenhagen Accord highlighted the following principles :

(*i*) The principle of common but differentiated respouribilites and respective capabilites among the developing countries;

(*ii*) Climate change is to be so addressed, that the temperature rise is held below 2° Celsis by 2050;

(*iii*) Developed countries shall submit their reduction commitment in Appendix I form and Developing countries shall submit then nationalty appropriate mitigation in Appendix II form.

The above national undertaking should be subject to intanational consultation one analysis.

(*iv*) Developed countries shall commit mitigation funding of USD 30 billion for the period 2010-2012, which fund should be impressed to USD 100 billion by 2020.

❑❑❑

Part II

References

Ahmed Feroz 2006, Report: *"The Scary Story of Climate Change"*, Business World, 18 September 2006, Vol.26, Issue 17, New Delhi.

Gilbert M Masters 1991, *Introduction to Environmental Engineering and Science*, Prentice Hall (USA1991)/ (India 2000).

Hamit J K , Wolf K A *et al* 1986 *Product Uses and Market for Potential Ozone-depleting Substances in Assessing the TRACE Gases*, USEPA .

Hansen J & Leberdoff S 1988, *Global Surface Air Temperature Update through* 187, Geophysical Research Letters 15:323-326.

Kalpagam U 2007, *India's Business Prospects in the Global Emission Market*, Global Business Review, 8:2(20070; p237-250, Sage Publication, New Delhi.

Keeling C D & Bacasta 1989, *Impact of Industrial Gases on Climate, Energy and Climate*, Natural Research Council, Washington D C.

Krishnamoorthy Bala 2005, *Environmental Management*, Prentice Hall India, New Delhi.

Mahajan S P 1985, *Pollution Control in the Process Industries*, Tata- McGraw hill, New Delhi.

Molina M J & Rowland F S 1974, *Stratospheric Sink for Chlorofluoromethane – Chlorine Atom Catalyzed destruction of Ozone*, Nature 249: 810-812.

Shankar Ulganathan ed. 2001, *Environmental Economics*, Oxford University Press, New Delhi.

Shea C P 1989, Global Warming, Sierra Club Books, San Francisco.

Sandler Todd 1997, *Global Challenges* – An Approach to *Environmental, Political & Economic Problems*, Cambridge University Press.

Siegelhalter U *et al* 1987, *Biosphere CO_2 Emission during past 200 years reconstructed by deconvolution of ice core date,* Tellus 39B:140-154.

Sincero A P & Sincero Gregoria A 1996, *Environmental Engineering – A Design Approach*, Prenctoce Hall,(USA1996)/(India1999).

Sonnenfeld D A & Mol P J 2006, "*Environmental Reform in Asia*", The Journal of Environment & Development, vol. 15,Number 2, June 06, Sage Publication.

Sunita Satyapal et al 2007, *Gassing up with Hydrogen*, Scientific American India 2007, p61-67, New Delhi.

Swamy N M 2003, *Textbook on Environmental Law*, Asia Publishing House, Hyderabad.

Walsh M P 2002, *Air Pollution: Automobile*, Wiley Encyclopedia, Energy & Environment Vol 1, Ed. Bissio.

Zweibel K *et al* 2008, *Solar Grand Plan*, Scientific American Jan 2008, p38-43, New Delhi.

Part III
Challenges of Energy

12
CHAPTER

Global Energy Challenges

Modern life thrives on energy. When we drive a car on roads, ride on railways or fly in the sky, motive engine consumes energy all the time. From the kitchen to the drawing room to the office and factories, the modern civilization consumes energy.

Global energy consumption (2002) is about 420 EJ (420x10**18 J), of which 25% is consumed by USA alone. Table T12.1 shows the energy consumption per capita per year at 173 GJ (173x10**9 J) in the developed countries and that for the developing countries is 34 GJ per capita per year. In other words, the people of developed countries use about 5.1 times more energy than that of developing countries.

Table T12.1: Global energy Consumption (2002)

countries %	World countries %	Industrial countries %	Developing
Oil	35	38.4	25.8
Coal	24	24.5	23.4
Biomass	11	2.8	38.2
Hydro	2+1	5.7	5.1
Nuclear	7	5.9	0.6
Natural Gas	21	22.7	7.1
Total	**100**	**100**	**100**
World : Total Energy		420 exajouls	
: Population		6.2 billion	
: Energy per cap		77 gigajouls	
Industrial Countries			
: Total Energy		260 exajouls; 66%	
: Population		1.5 billion; 25%	
: Energy per cap.		173 gigajouls	
Developing Countries			
: Total Energy		160 exajouls; 34%	
: Population		4.75 billions; 75%	
: Energy per cap		34 gigajouls	

Source: World Resource Report based on CAIT and IEA, 2002

In Table T12.2, we show country wise energy consumption (2002) for 25 countries, which are also top emitters of GHG (greenhouse gases) for 2002. From this Table, we see the obvious truth that higher the energy consumption, more is the emission of carbon dioxide and other GHG leading to the very undesirable climate change and the consequent global warming. In other words, the energy-intensive western countries are generally the prime polluter of the environment, as we will see in this book in the Part Environmental Challenges.

We also show in the two Pie Diagrams of Fig. F12.1 that fossil fuels oil, Gas and coal contribute the major parts of all primary energy needs of the globe. Table T12.3 summarizes the data.

Table T12.2: Energy Consumption of Top 25 GHG Emitting Countries (2002)

Countries	Total Energy million TOE	TOE / capita	Rank (TOE/cap)
USA	2300	8.0	1
E U	1700	3.7	11
China	1200	1.0	22
Russia	700	4.4	6
India	510	0.6	24
Japan	500	42	9
Germany	350	4.3	8
France	250	4.6	5
Canada	225	8.0	1
U K	200	3.8	10
S Korea	190	4.4	6
Brazil	175	1.2	20
Italy	160	3.0	13
Mexico	150	1.7	18
Indonesia	150	0.8	23
Iran	140	2.0	17
Spain	140	3.3	12
Ukraine	140	2.8	14
S Arabia	140	5.8	3
S Africa	130	2.6	15
Australia	130	5.7	4
Poland	110	2.3	16
Turkey	100	1.2	20
Pakistan	95	0.5	25
Argentina	90	1.5	19

Source : WRI 2002

Table T12.3 : Fossil Fuels & GHG Emissions 2002

Energy Source	Energy Contribution %	GHG Emissions %
Oil	35	37
Coal	24	37
Gases	21	20
Total	80%	94%

Source : WRI, 2002

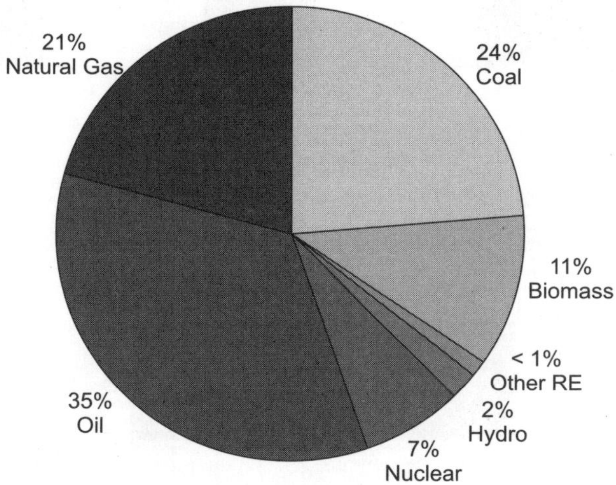

Fig. 12.1A: Primary Energy Source (2002)

Source: WRI/CAIT, IEA(2004B)

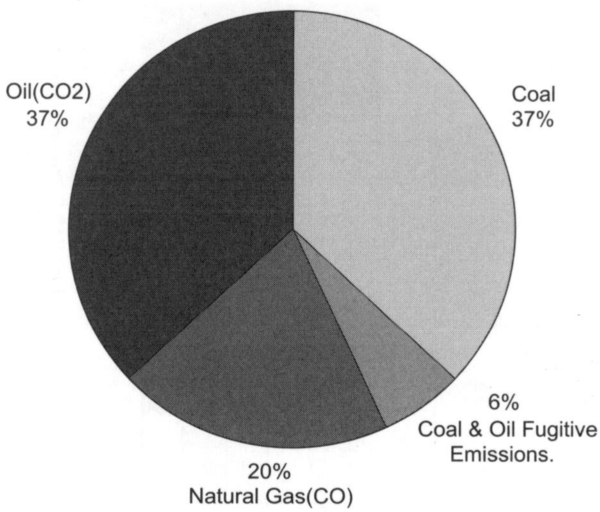

Fig.12.1B: Primary Energy GHG Emission (2002)

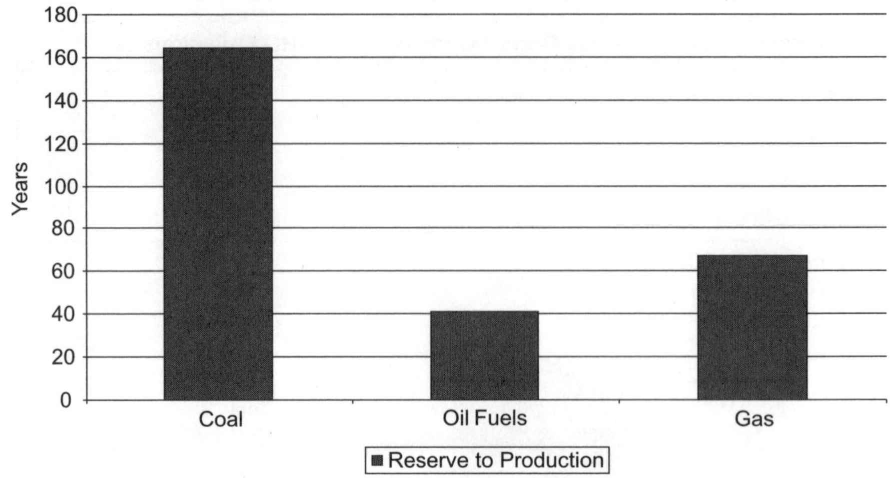

Fig.12.2 Reserve to Production

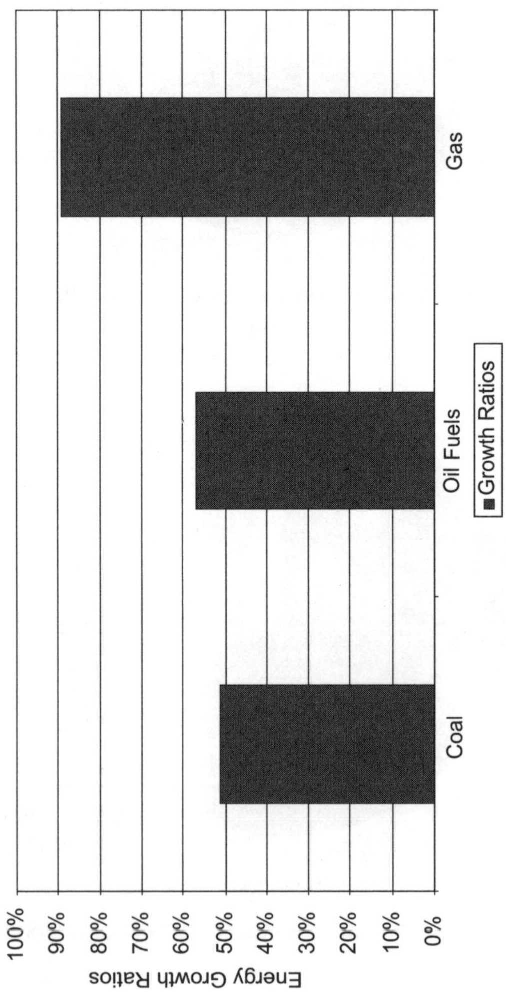

Fig. F12.3 Projected Growth in Energy Demand

We note from Table T12.3 that fossil fuels, while supplying 80 percent of world energy needs, contribute almost all the global warming gases at 94 percent. This is a strong indictment against the fossils fuels pointing to the need of alternative non-polluting fuel. Fortunately, global stocks of these three fossil fuels are quite limited, as seen from Fig.F12.2. Provided the energy consumption remains at the current level, coal will last the longest for 164 years, while petroleum will last just 41 years. The availability scenario will look grimmer, if we notice that the energy demand in 2030 will substantially increase as predicted in Fig. F12.3. Gas demand is likely to increase by 90%. Even assuming that there will be new finds of fossil fuels, it is safe to predict that the days of fossil fuels, especially oil and gas are numbered. In fact, the increase in energy consumption should be expected as all less developed countries are undergoing vast infrastructure developments, particularly China and India.

Now fossil fuels are almost exclusively responsible for GHG emissions and the eventual global warming. The supplies of fossil fuels are reaching the peak and in the coming decades, their availability will start waning. Hence the hunt for alternative fuels which are eco-friendly. We refer to the energy Spectrum in Fig. F12.4. From the energy spectrum, we may infer, among the solid fuels, coal may last till the end of the current century. But the nuclear energy, having effectively overcome the problem of radio-active waste disposal and having reached the economic viability through continuous technological developments including fuel reprocessing and Breeder Reactor Technology, can support the world energy demand from 2030 onwards.

Among the renewable energy sources, solar thermal and photovoltaic energy is finding increasing applications and undergoing technological maturity and economic viability.

The other renewable wind energy source, which is already supplying considerable electrical power in Germany, USA, Denmark, Spain and even, in India. Large wind farms of capacities up to 160 MWe or more are already commissioned.

Another real renewable energy is Bio-energy producing both biogas and liquid bio-fuels like Ethanol, fermented from sugars are being produced in large measure in Brazil and USA for automotive engines. The latest bio-fuel is Bio-diesel, which has been successfully developed from transesterification of vegetable fatty oils (also from animal fats in small measure). Bio-diesel compares very favaourably with petro-diesel as per thermodynamic properties for use in the internal combustion engines. The Bio-diesels also emit GHGs but in much smaller scale than petroleum fuels. As long as the annual consumption of bio-energy is within the annual growth rate of the energy crops of all types, this energy source will be continuously available, hence called renewable energy.

Alternative Energy Programmes

In the light of the broad discussion on the various environment-friendly potential energy resources, we may summarize possible energy policies and programmes for the medium term (for a few decades) and long-term, say, up to 2100 or even 2150.

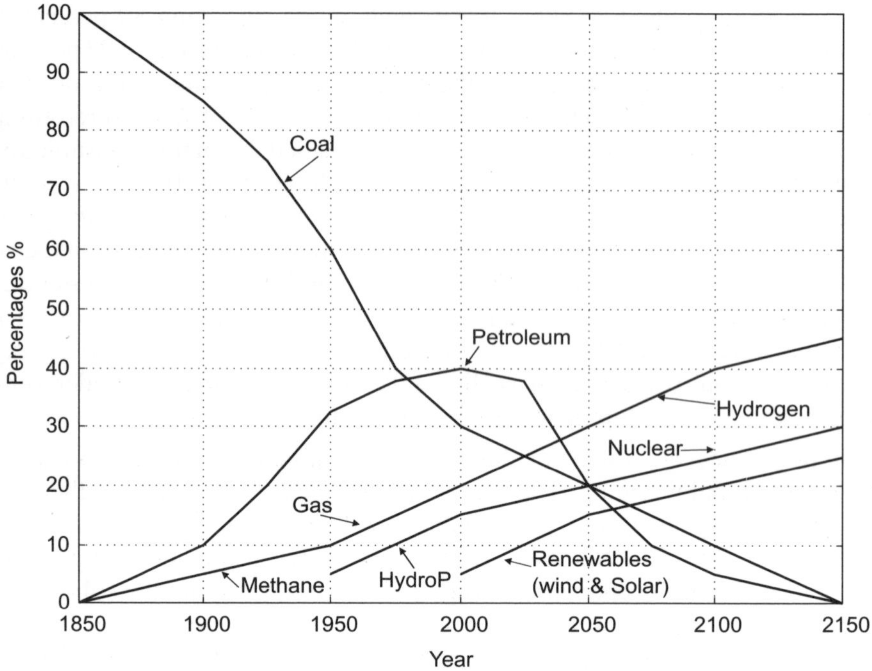

Fig. F12.4 Long-Term Energy Availablity (Adapted from Dunn)

Firstly, fuel cells with hydrogen energy will replace fossil fuel for the automobile industry in the next few decades. The vehicular transport sector accounts for about 30% of the energy consumed. Now, hydrogen energy is completely devoid of GHG emission. Hence the world will see significant drop in the environmental pollution.

Secondly, electric power generation will continue to use highly polluting coal/gas energy in the next several decades. But it will be overtaken by the nuclear fission energy within these two decades even in India. This option for the nuclear energy is prompted by the very significant improvement in the safety technologies of radioactive waste-processing ad even in the operation of the nuclear power plant. Hence, another 30% of energy consumption will get freed from GHG emission and other pollution hazards.

Thirdly, industry processes will gradually veer to more soft technology like biotechnology, where fermentation process may gradually but steadily replace many of the energy-guzzling combustion process, thus saving the world further of polluting industrial processes.

Fourthly, in the domestic sector, the use of solar, both thermal and photovoltaic energies will also lead to much reduction of GHG and other pollution emission.

Finally, in the long term by the end of the 21st century or about thereafter, the global energy supply will see the exit of the fossil fuel more or less completely and the emergence and domination of the hydrogen and nuclear energy of both fission and also possibly fusion types. Also there will be no GHG and other pollution nor waste radioactivity. Hopefully, given higher socio-economic knowledge and responsibility, the world will have nurtured the environment for the long-term healthy survival of the humans, other living creatures and the ecosystems in general.

❑❑❑

13
CHAPTER

Nuclear Energy

13.1 Nuclear Energy Scenario

Sir Arthur Edington first indicated the vast amount of sub-atomic energy at the 1930 World Power Conference in Berlin. How to access, control and use the energy stored in the atom per Einstein's famous mass-energy equation $E=m.C^2$ was the challenge.

On December 2, 1942 Enrico Fermi produced the world's first Nuclear Reactor named Chicago Pile I. Fermi's method to design, construct, operate and control the first nuclear reactor laid the foundation of nuclear engineering.

It will be seen from the Pie diagram Fig.F13.1 that coal at 40% is still the major power generation source, whereas nuclear power accounts for 16% of global electricity generation.

The uneven development of nuclear electrical power globally will be observed from Table T13.1. While France produces 78% of its electrical power from nuclear source, US produce only 20% of electricity from nuclear reactors.

The slow growth of the nuclear energy is much due to fear of nuclear accident and its radiation hazards and only partly due to higher economic cost at the initial stage of development. In the past 50 years there has been only two major nuclear accidents, *e.g.,* Chernobyl Nuclear Plant in Russia and the Three-Mile Power Plant in US. As a result, the building of nuclear plants was stopped in US due to public pressure from the eighties. This is the reason of lower percentage of nuclear power generation in US, but it has still the largest number of nuclear reactors in the world, 101 of 441 numbers in the whole world.

However, there have been no accidents in France and nuclear power developments in France have been progressing unabatedly. There has been considerable progress in the nuclear safety, including the containment system and also in the waste disposal. In the future Indian programme of high temperature Nuclear Reactor, there will be no moving parts, *e.g.*, pumps, which will further minimize the radiation contamination risks. There is also provision for cooling off of the nuclear core by a surge of automatic water flooding, should the temperature of the reactor rises above the permissible limit.

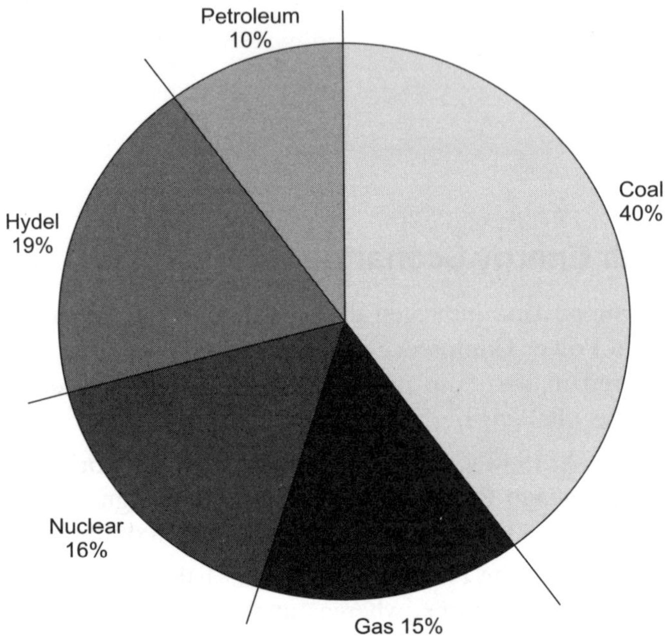

Fig. F13.1: Share Of Nuclear Power

Table T13.1: Nuclear Power (Reyce et al 2004)

Countries	Percentage Nuclear Power
France	78
Japan	35
Germany	32
United Kingdom	27
U.S. (98.2 Gwe)	20
Canada	15

Source: Reyce et al 2004

13.2 Basic Nuclear Physics

Nucleus

The atoms comprise of a densely packed central nucleus containing several positively charged protons and several uncharged neutrons and a peripheral cloud of negatively charged electrons, obviously of same number and same charge as of the protons. As a neutron or proton is 1800 times heavier than an electron, the atomic mass is assumed same as the mass of the nucleus. The following definitions should be noted in this respect.

Atomic Number: The number of protons is known as the atomic number representing the electrical property of an atom.

Mass Number: The total number of protons and neutrons together are known as the mass number.

Isotope: Elements having the same atomic number but different mass number are called Isotopes.

For example, the lightest atom Hydrogen has one proton and one electron and has also two isotopes. The first isotope Deuterium has one proton, one neutron and one electron and is symbolized as 2H (*i.e.,* mass number is 2 and the atomic number is 1). The second Isotope Tritium has one proton, two neutrons and one electron is symbolized as 3H (*i.e.,* mass number is 3 and again atomic number 1). Please see Fig. F13.2 for the three isotopes.

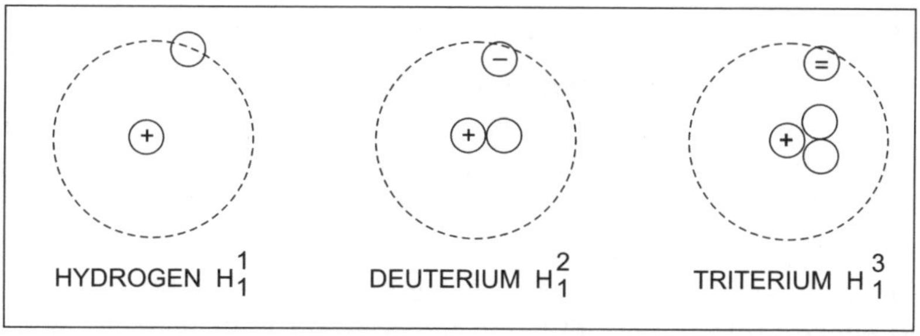

Fig. F13.2: Isotopes of Hydrogen

Now Uranium is one of the heaviest elements having atomic number 92 but it has several important isotopes having mass numbers 233, 235 and 238 written as

uranium-233	uranium-235	uranium-238
(U-233)	(U-235)	(U-238)

Usually, atomic number being a common number is not written and the isotopes are symbolically known as U-233, U-235, U-238.

Radioactive elements

Elements having atomic numbers above 83, are generally radioactive, *i.e.*, they emit radiation because of continuous changes in the nucleus. These radiations are hazardous to human beings. There are principally three types of nuclear radiation.

(*i*) Alpha radiation: It consists of emission from nucleus of two protons and two neutrons. Thus, when an unstable nucleus emits alpha radiation, its atomic number reduces by 2 units and its mass number reduces by 4 units. For plutonium,

239 Pu 94 ——235 Pu 92 + 4 A 2 + G .... (E13.1)

where G stands for a type of electromagnetic radiation and A for Alpha

(*ii*) Beta radiation: It consists of emission of one electron from one of the neutrons in the unstable nucleus, converting the neutron into a proton. As a result, the atomic number increases by one unit and mass number remaining the same. For example, for Strontium,

90 Sr 38 —— 90 Y 39 + B, (E13.2)

where Y stands for Yttrium and B for Beta radiation.

(*iii*) Gama radiation: It has no charge or mass, being simply a wave of electromagnetic radiation.

Nuclear Binding Energy, Fission & Fusion Energy

The mass of an atom is approximately equal to the nuclear mass, as the electron mass is almost negligible. However, if the nuclear mass is measured, it is found that the actual nuclear mass is less than the total of the masses of its individual particles. The reason of this mass deficiency is that, in forming a nucleus, the particles achieve a lower energy state together than they would separately. The absolute value of this energy difference is called **binding energy** ($dE=dm.C^2$, C being the constant light velocity, the energy difference dE is due to mass difference dm.).

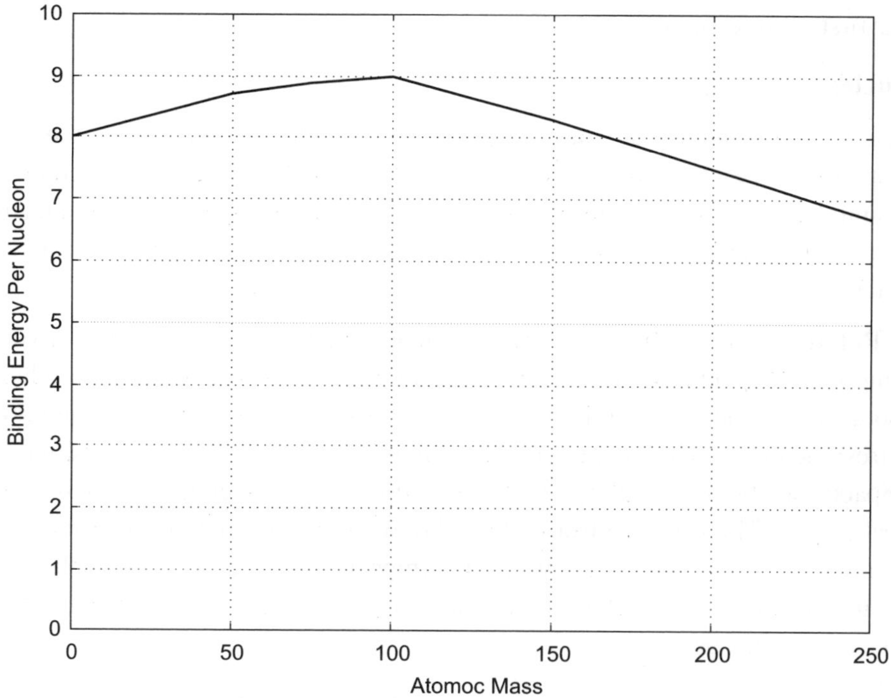

Fig. F13.3: Binding Energy of Isotopes

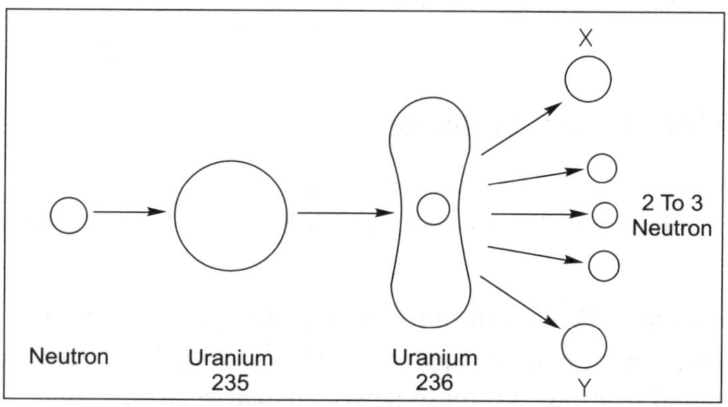

Fig. F13.4: Nuclear Fission

In Fig. F13.3 is shown the binding energy (BE) per nucleon (i.e. per nuclear particle). It is observed that the BE per nucleon is maximum at around the nuclei mass number 60 (A = 60). Thus for light nuclei, energy can be liberated by binding nuclei together and for the heavier nuclei, energy can be released by splitting them apart.

The former process known as **fusion** is operative in the sun and in the explosion of hydrogen bomb.

The second process, by contrast, is known as **fission** and was utilized in the first atomic bomb.

Nuclear Fission

Heavy atomic nuclei are not as stable as the light nuclei, because of the repulsive forces caused by the presence of the large number of protons in the nucleus. For this reason, it is possible to cause fission of a heavy nucleus of a fissile element uranium-235 by bombarding by a neutron in a nuclear reactor.

Referring to Fig. F13.4, a U-235 neutron absorbs a neutron, converts temporally into an unstable U-236 and splits into high velocity fission products (notably, radioactive strontium, cesium-137, Iodine- 131, xenon etc) and 2 or 3 fresh neutrons. The kinetic energy of high velocity fission products, which impacts on the surroundings and slows down, is converted into enormous heat energy. The fresh neutrons, slowed down in a chosen moderator medium like water, heavy water, graphite etc, impacts on other U-235 nuclei and create a chain reaction releasing an enormous energy of 200MeV per reaction equation E13.3 below:

$$235\ U + n \text{ —— } X + Y + 2.5\ n + \text{energy (200MeV)} \quad \ldots\ldots\ . \qquad (E13.3)$$

where X and Y represent fission products and n represents neutron.

The quantum of energy released amounts to 82 gigajoules per gm of U-235.

13.3 Nuclear Power Reactor

Having discussed the basic nuclear science and nuclear fission, it is now considered appropriate to consider application of nuclear fission as power source.

Now, uranium-235 fissions through the absorption of thermal neutrons. Each fission, in turn, releases approximately 200 MeV of energy and two or three additional neutrons. Of these newly created neutrons, some will be lost in leakage and absorption in non-fissile materials. Others will be absorbed in the uranium-235 nuclei to generate additional fissions and neutrons. Accordingly, it is possible to sustain the fission process *via* a chain reaction of following three types:

Supercritical fission: If one fission induces more than one fissions on average, the system is supercritical and the fission rate will increase.

Sub critical fission: On the other hand, if one fission induces less than one fission on average, the system is sub-critical and the fission rate will decrease.

Critical fission: But if one fission induces exactly another fission, the system is critical and the fission rate will remain constant.

In nuclear reactor, therefore, the design task is to control the chain reaction to maintain a constant power level. In practice, the control is executed by adjustment of control rods, which contain a neutron-absorbing material, such as boron. By adjusting the position of the control rods, the fission rate and the corresponding heat generation rate may be increased, decreased or held constant.

Nuclear engineers are required to calculate the population of neutrons at each position at a time and also the flux representing the number of neutrons per unit area per unit time at a given position and time inside the reactor. The integral-differential equation is quite complex mathematically. Instead of solving the neutron transport equation directly, the nuclear engineer uses the diffusion theory, which gives reasonably satisfactory and accurate results.

The basic components and sub-systems of a typical Fission Reactor may be identified with reference to Fig, F13.5:

(*a*) Heavily-built concrete building around the pressure vessel;

(*b*) Pressure vessel;

(*c*) Nuclear fuel bundle;

(*d*) Control rods;

(*e*) Moderating medium (shown in Fig. F13.6);

(*f*) Cooling medium (shown in Fig.F13.6)

Types of Nuclear Reactors & Systems

Table T13.2 shows the principal types of nuclear reactors and their main subsystems in use in the world today, including Pressure Water Reactor (PWR), Pressurized Heavy Water Reactor (PHWR), Boiling Water Reactor (BWR), Russian RBMK, Gas Cooled Reactor (GCR), Liquid Metal-cooled Fast Breeder Reactor (LMFBR), Pressure Tube Reactor (German AKB) etc.

(a) Pressure Vessel
(b) Core-supporting Framework
(c) Core Cage
(d) Top Supporting Plate
(e) Bundle of fuel elements
(f) Control-rod drive
(g) Control-rod
(h) Bottom Supporting Plate
(I) Thermal Shield

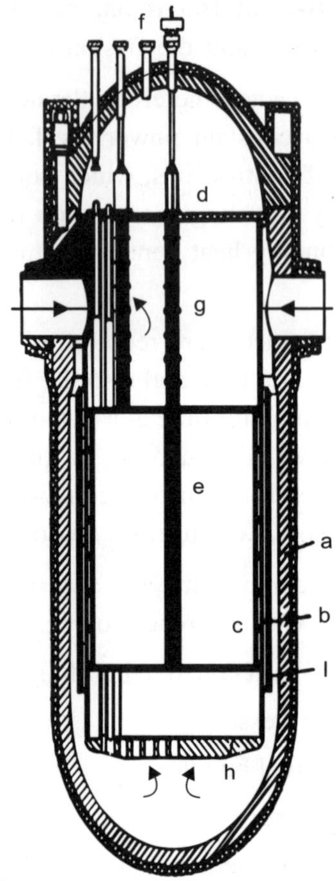

Fig. F13.5: Layout of a Pressure Water Tube Reactor

Construction

The most common reactor types PWR, PHWR, and BWR etc have similar construction as shown in Fig.F13.5, but differing in functional systems. But the Pressure Tube Reactor (PTR) of Fig.F13.6 does not have the Pressure vessel, though Pressure vessel is considered essential today for safety measure. In PTR of Fig.F13.6 the fuel elements are stacked inside the Pressure Tubes, through which also flows the cooling medium pressurized by an independent pump. Pressure tubes are immersed in the moderator medium, which in turn is again worked by another independent Pump system. Table T13.2 gives a broad outline of various materials used in a typical nuclear reactor.

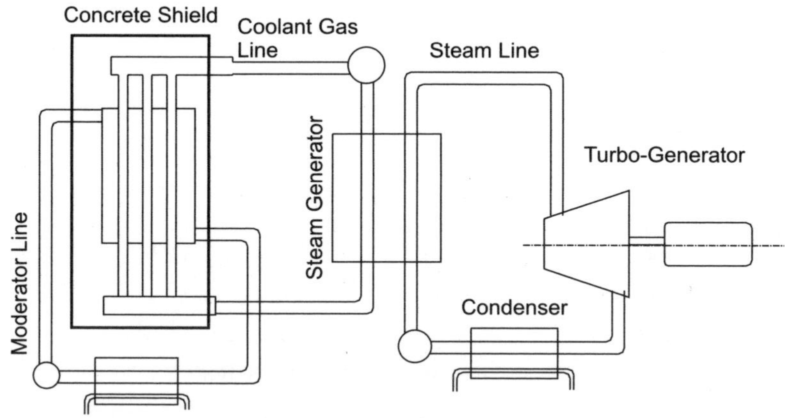

Fig. F13.6: Assembly of A Pressure-Tube Reactor

Nuclear Fuel System

A typical reactor uses fuel pellets of uranium oxide enriched to about 3.5% uranium-235.Ceramic matrix is used in its construction, so as to withstand very high temperature. The pellet ends are dished to allow thermal expansion and also made somewhat porous to trap fission gases. The pellets are stacked end to end in a long metal tube as cladding, which provides structural support to the fuel and to prevent radioactive fission products to enter the reactor coolant. For these reasons, the cladding is made of zirconium alloy, which is heat and corrosion-resistant and also very poor absorbent of neutron. The fuel rod with the cladding is known as the Fuel Rod (see Fig.F13.7A).

The fuel rods are assembled and aligned in a rectangular array 17x17 container including the control rods (see Fig. F13.7B)

These rectangular assemblies are then grouped together to approximate a right circular cylinder to form what is known as Core Barrel (see Fig. F13.7C).

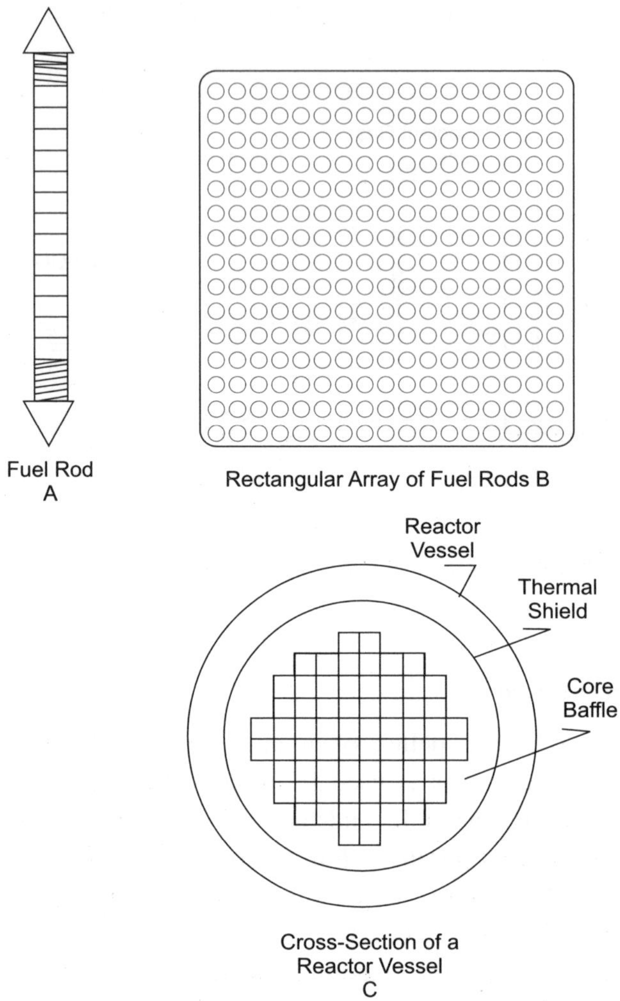

Fig. F13.7 General Arrangement of a Fuel Core

Operation

The different cooling and operating media are the main reasons of somewhat different operating principles of nuclear power plants.

In the Light Water Reactors (LWR), water flows through the core to remove heat from the fuel rods. Moreover, in addition to serving as coolant, the water also acts as the Moderator, *i.e.,* it also acts to slow down or moderate the high speed neutrons released from the fission.

With the LWR, there are two basic designs –

- Pressure Water Reactor (PWR), see Fig.F13.8;
- Boiling Water Reactor (BWR), see Fig.F13.9;

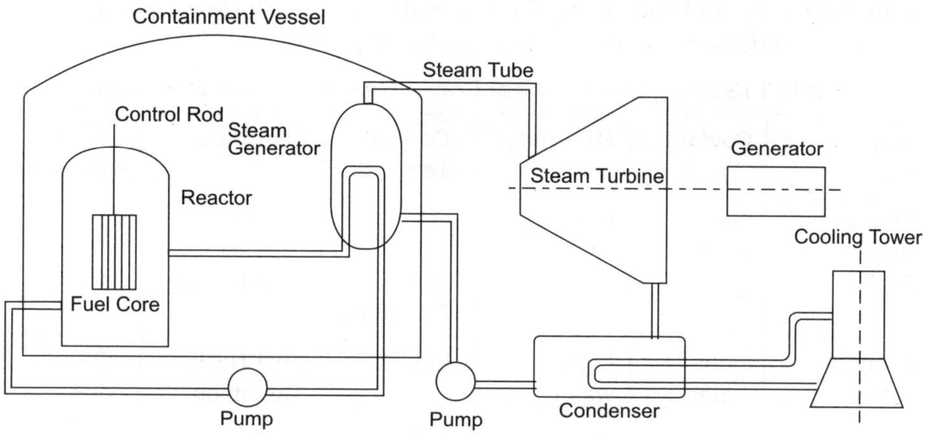

Fig. F13.8 Schematic View of a Pressurized Water Reactor

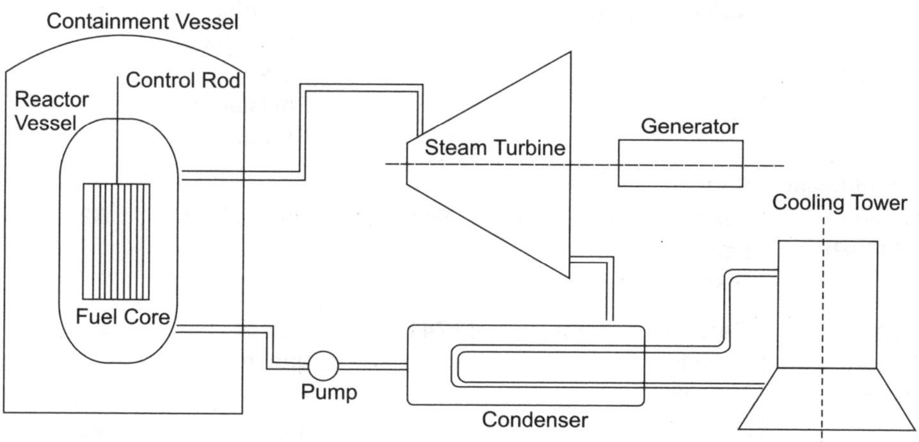

Fig. F13.9 Schematic View of a Boiling Water Reactor

PWR has three flow loops –

(*i*) Primary cooling, where water directly cools the fuel core under high pressure 15.5 MPa (2250 psi) and in turn heats up the water in the steam generator.

(*ii*) Secondary fluid gets heated up into steam to run the turbo-generator.

(*iii*) Liquid and low-pressure steam is then cooled to water by the third water flow. The secondary water is then pumped back to the fuel core.

In BWR, there is one less flow system. The primary coolant itself boils in the fuel core and the steam then directly drives a turbine. There is , of course, a condenser cooling system as for the PWR.

Table T13.2: Types of Nuclear Power Reactors in use worldwide

Type	Coolant	Moderator	Coolant Temp. C	Countries	Total Number
Pressurized Water (PWR)	Light water	Light water	267 Countries	Most Nuclear	259
German RBMK	Light water	Light water	270	Most Nuclear Countries	92
Boiling Water (BWR)	Light water	Graphite	288	Former Soviet Union	13
Pressurized Heavy Water (PHWR)	Heavy water	Heavy water	330	Argentina, Canada, China, India, Korea, Pakistan, Russia	43
Liquid-Metal Cooled (LMFBR)	Sodium, Lead, Lead-Bismuth	Graphite	545	UK, Russia	32
Gas-Cooled (GCR)	Carbon Dioxide, Helium	none	741	France, Japan, Russia, India	2
				Total	441

Source : Teste ret at (2005)

In PHWR type, originally developed by Canada, the moderator is the heavy water (*i.e.*, Deuterium D_2O, where D is actually $2H_1$ instead of ordinary water 1H1), which has higher capability of slowing down the neutrons and allowing the use of natural uranium instead of the enriched uranium. In the CANDU reactor of Canada, the cooling medium is also heavy water (Fig. F13.10). Like in PTR of German AKB models (see Fig.F13.6), it has both the fuel elements and the cooling medium inside pressure tubes, but unlike the PTR, it has a pressure vessel called the calendar containing the moderating medium.

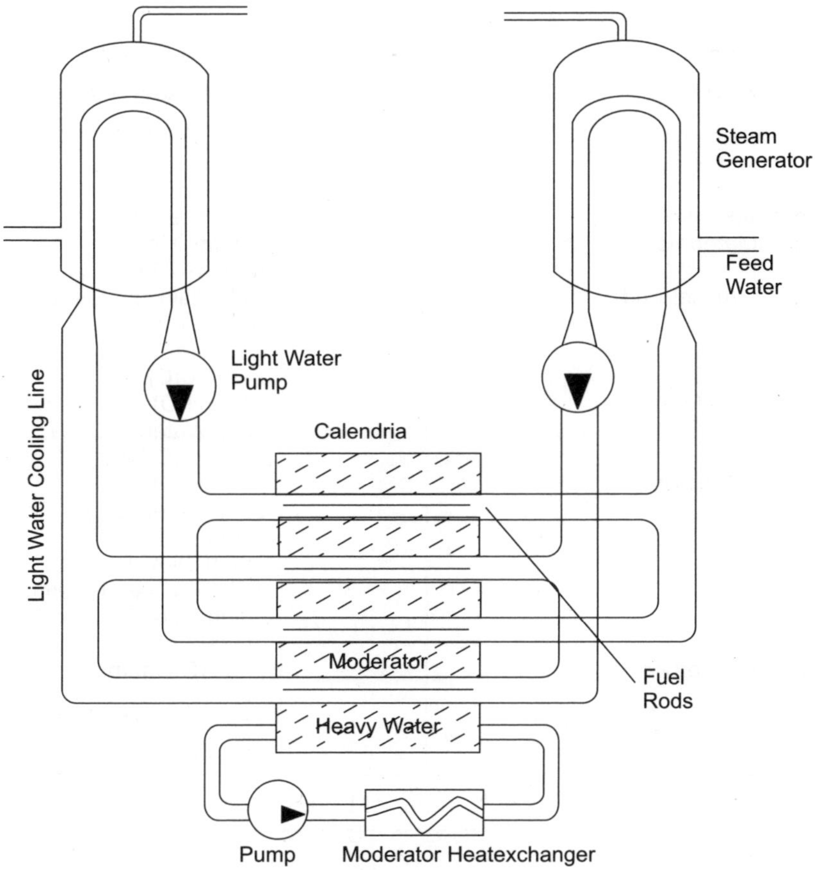

Fig. F13.10 Candu Light Water Pressure Tube Reactor

In the LMFBR Breeder Reactor, uranium-238 is converted into Plutonium-239 by actions of fast speed neutrons. Hence the moderating medium is omitted and cooling is done by liquid metal sodium Na, which has no moderating properties. Gases could also be used instead. In GCR of UK, Carbon Dioxide was used as Coolant and Graphite as moderator.

World Population of Nuclear Reactors

Table T13.2 from Tester et al (2005) gives a total of 351 LWRs but Table T13.4 from Voigt et al (2002) mentions a countrywise population of LWRs, which total to 354. However, Table T13.2 shows a total of 441 nuclear reactors all over the world. Table T13.4 shows total LWRs in USA at 104, the highest number in one country.

Technical Specification of a Nuclear Reactor

Table T13.5 from Voigt et al (2002) presents the principal parameters of a Power PWR of capacity 1100 MW electrical power. Table T13.3 shows some standard materials used in construction of fission reactors.

Table T13.3 Some Standard Materials of Fission Reactors

Sub-Systems	Typical Materials
Self-sustaining Chain Reaction	Fuels:URANIUM-233 Uranium-235 Plutonium-239
Conversion into Fissile Materials	Fertile Materials: Thorium 232 Uranium 238
Reactor Vessel and Parts	Steel, Stainless Steel Zircalloy (Zr) Aluminium (Al) Graphite (C)
Neutron Moderator	Water (H_2O), Heavy Water (D_2O) Graphite (C)BeO_2
Reactor Coolant	Water (H_2O) Heavy Water (D_2O) Helium (He) Carbon Dioxide (CO_2) Lead (Pb)/ Lead-Bismuth Sodium (Na)Sodium Potassium(Na-K)
Neutron Absorber	Boron (B)Cadmium (Cd) Gadolenium (Gd) Hafnium (Hf)

Table T13.4: Number of LWRs in World.

Countries	Number
USA	104
France	58
Japan	52
Germany	19
Russia	16
Ukraine	13
South Korea	12
Sweden	11
Spain	9
Belgium	7
Bulgaria	6
Taiwan	6
Slovakia	6
Switzerland	5
Czeckosvlakia	5
Hungrary	4

Contd...

Finland	4
China	3
India	2
South Africa	2
Mexico	2
Brazil	2
UK	2
Pakistan	1
Slovenia	1
Netherlands	1
Armenia	1
Total	354

Source : Vogt (2004)

Table T13.5: Pressurized Water Reactor:
Principal Specifications of a 11000 MWe Power Reactor

Design Parameters	Typical Values
Thermal Power	3411 MWt
Electrical Power	1100 MWe
Reactor Pressure	2250 psia
Reactor Coolant Inlet Temp.	560 F
Reactor Coolant Outlet Temp.	620 F
Reactant Coolant Flow Rate	
Number of Fuel Assemblies in Core	193
Number of Stem Generators	4
Steam Turbine Inlet Temp.	541 F
Steam Turbine Inlet Pressure	970 pia
Fuel Rod Array	17x17
Fuel Rods per Assembly	264
Control Rod Locations per Assembly	25
Fuel Assembly cross Section	8.426"x8.426"
Fuel Assembly Length	160"
Fuel Rod Outer diameter.	0.36"
Fuel Rod Clad thickness	0.0225"
Fuel Pellet outer diameter.	0.3088
Fuel Pellet Length 0.53	
Fuel Assembly Weight	535 kg.
Fuel Assembly Uranium Weight	423 kg

Source: Vogt et al (2004)

13.4 Nuclear Fuel Cycle & Reprocessing

Nuclear Fuel belongs to one of the following three categories:

Fertile Isotopes – U-238, Th-232, which are convertible to fissionable isotopes;

Fissionable Isotopes – U-235, U-233, & Pu-239;

Natural Uranium - 0.7% U-235 and 99.3% U-238 as mined before enrichment.

Nuclear Fuel Technology

Economically recoverable uranium in the world is estimated at 4 million tons. The reserves of thorium in the earth are far larger than that of uranium. Thus if all the recoverable uranium and thorium are converted into fissionable isotopes, there would be an ample energy supply for the mankind several millennia. In fact, each gram of U-235 releases 1 MWday energy by reaction E13.4.

The natural uranium, which is generally mined, consists of 0.7% U-235 and the balance is U-238.For operation in the nuclear reactor, natural uranium is enriched to 3 -5% U-235.

Uranium that remains after the enrichment process is called is generally known as the depleted uranium (consisting mostly U-238), because U-235 concentration has been reduced to 0.3% or less. This depleted uranium can be converted to a fissionable isotope Pu-239 by a process called breeding. Thorium Th-232 can also be breeded to fissionable isotope U-233. The process involves absorption of a neutron by the nucleus of U-238 and Th-232 in a Breeder Reactor.

The reactor driver fuel consists of the elements or compounds of fissionable isotopes U-235, U-233 and Pu-239 along with fertile isotopes U-238 and Th-232.

The important nuclear processes that take place in the reactor are the followings:

U-235 + n = X+Y + 2.5n + energy (200 mev) E13.4;

U-238 + n = Pu-239 .... E13.5;

Th-232 + n = U-233 .... E13.6;

Where X and Y are called fission products, which are also other radioactive elements of the Periodic Table.

The general reactions in the nuclear reactor generally follows E13.4, but as U-235 is fissioned, some U-238 also gets converted to Pu-239 and in fact some 40 per cent of the energy generated in the reactor is from the fission of U-238 in the 4-year life cycle of the fuel (E 13.5). Some Th-232, if present, can also get converted into uranium – 233 by neutron absorption per E13.6. Another series of reaction by successive neutron capture similar to reactions E13.5 and E13.6 produces radioactive Actinide, Americium, Curium and Neptunium.

Nuclear Fuel Processing & Reprocessing

The deployment of nuclear power generation plant will depend on having in place a fully functional and highly safe nuclear waste depository. To understand the problem of nuclear waste disposal, we have to follow up the nuclear fuel cycle:

(*i*) Uranium ore is mined, separated and milled to produce yellow cake U_3O_8 oxide.

(*ii*) The yellow cake is then chemically converted into a gas Uranium hexafluoride UF_6.

(*iii*) Enrichment through a process of gaseous diffusion or centrifuging: UF6 is enriched to its form 235-UF_6, which being lighter is separated out from 238-UF6.

(*iv*) Once the permeate UF_6 is sufficiently enriched to 235-UF_6 (about 3%), the gas is oxidized to uranium dioxide (UO_2) pellets in the fuel fabrication plant.

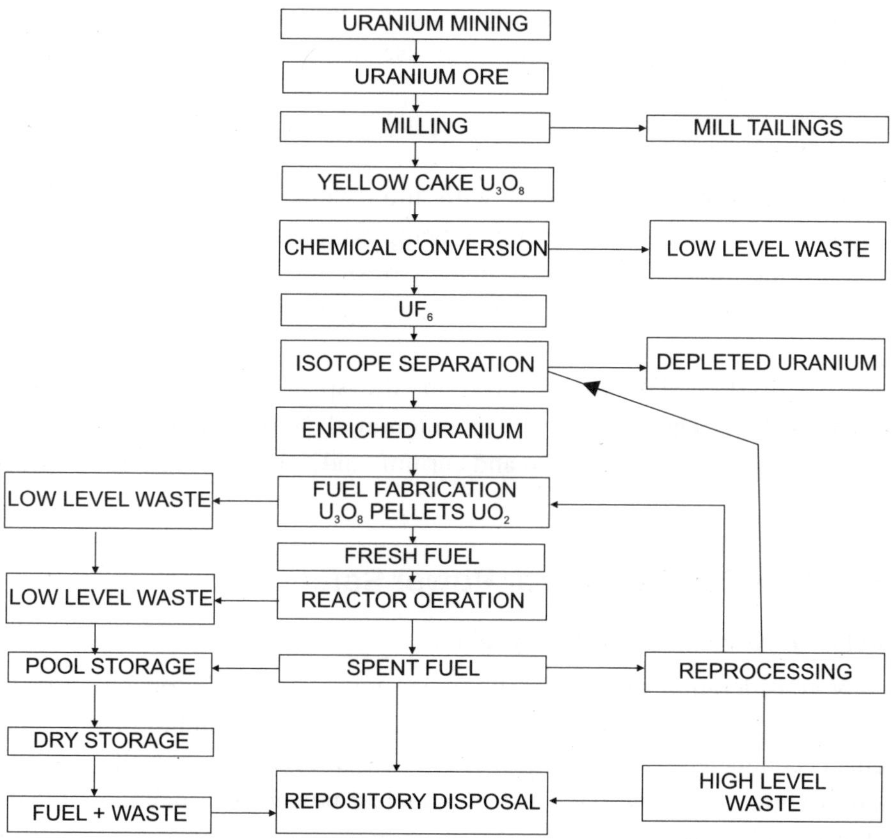

Fig. F13.11: Nuclear Fuel Processing & Reprocessing

(*v*) Fuel pellets in the fuel rod is used in the nuclear plant for three years.

(*vi*) In USA, the spent fuel rods are stored in water storage to allow most intensely radioactive fission products to decay. Once the fuel rods are sufficiently decayed, they are then stored in a dry storage awaiting burial in a 300 m deep tunnel in a geologic medium.

(*vii*) Alternatively, the spent fuel which consists of 95.5 per cent uranium-238 and small amount of plutonium, are separated out from 4.5 per cent actinide and other industrial isotopes by dissolving in nitric acid. The uranium is finally separated out from plutonium and then reprocessed into Uranium-235.

Aim of Reprocessing

Reprocessing is essential to closing the nuclear fuel cycle. It takes more than 7 tonnes of uranium ore feed to produce 1 tonne of 3 per cent enriched uranium. Nuclear fuel discharged at the end of its economic lifetime contains less than 1 percent U-235 but still more than that of the natural uranium contains. The other naturally occurring isotope U-238 transmutes into Pu-239 and produces 40 per cent of the fission energy. It is the recovery of plutonium and to a lesser extent the uranium for use in the recycled fuel that is the focus of reprocessing.

Spent Nuclear Fuel

Average burn-up has increased with improvements in fuel technology, amounting to 40,000 MW-days/tonne being typical of spent fuel discharged. The mass of uranium contained in two fresh fuel assemblies for a PWR reactor is approximately 1 tonne. Each spent fuel assembly includes approximately 150 Kg of radioactive hardware that holds the fuel pellets in place. At 40,000 MW-d/tonne burn-up, the spent fuel assembly contains approx. 94.5 per cent unused U-238 uranium, 1 % plutonium, 0.2 % minor actinides (neptunium, americium and curium) and 4.3 per cent fission products by weight.

13.5 Overview of Nuclear Power Safety

Safety of Nuclear Reactor Systems

In any consideration of nuclear safety, we must review the operation of the nuclear power reactor, in which the nuclear fission of uranium or plutonium atoms is conducted so that the energy is released in a duly controlled process. To start with, a nuclear explosion in a nuclear power plant is impossible due to low enrichment.

The worst possible accident is a severe loss-of-coolant accident (LOCA), leading to core-meltdown, for which there are many precautions. Firstly, the control will immediately act to stop the nuclear fission. Secondly the zirconium cladding will stop the escape of neutron and radioactivity and also the rector wall, 25.4 mm thick steel with stainless cladding is the next line of containment. Thereafter, the very heavy concrete containment structure is the last barrier to prevent any radioactive leakage.

Finally, the cooling system is so designed that in emergency it will fill up the rector vessel to meet worst pessimistic condition.

We now need to examine following other aspects which have direct bearing on the overall safety of the nuclear operation.

Radiation from Nuclear Power Plant

The radiation and other harmful emissions from a nuclear power plant are, in fact, almost non-existent. The neighbouring populations typically receive less than 0.001 rem of radiation per year. This is insignificant compared with natural background radiation of 0.3 rem per year.

In fact, most people are not aware that a coal burning power plant emits more radiation from the plant stack due to the presence of radium and the other radioactive isotopes in coal.

In conclusion, a normally operating nuclear power plant emits no air pollutants and typically less radiation than a coal-burning power plant.(Note: The defnition of the radiation measure "rem" is explained in Box B13.1)

Transport of Radioactive Materials

Radioactive materials must often be shipped over sea or land.. The main concern is relating transport of spent radioactive fuel from the nuclear plant. Fortunately, having regard to the safety packaging, there has never been any report of any radioactive leakage.

Radioactive Nuclear Waste

The problems are complex, but as on date, the problems of waste management and disposal have been largely solved in the technical sense. However, we will briefly review the dimensions of the radioactivity of nuclear wastes. The major concern is, of course, with high level wastes, which include spent fuel rods from reactors which contain many radioactive isotopes. The dangers to human lives depend upon the "half-life" of these isotopes (see Table T13.6).

(*i*) *Radioisotopes with very long life (million years): uranium-235 and neptunium-237*:

These are not very dangerous as very long life means very low level of radioactivity.

(*ii*) *Radioisotopes with short half-lives tens of years: Strontium-90, cesium-137 & 134*:

Being of short half-lives, they burn themselves quickly and may not pose long-term problem.

(*iii*) *Radioisotopes with intermediate half-lives (thousands of years): plutonium-239, americium-243:* In addition to uranium and plutonium, spent fuel rods contain some tansuranic elements (elements, which are heavier than uranium and which do not exist in nature). These isotopes are much more radioactive and can be very dangerous and must be sequestered for very long time.

Safety and Reprocessing of Spent Fuels

US Nuclear Waste Policy Act of 1983 stipulate that untreated spent fuel, which contains high levels of transuranic isotopes with intermediate halve-lives, be isolated for 10,000 years. However, to dispose of such untreated spent fuel, as US proposes to do, is really an unscientific approach and unnecessarily increases the hazards. If one reprocesses the spent fuel, 99 percent of the uranium and plutonium can be separated and can be reused , extending the useful lifetime of fuel many times. Equally important is the great reduction in the waste volume and the time for which the waste fuel must be sequestered.

Safe Sequestering of Nuclear Waste

One of the preferred options is deep land burial or geologic disposal. The waste is concentrated by dissolving in nitric acid and distilling and by converting into dry particles and by vitrifying with molten glass and pouring the same in stainless steel canister, duly sealed. Six vitrification plants are now in operation around the world but not in USA.

Accidents

The nuclear accidents are uncommon even globally and no body has so far died resulting from nuclear accidents except in former USSR.

The only accident to have occurred in USA was at Three Mile Island in 1979 where there was a partial meltdown of the fuel core. But the fail-safe systems worked and no one in the plant was harmed. This was a clear case of human error which is unlikely to repeat in view of lesson learnt and new knowledge gained.

The Chernobyl Nuclear plant accident was much graver resulting in the direct loss of 31 lives arising out of a steam explosion and furious burning of the graphite moderator. The accident did not occur during the normal operation of the plant but during an unauthorized experiment the operators were trying to carry out. The experiment was to test if, in a station blackout, the "coasting-down" generators would generate enough electricity to run the cooling pumps until the standby diesel generators could take over.

Table T13.6: Half-Lives of Some Typical Radio-Isotopes in Spent Fuel

Radioisotopes	Half-Life (Years)
Very Long Life	
Uranium-235	700,000
Neptunium-237	200,000
Intermediate Life	
Plutonium-239	24,000
Plutonium-240	7,000
Americium-241	500
Americium-243	8,000
Short Life	
Strontium-90	30
Cesium-137	30
Cesium-134	2

Source: Jagger 1994

The Chernobyl reactor permitted coolant water to boil sooner, making steam explosion likely, that is what happened. The Reactor was inherently unstable at low power, where it was during the accident. The reactor had graphite moderator and the graphite material burned for ten days, creating half of the radiation release. Finally, the Chernobyl Reactor did not have the concrete containment building, which could have prevented the explosion and loss of lives.

The main concern was the radioactive cloud it spread, helped by the wind to and beyond Belarus releasing 50 million curies of radiation

Box B13.1: Measures of Radiation

In considering the effect of radiation on human body, there are two aspects:

Energy level of radiation: measured by "radiation absorbed dose" *i.e.,* 1 rad = 0.01 joule per kg. of body mass.

Energy concentration level: higher penetrating Gamma (G) rays transfer less effect per linear body than a less penetrating alpha ray transfer higher effect per linear body for the same rad.

The overall effect is given by the dose equivalent "roentgen equivalent man" (rem). Thus, referring to Table T13.7A, an absorbed dose of alpha ray of 1 rad will have an effect 1 rad x 10 bqf = 10 rem.

Table T13.7A: Biological Quality Factor for varying Types of Radiation

Types of Radiation	Biological Quality Factor (bqf)
X-ray & G-ray	1
B-ray	1 to 1.7
Neurons	2 to 11
A-ray	10
Heavy Nuclei	20

Table T13.7B: Radiation absorbed in USA

Types of Radiation	Total Radiation absorbed in a year
Inhalation of Radon	200
Gas from soil/rock	28
Cosmic rays	27
Own body	40
Man-made sources	65
Total	360

As shown in Table T13.7B, the average American receives a radiation dose of about 360 mrem (or 0.36 rem) per year. Of this exposure, the inhalation of radon gas accounts for 200 mrem, with another 28 mrem coming from rocks and soil.

Radon is a decay product of uranium naturally present in rocks and soil, as a noble gas, radon easily seeps into buildings. Further 27 mrem results from cosmic rays, which consist of high energy radiation emanating from outer space. Own body exposes to 40 mrem and man-made sources (such as X-ray) 65 mrem.

The above environmental radiation of 0.36 rem compares with very unfavourably with the radiation from nuclear power, which amount to mere 0.001 rem per year per person. Thus nuclear power is safer than the environmental background radiation.

13.6 Breeder Reactor

We have seen from equation E13.4 that in nuclear fission, each neutron causing nuclear fission produces 2.5 neutrons, which make the fission chain not only to sustain but to grow. The growth is regulated by a moderator medium so that the chain reaction is made just sustainable by slowing down the neutron.

In the Breeder Reactor, the moderator medium is dispensed with and the neutron from the neutron stream from U-235, in addition to fissioning other U-235 nuclei, also impacts the fertile U-238 isotope of uranium and also Thorium isotope Th-232. Thus, two fissile isotopes Pu-239 and U-233 are breeded in addition to the initial fissile uranium-235.

In the Fast Breeder Reactor, the reactor vessel is quite compact as the moderator is dispensed with. As indicated in the earlier paragraph, there are actually two breeding cycles:

Uranium-235 → Uranium-238 → Plutonium-239 (U-Pu cycle)

Uranium-235 → Thorium-232 → Uraium-233 (Th-U cycle)

However, in the Breeder Reactor, the U-Pu cycle is mainly employed with Plutonium charge. The Thorium-Uranium cycle operated more favourably in the thermal energy range and employed in the advanced type of Thermal Reactor.

As compared with the ordinary reactor, fast breeder reactor is capable of utilizing the uranium resource fifty times more when U-Pu cycle is used. The uranium feed contains more than 90% U-238 isotope that is breeded to fissile Pu-239 in the reactor and takes part in fission and heat energy generation. Once the Breeder Technology is fully developed, the nuclear power will be able to sustain global energy requirement for several centuries.

The fast breeder reactors have already been built in the United States, Russia and Great Britain. All use fuel elements as rods and the coolant as liquid sodium. Fig.F13.12A shows the cut-out of the American EBR II fast breeder reactor. Fig.F13.12B shows the fuel element of the same Reactor, where the core zone at mid-height contains the sheathed fuel elements and the breeding zones are below and above the core zones. The breeding zones likewise contain sheathed U-238 fertile elements, where the escaping neutrons from the core zone cause transformation of U-238 to Pu-239.

India has limited deposits of Uranium. Hence the country has developed U-Pu cycle breeding technology in a big way and is planning to develop 150,000 MWe of nuclear power through the breeder technology route. India has also embarked upon the Th-U cycle of better utilization of large Thorium deposits with plan to produce 300,000 MWe power from Thorium fuel. Please see the last Section of this Chapter on Indian programme for more details.

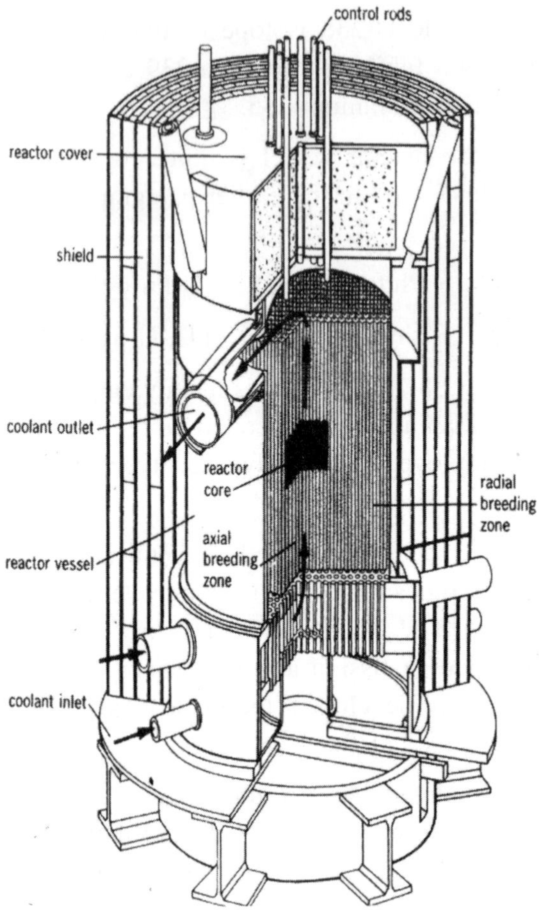

Fig. F13.12A: Breeder Type EBR II Reactor

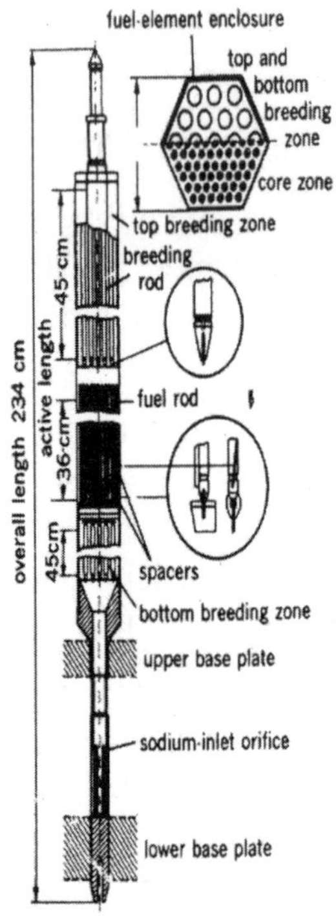

Fig. F13.12B: Fuel Elements EBR II

13.7 Nuclear Fusion Reactors

In suns and stars, light nuclei combine or fuse to create heavier nuclei. The process releases a very high quantity of energy. Our sun has been shining for 4 billion years and according to astro-physicists, is likely to shine another billion years.

A Fusion Reactor works on the principle of nuclear fusion of the solar system. A fusion reactor gives out no radioactivity nor any pollution. The light elements like isotopes of Hydrogen are available in unlimited quantities from the sea.

Fusion Reaction

To obtain a fusion reaction, two nuclei that have a natural tendency to repel each other, both being positively charged, must be brought closer together. Therefore certain amount of energy is necessary to overcome the repulsion forces of like charges and reach the zone close to the nucleus, where the nucleus forces are capable of overcoming the barrier. The possibility of crossing the barrier may be quantified by the cross-section of the reaction. The cross-section of the fusion reaction is on the order of 1 barn (1 barn = 10**(-28) m2), compared with 600 barns for fission reaction of U-235 obtained by thermal or slow neutron bombardment.

Hydrogen (H-1) is the lightest element having only one proton in its nucleus. Its two isotopes, Deuterium D (H-2) having one proton and neutron in its nucleus and Tritium T (H-3) having one proton and two neutrons in the nucleus, are the next lightest elements, which are used in nuclear fusion as shown below:

D + T = [He-4 + n]	[17.59 MeV]	..... E13.7
D + D = 50% [He-3 + n]	[3.27 MeV]	..... E13.8
+ 50% [T + p]	[4.03 MeV]	
D + He-3 = [He-4 + p]	[18.35MeV]	..... E13.8

Production of Tritium

Whereas Deuterium is naturally available from seawater, Tritium is produced from neutron bombardment of Lithium

n + Li-6 = T + He-4	....	E13.10
n + Li-7 = T + He-4 + n	....	E13.11

Fusion demands high temperature, typically more than 100 million Celsius (10 KeV). At such high temperature, electrons are separated from nuclei; hence, plasma the fourth state of matter is created. But plasma is the most common state of the matter in the universe and has already found numerous applications on our earth. (neon tube, plasma torch for metal welding and cutting).

The most accessible fusion reaction is that involving deuterium and tritium, two isotopes of hydrogen. Though Deuterium is abundant, Tritium exists in trace amounts. Tritium, which is radioactive, is produced as mentioned by the nuclear absorption of a neutron with Lithium (see equations E13.10 and E13.11).

Plasma is electrically conducting fluid, in which ions and electrons flow practically independently of each other. Submerged in a magnetic field, ions and electrons will follow helical trajectories winding around field lines and will be forced to move along the field. This is the magnetic confinement principle. To avoid that the plasma does not escape at extremities, the plasma cylinder was closed on itself, resulting in torrid configuration.

Fusion Reactor

A schematic diagram of a magnetic Fusion Reactor, a donut shaped toroid is shown in Fig. F13.13. The deuterium/tritium mixture is injected in a gas or frozen ice-pellet form into a vacuumed torus chamber., where by means of confinement and heating system, it passes to the plasma state and burns continuously. The plasma produces ash (helium atoms for a D-T plasma) and energy in the form of particles and the radiation. Ashes are extracted in gaseous form and are processed outside the reactor.

Charged particles and radiation are absorbed, losing their energy in the first physical boundary surrounding the plasma (so-called "first wall"). Neutron kinetic energy (about 80% of the total energy for a D-T plasma) is converted into heat in the so-called "blanket" region located after the first wall but still within the vacuum chamber.

The vacuum chamber encloses the space where the fusion reaction takes place. The first wall, blanket and the vacuum chamber are cooled down by a heat extraction system. The heat is used to generate steam that supplies a conventional turbine and alternator electricity generating system.

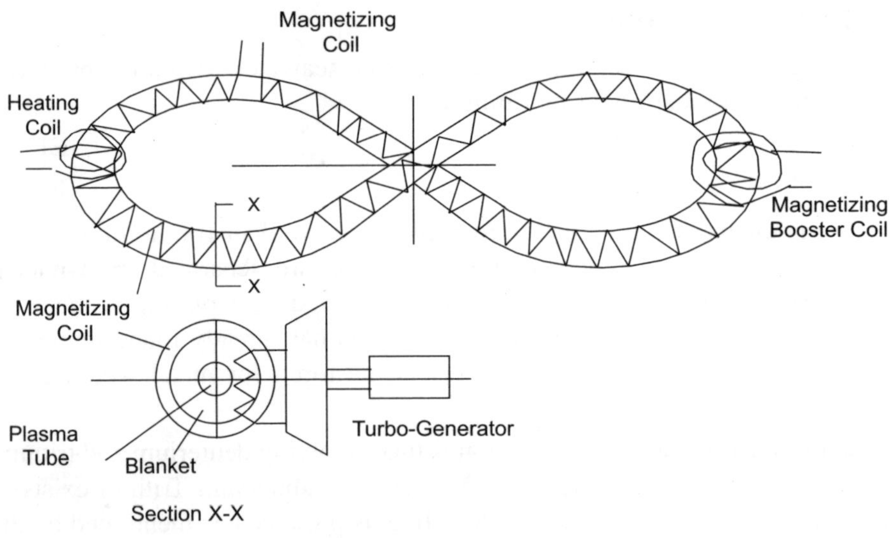

Fig. F13.13 Fusion Reactor Scheme Toroidal Plasma

Reactor Fuels

The fusion reaction involving deuterium and tritium is the most accessible and obviously, is the reaction on which the most creditable concepts are based. Tritium is a radioactive element with a relatively short half-life (12.3 years) that will be produced on-site by neutron capture in lithium. Therefore, the primary fusion fuels are deuterium and lithium, two non-radioactive elements.

Deuterium is a stable isotope of hydrogen. It is very abundant and may be extracted from seawater (33 gm of deuterium per cubic metre seawater) by conventional industrial processes (distillation, electrolysis and isotope exchange). The estimated resource in oceans is of the order of 4.5 x 10E13 tonnes, corresponding to an energy potential of 5 x 10E11 TWe per year. Bearing in mind that the current global energy consumption is of the order of approx. 12 TWe per year (1990), the deuterium energy source would exceed the lifetime of the sun (5 billion yeas).

The Lithium content of the earth's crust is approx. 50 ppm (0.05 gm/kg). It is more abundant than tin or lead and is even 10 times more abundant than uranium (3-4 ppm). Lithium occurs naturally in a mixture of two isotopes: lithium-6 (7.5 % of naturally occurring lithium) and lithium-7 (92.5 % of naturally occurring lithium).

Lithium may also be extracted from seawater that contains 0.17gm/cbm, constituting a potential reserve of 230,000 million tonnes. In a Fusion Reactor producing I GWe per year of electricity, naturally occurring lithium will be used.

In Table T13.8 are shown the ITER (International Thermonuclear Experimental Reactor) at the Zurich Research Station and also two other extrapolated designs of Fusion Reactors.

Table T13.8: Main Parametric Characteristics of Fusion Reactor Experimental Projects

Parameter	ITER System	Reactor Type I	Reactor Type II
Fusion Power(MW)	500	3500	2500
Electrical Power	—	1500	1500
Reactor Efficiency (%)	—	35 - 45	60 – 65
Q	10	20	35
Plasma duration (Sec,)	400 s	continuous	Continuous
Major radius (m)	6.2	99 - 10	6
Averg Temp. (Kev)	10	20	12

Source: Magaud 2005

13.8 Indian Nuclear Programme

Background

The programme of nuclear energy development started with the establishment of the Tata Institute of Fundamental Research (known popularly as TIFR), a non-profit private organization with an endowment of Dorab Tata Trust on

June1, 1945 on the campus of the Indian Institute of Science, Bangalore. In September, 1949, it shifted to its own premise at Bombay.

The doyen of Indian nuclear science, Dr. Homi Jehangir Bhaba, Scientist par excellence, institution-builder, piano-player, painter and an administrator of astonishing ability was naturally chosen the Director of TIFR.

The Department of Atomic energy of the Government of India was set up naturally with Dr. Homi Bhaba as its first Secretary and the Administrative Head. Two other Institutions – the Atomic Energy Commission (AEC) to formulate, nourish and regulate the Indian nuclear policy and the Atomic Energy Establishment, Trombay (AEET) off Bombay with actual infrastructures for nuclear research and nuclear power development work were established. The AEET was renamed as Bhaba Atomic Research Centre (BARC) on January 24, 1966.

Bhaba was convinced that as far as India was concerned, there were serious limitations in respect of fossil fuels in magnitude and locations, and it would be vital for India to develop the atomic energy.

India achieved the distinction of commissioning Asia's first research nuclear reactor Apsara in 1956 on her own but with fuel elements uranium-235 supplied by the United Kingdom.

Three-stage Development Programme

Considering that India, so far, does not have enough uranium ore, but there is more plentiful thorium, a three stage nuclear development programme has been as drawn up:

In the first stage, nuclear reactor using natural uranium with heavy water cooling and moderation is to be used. It was concluded that the country could sustain a nuclear power development programme of 10,000 MW under this programme. Accordingly, today there are 15 PHWR (Pressurized Heavy Water Reactor) are being operated by the Nuclear Power Corporation and 5 more PHWR are under construction. India today has complete mastery of the nuclear fuels, right from the mining of natural uranium ore to finally disposing the waste in safe repositories.

In the second stage, India will first develop prototype Fast Breeder Reactor Technology to convert uranium-238 and thorium-232 to fissile Plutonium-239 and uranium-233 respectively. Now, India has actually entered the second stage of its nuclear programme with 500 MWe Prototype Fast Breeder Reactor (PFBR), which is under construction at Kalpakam in Tamilnadu. This FBR will use plutonium-uranium mixed oxide fuel. Four more FBR with a capacity of 500 MWe are proposed to be built before 2020.

In the third stage, the thorium-uranium233 fuel is to be used in the 300 MWe AHWR (Advanced Heavy Water Reactor). This innovative Reactor system, which has now been designed by BARC and duly vetted for safety by the Atomic Energy Regulatory Board (AERB) will have many pioneering features. The AHWR will employ what are called positive safety features, *e.g.,* there will be no pumps and no moving parts. It neither depends on help from instrumentations nor relies on control mechanism, employing natural circulation. In the case of an accident, water stored in a huge overhead tank inside the reactor building will ensure core cooling for three full days, without any human intervention. The AHWR will use thorium as feed and convert it into fissile uranium-233, which will undergo fission in situ to generate electricity. The AHWR will also use a small amount of plutonium. The construction of AHWR was to start in 2007.

India's Futuristic Programmmes in Nuclear Power

There are a number of very important areas of nuclear development programmes that are being followed in India.

1. **AHWR Programmes:** The construction of the Advanced Heavy Water Reactor will enable India to proven technology of large-scale utilization of thorium in a reactor and also demonstrate the efficiency of a number of pressure safety features.
2. **High-Flux Research Reactor:** India has already a high-flux Research Reactor 100 MWe "Dhruv" which went critical in August 1985. Constructing another high flux reactor for testing materials and producing high specific radioactive isotopes is an important thrust goal.
3. **Augmenting Reprocessing:** India has been reprocessing spent fuel and recovering plutonium, which is a fuel for the FBR. To meet the requirement of India's expanding FBR programme, it is essential to augment the scale of its operation. The overall growth of India's nuclear programme will depend on how fast India keeps building more FBRs. The augmentation of reprocessing is thus a thrust area.
4. **Enrichment of Uranium:** India is working on uranium enrichment for the growing number of Indian Light Water Reactors. The target is now to improve the capacities and efficiencies of the isotopes separation processes and to deliver products at the internationally competitive prices.
5. **Nuclear Waste Management:** This is an important thrust area of growth of nuclear power. To handle high-level waste, that is, radioactive material, BARC has already developed Waste Immobilization Plant vitrifying them into liquid glass and pouring and solidifying them into canister, to be finally placed in sealed stainless steel container in deep repositories.
6. **Accelerator-driven systems (ADS):** It is required for two purposes. Firstly, ADS produce neutrons by spallation process, which means boiling. Neutrons in a reactor are produced by fission. When high energy protons (from an accelerator) are bombarded on heavy nuclei

lead, the target nuclei get excited to level that some of the neutrons are liberated from the nucleus. This is known as spallation. Spallation neurons can augment the neutron supply, which can again be used for converting thorium to uranium-233. The second reason for development of ADS is that they are eminently suitable for transmuting the highly radioactive waste from the conventional nuclear reactors to short-lived radio-nuclides, which do not require long-term storage under surveillance.

However, ADS development will take time, because this involves a series of steps and each of them are technically challenging.

Development of Compact High Temperature Reactor

The present generation of PHWR reactors uses heavy water as coolant and the maximum allowable temperature is 300° Celsius only. In FBR, it is 550° Celsius, hence coolant is liquid sodium. Now, India wants to develop a futuristic reactor where the core temperature can reach 950° Celsius, which is required to generate hydrogen from water. Hydrogen is the new fuel of the future hydrogen economy, when fossil fuel will be all exhausted. The objective of the future Compact High Temperature Reactor (CHTR) is to develop a reactor system that will produce both electricity and hydrogen. When fossil fuel will not be available, liquid hydrogen can be bottled in a car as automobile fuel either for fuel cell or even for the possible use in the internal combustion engine.

The production of hydrogen using sulphur-iodine process requires high temperature of 950° Celsius. The CHTR programme will require development of a variety of materials, because the reactor will become white-hot object at 950° Celsius. The coolant has to be liquid lead or helium, preferably lead. The third option is to use molten salt which can extract heat at that high temperature. The entire structural material inside the reactor will be ceramic or carbon-based material. Very few materials can withstand CTHR temperature and retain their properties. India will need to develop special kind of fuel for CHTR.

India's Prospective Nuclear Power Generation Programme

India has already several nuclear power plants using natural uranium-235 and the present generating capacity is about 3000 MWe. But as outlined above, there are options for three nuclear routes *i.e.,* PHWR Uranium-235, BFR U-238—Pu-239 and AHWR Th-232—U233 technologies with the availability of uranium and thorium in the country. Based on these studies, the long-term projections of electric power generation have been projected in Table T13.9:

Table T13.9: Projected Indian Nuclear Power Programmes

Nuclear Power Routes	Annualized Power Capacities
Natural U-235 PHWR	10,000 MWe per annum
Pu-239 based FBR (U-235-U-238-pu-239)	150,000 MWe per annum
Th-232 based AHWR (U235 – Th232 – U233	300,000 MWe per annum

The above total projected capacities are over 3 times the present total installed electric power generation capacities.

13.9 Economics of Nuclear Power

The advent of nuclear energy for generation of electric power is being greatly advantaged for several extenuating circumstances, namely, to ever increasing hunger of the modern civilization, dwindling stock of fossil fuels, severe pollution hazards of fossil fuels, highly improved safety of nuclear energy production and finally competitive pricing of nuclear power today in comparison to steeply rising cost of fossil fuels. In this section, we look at the economics of nuclear power.

As on today, the capital plant costs of nuclear fission power is the highest and fuel cost is the lowest compared to other forms of energy. The US data in this respect are reflected in Table T13.10.

However, the nuclear technology is continuing to develop very fast and in the near future, the plant cost and the operation cost of the nuclear power are coming down. It is believed that in Europe the nuclear power is already fully competitive. In US, of course, the nuclear is still commercially somewhat dearer as in Table T13.11

Table T13.10: Cost Breakdown of Electric Power by different Technologies

Technology	Capital %	Operation %	Fuel %	Total %
Gas	35	30	35	100
Oil	40	20	40	100
Coal	45	15	40	100
Nuclear	60	25	15	100

Source: Tester et al (2005) & Ghoshroy (2007)

It may be noted that the global oil price has shot up to 90 plus per barrel compared to $ 50 / bbl assumed in Table T3.13, which will make the oil power the costliest power over nuclear power today.

Table T13.11: Grid cost of various types of electric Power (US Price 2003)

Fuel Type	Installed Plant cost $/kw	Annualized Plant cost $/kwh	O & M cost Cent/kwh	Fuel cost Cent/kwh	Busbar cost cent /kwh
Gas	600	1.5	0.3	0.8 – 3.1	2.6 – 4.9
Oil	800	2.0	0.3	3.4 -8.5	5.7 – 10.8
Coal	1200	3.0	0.3	0.6 – 4.0	3.9 -7.3
Nuclear	2400	5.9	0.4	1.0	7.3

Source: Tester et al (2005)

Capital Cost of a 1400 MWe Nuclear Power Plant (1999 US Price)

In Table T13.12, we present the capital cost of a 1400 MWe Nuclear Power Plant in US. At USD 1812 (@5% interest cost of capital), the costing per kwh is very much lower than the estimate of Table T13.11 but still higher than the capital cost of coal-based Power Plant at USD1200.00 per kwh.

Table T13.12: Nuclear Power Economics (Rothwell,2004)
Total Plant Costs for a 1400 MWe Advanced Boiling Water Reactor at Two Unit Site

Costs	Amount million USD	Total million USD
Direct costs –		
Structurals	1130	1420
Reactor Plant	520	
Turbine Plant	230	
Electrical Plant	150	
Miscellaneous Plant	45	
Main Heat Rejection System	45	
Indirect Costs –		
Construction Services	250	710
Engineering Services	70	
Field Office Services	190	
Owner's cost	200	
Total Base Construction cost–	2130	
Contingency	–	125
Interest during construction@ 5% /10%, lead time 5 years	282 (524)	
Total One Unit (5%/10%)		USD 2537/2819
Plant Cost per KWe (gross) @ 5%/10% interest		USD 1812/2013

Source: Rothwell (2004)

Note: Based on data from Nuclear Energy Agency's Reduction of Capital Costs of Nuclear Power Plants.

❑❑❑

14

CHAPTER

Biomass Energy–An Outline of Futuristic Biofuels

14.1 Overview

The global energy potential of virgin biomass is very large. It is estimated the world's standing terrestrial biomass carbon (*i.e.,* the renewable above ground biomass that could be harvested and used as an energy source) is approximately 100 times the total annual energy consumption. Table T14.1 shows the areas covered by forests, grasslands, swamps and marsh and other terrestrial plants growing areas. It also shows annual growth of biomass carbon and the Biomass carbon stocks globally.

Table T14.1: Global Biomass Distribution (Estimated)

Description	Forrest	Savanna & Grasslands	Swamp & Marsh	Remaining Terrestrials	Marine
Area (Million sqKm)	48.5	24.0	2.0	74.5	361
Percentage	9.5	4.7	0.4	14.6	70.8
Net Carbon Production (GT/Yr)	33.26	8.51	2.7	8.4	24.62
Percentage	42.9	11.0	3.5	10.8	31.8
Standing Carbon (GT)	744	33.5	14.0	37.5	4.5
Percentage	89.3	4.0	1.7	4.5	0.5

Source: Klass 2004

14.2 Photosynthesis & Biomass Energy Spectrum

Photosynthesis enables biomass to be created. The carbon in the atmospheric carbon dioxide (CO_2) is captured through the absorption of solar energy into organic carbohydrate. The chemical formulation is given in E14.1 below.

$$CO_2 + H_2O + \text{sunlight} = (CH_2O) + O_2 \qquad \text{..... E14.1}$$

The organic building block (CH_2O) is the source carbohydrate of all cellulose, sugar, and starch, which constitute the plants, trees and all biomass. Thus, for each mole-gram of carbon is fixed, about 470 KJ (112Kcal) of energy is absorbed.

Fig.F14.1 shows the overall spectrum of general biomass, its solar genesis, its application in producing energy as hydrocarbon and synthetic fuels, foods and feeds and recycling of its waste.

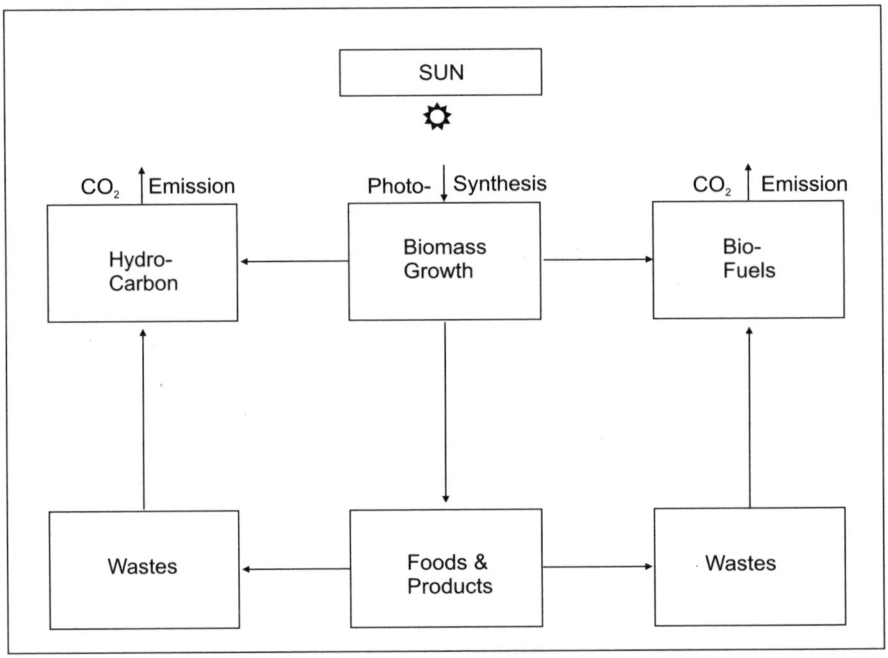

Fig. F14.1 An Overview of Biomass Energy Spectrum

14.3 Biomass Processing & Treatment System

To obtain energy and other byproducts, biomass is treated by any of the following processes –

(*i*) Combustion

(*ii*) Gasification

(*iii*) Liquefaction

(*iv*) Fermentation,

(*v*) Microbial Conversion.

Wood is directly combusted to generate heat for domestic cooking in most of the developing countries. Besides, wood is combusted for many of the industrial applications, namely to generate steam for driving turbo-generator to produce electrical energy.

Gasification or liquefaction is effected by rapid pyrolysis or thermal decomposition of wood and wood waste to yield fuel gas, fuel oil, and biochemical.

Fermentation is a process of microbial conversion in presence of enzyme catalyst. This is the way to produce alcohols as a source of bio-energy.

Other microbial conversion is by anaerobic digestion of bio-solids to obtain high methane content biogas of medium energy.

Fig. F14.2 shows the basic biomass processing leading to the output of energy fuels and other chemicals.

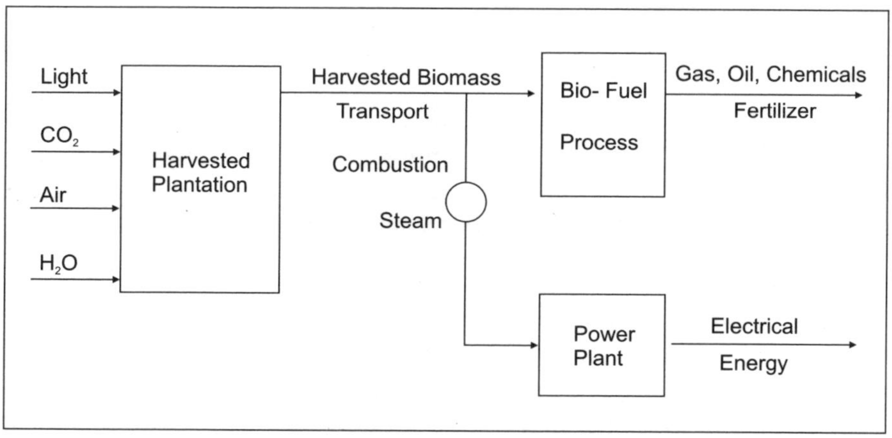

Fig. F 14.2 : Integrated Biomass Conversion Scheme

Combustion Technology

The calorific value of wood depends upon the moisture content which can vary from 40% to 75% . The higher heating value (HHV) is the calorific value of dry wood while Lower Heating value (LHV) is the calorific value of wet wood, the relationship being given by the equation.

$$LHV = HHV\,(1-W) - E\,(W + Hw) \quad \ldots \qquad \ldots \text{E14.2}$$

Where E = 2.26 MJ/kg water, latent heat of evaporation,

Hw = water equivalent of Hydrogen, 8.94 kg of water/kg Hydrogen,

W = water content in kg.

Assuming carbon content of wood at 50%, HHV will be 25 MJ per Kg carbon approx. The heating efficiency of wood in a traditional fixed bed grating type furnace is quite low. Bubbling Fluidized bed, where wood chips as fuel are fed and where a strong vertical air flow makes a bed of sand to bubble up and down, has higher efficiency, heating the flue gases to higher temperature. In the circulating Fluidized bed furnace, still higher velocity is used so that the sand particles are carried up to some distance above and the sand is returned to the furnace bed through a collection chute. This type of furnace has the highest combustion efficiency. The above three basic combustion concepts are shown schematically in Fig. F14.3.

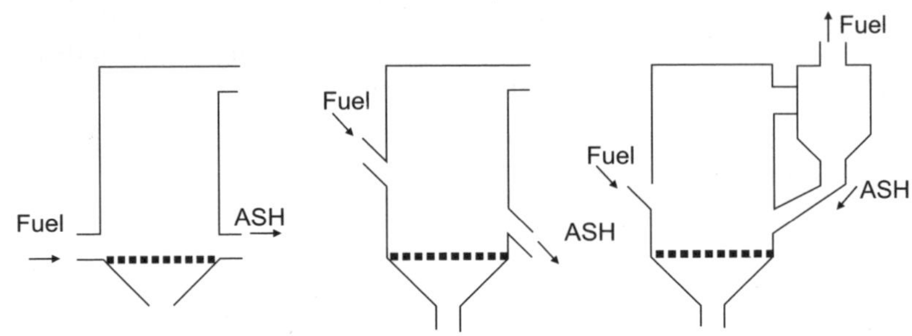

Fig. F 14.3 : Three Basic Combustion Furnaces

From the Fig. F14.3 it will be observed the fluidized bed furnace leads to fuller combustion and generate minimum amount of pollution. In the fixed grating type of furnace or stove, where the efficiency is rather low, the particulate and gaseous pollution is quite high. Table T14.2 shows the pollution data in respect of Nitrogen oxide (NOx), particulate matter (PM), Methane (CH_4), Carbon Monoxide (CO) and Polycyclic Aromatic Hydrogen (PAH).

Table T14.2: Pollution Characteristic of Different Furnaces.

Combustion Type	NOx mg/MJ	PM mg/MJ	Tar mg/MJ	CH_4 mg/MJ	PAH mg/MJ	CO mg/MJ
F.B Boiler	1.70	2.0	-	1.0	4.0	0.0
G.F.B.	110	120	-	70	4040	1850
Modern Wood stove	60	100	70	200	30	1730
Traditional Wood Stove	30	1920	1800	17500	3450	960
Fireplace	-	6050	4200	-	110	6720

Source: Faiji 2004

It will be seen from the Table T14.2 that the traditional fire place generates the maximum pollution. Similarly the traditional wood stove generates more pollution than the modern wood stove. In the case of the Fluidized bed boiler, because of high temperature attained, high volume of NOx is generated, but otherwise the other pollutant is very minimal.

Table T14.3: Total Energy & Biomass Energy Consumption (USA & Top 10 Energy Consuming Countries) 2002

Country	Total Energy (EJ/Y)	Total Biomass Energy(EJ/Y)	% Total
USA	99.85	3.373	3.37
China*	47.25	9.244	19.56
Russia*	26.18	0.326	1.24
Japan	22.78	0.242	1.06
India*	20.84	8.596	41.25
Germany	14.74	0.366	2.48
France	11.16	0.496	4.44
Canada	10.90	0.487	4.47
U.K.	10.10	0.093	0.92
South Korea	8.41	0.092	1.09
Brazil	7.80	1.862	22.14
World	400.00	44.000	100.00

Source: Klass 2004 *Data 1999

14.4 Bio-energy

The modern Commercial Energy production from biomass in the industry and in the Power Generation makes a low but significant contribution. Nevertheless, biomass has the longest history of any energy source and still provide 44 EJ/Y amounting to 11% of world's primary energy needs (See Table T14.3).

Of the 44 EJ of bio-energy, about 50% is used in the developing countries, USA uses 3%, Japan 1%, England 1%.

Today the use of Biomass for energy production in the world is dominated by the combustion-steam Generator system. In term of net energy produced, gasification, pyrolysis, fermentation of sugar for ethanol production, extraction of liquid fuels for bio-diesel and other options play a marginal role. Although, various technological alternatives mentioned above may have better economics and efficiency perspectives in the long term in general, it is likely that biomass combustion will have a major role to play for a considerable time to come.

The above facts is corroborated by the pattern of uses of various bio-fuels in USA is reflected in the Table T14.4. Wood is the primary biomass used @ 2.5% included in the 3.4% bio-energy, while synthetic fuels like ethanol and bio-diesel constitute a very small percentage.

Table T14.4: Short Profile of Biomass Energy Consumption in USA

Biomass Resource	Energy Consumption in 1990 (EJ/Y)	Energy Consumption in 2000 (EJ/Y)
Wood	2.510	2.737
Waste	0.381	0.570
Alcohol fuels	0.063	0.147
Total	2.954	3.454
% Primary Energy Consumption	3.3	3.46

Source: Klass 2004

Power Generation System

Biomass combusted steam boiler and turbo-generator sets are in use in several countries including USA, Scandinavian countries and even some South-East Asian countries. All smaller Power plants do have exclusive biomass feed, larger plants have coal combustion furnace, that is, they can burn either coal or biomass. Boiler capacities may use different combustion technologies below.

Fixed Bed Combustion Boiler — below 10 MW

Bubbling Fluidized Bed Combustion Boiler — below 100 MW.

Circulating Fluidized Bed Combustion Boiler — above 100 MW

In Finland, there is already 500 Mwe Power Plant co-fired with a portfolio of biomass fuel. Malaysia and Thailand are two countries in Asia that are active in deploying state of art combustion technology for 20 to 40 Mwe utilities using forest, plantation and agricultural residues.

The schematic Steam-Turbine Power generation system is displayed in Fig. F14.4. TableT14.5 shows both the biomass-based electrical power generation and total electrical power generation capacities of 10 important countries.

Energy from Liquid Bio-Fuels

In developed countries, about two-third of Petro-energy is consumed in the transportation sector of automobile, railways, shipping and aviation industries. These huge consumptions have led to unprecedented pollution of the global atmosphere. Fortunately two liquid bio-fuels –

(*i*) Ethanol from fermentation of cane sugar, corn, etc.,

(*ii*) Bio-diesel by transesterification of vegetable oil and animal fats have now been developed and found technologically and commercially viable as transportation fuel, especially automobile fuel. Ethanol (or ethyl alcohol) was a known product for a long time, but its application in auto-fuels as a substitute of Petrol or gasoline has now been proven comprehensively in Brazil. Similarly, Bio-diesel as a substitute of petro-diesel is proving successful in USA, Germany, and other European countries. Once these fuels find fuller application as auto-fuels, the world will witness quantum reduction in atmospheric pollution.

In the next two sections, we will deal with these two bio-fuels separately.

Table T14.5: Total Electricity generation & Biomass-based Electricity. (10 Top Energy Consuming Countries)

Countries	Total Electricity EnergyTWH/Y	Biomass-based Electricity TWH/Y	Percentage of Total%
U S A	4003.5	68.805	1.718
Russia	1239.3	1.963	0.158
China	845.3	2.075	0.245
Japan	1081.9	16.518	1.526
India	527.3	-	0
Germany	567.1	10.121	1.785
France	535.8	3.290	0.614
Canada	605.1	7.379	0.012
U K	372.2	4.360	0.012
South Korea	292.4	0.396	0.135
Brazil	332.3	8.519	0.026

IEA 2002

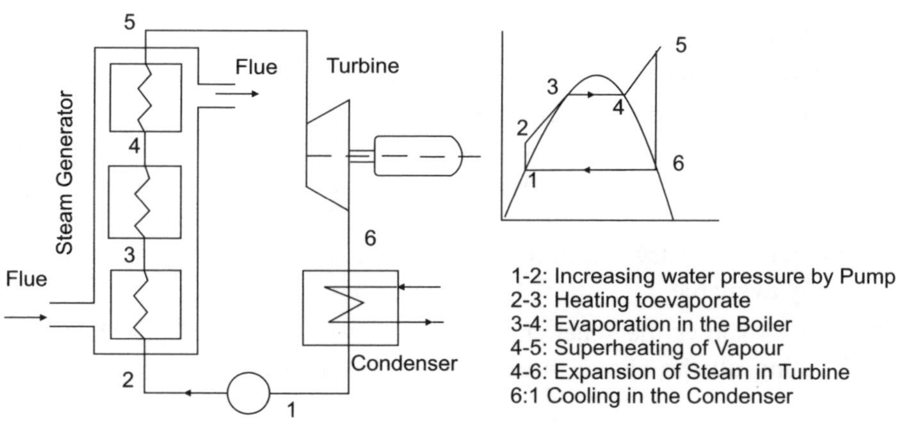

Fig. F14.4: Steam Turbine Generator Systems

14.5 Bio-Ethanol Fuels

Properties of Ethanol

Ethyl Alcohol or Ethanol C_2H_5OH is a clourless, clear liquid and looks like water. Bio-ethanol produced by fermentation from biomass is non-toxic and biodegradable. It has excellent fuel properties for spark ignition internal combustion engine and compares very favourably to the Petro-gasoline as seen from the Table 14.6:

Table T14.6: Fuel Properties of Ethanol

Properties	Ethanol	Gasoline
Octane number	99	88
Latent heat of vaporization (KJ/L)	664	250
Calorific value (MJ/L)	21.2	30
Flash point celsius	78	27
Reid vapour pressure (Kilopascal)	16	71
(Psi)	2.3	10.3

Source: Wyman 2004

From the calorific value above, it will be observed that for the same engine energy output, more ethanol will be required than gasoline, but using higher compression and hence higher efficiency of ethanol engine would reduce the difference.

Production methods of Bio-ethanol

The final process in the production of Bio-ethanol is the fermentation of sugar. But currently there could be three distinct biomass feed stock for the production of Bio-ethanol:

(*i*) Sugar cane crop

(*ii*) Corn/beat starch crop

(*iii*) Plant cellulose biomass

Brazil now is the largest producer of bio-ethanol and sugar cane is the main feedstock of producing ethanol in Brazil. In US, the next largest producer, corn starch is used. The cheapest source is the plant biomass conversion, for which the technology is presently under development. India cannot possibly use either sugar cane or corn because of high domestic demand as food. Hence India should try the third process of using plant biomass, which are abundantly available.

Fig. F14.5 shows the schematic Process Flow diagram of producing bio-ethanol using all the above three feed stock sources.

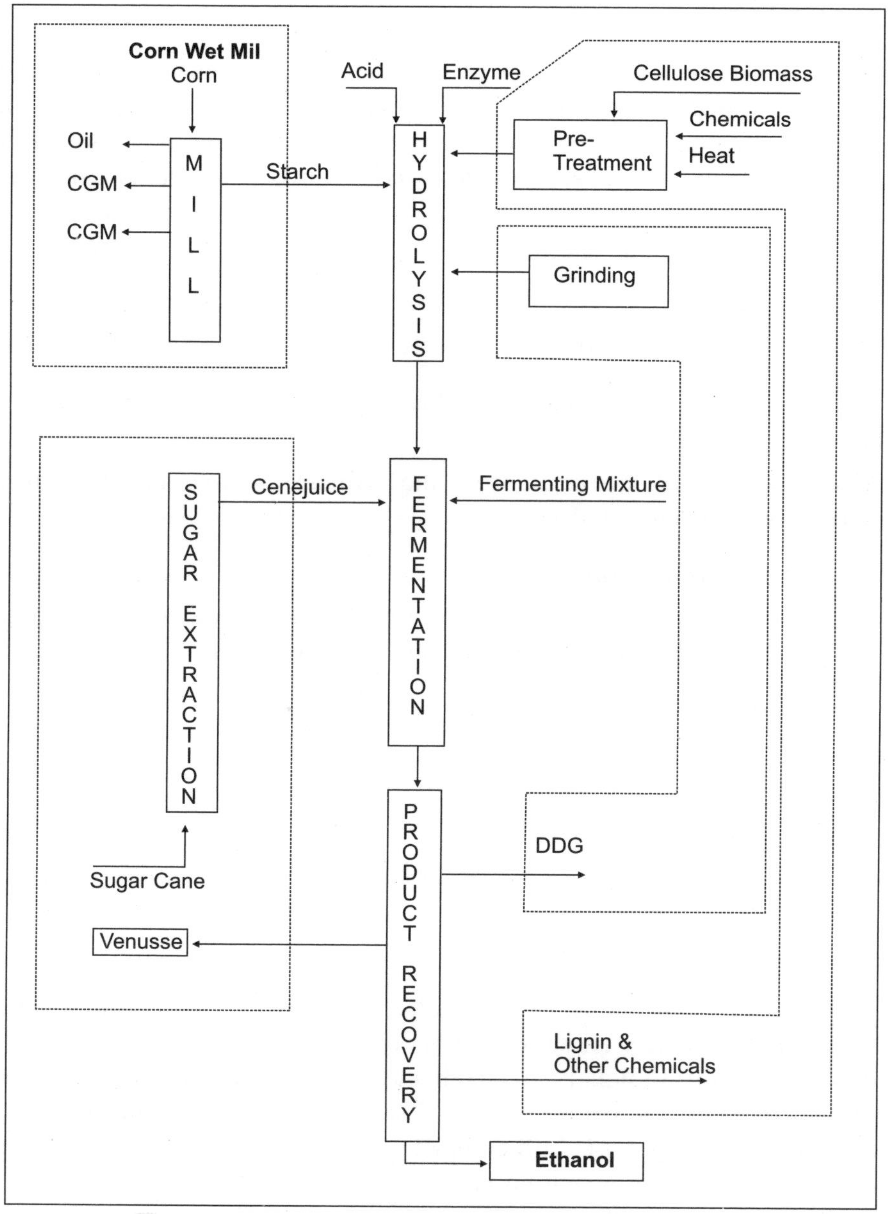

Fig. F14.5 : Process Flow Diagram for Production of Ethanol (Wyman 2004)

Production of Ethanol from Sugarcane

From Brazilian sugar cane

Brazil produced 272 million tons of sugar cane in 2001-2002, second only to India. From this 272 million tons of sugar cane, they produced 13.9 billion litres of ethanol and also 13.5 million tons of sugar. Brazil produced about

5100 litres of bio-ethanol from a hectare of land in 1996. Brazilian sugar generally contains 12 to 17% sugar (90% sucrose, 10% glucose), 68 to 72% moisture.

Process: In the dedicated ethanol plant, cane juice is heated to about 105° to 110°C to reduce microbial contamination and is then fermented. In the combined ethanol- sugar production facilities, sucrose is removed by crystallization and the remaining molasses containing 65% sugar is fermented.

In either process, sugar $C_6H_{12}O_6$ is readily fermented to C_2H_5OH ethanol in presence of common yeast (sacharomycos cerevisae.) by the following reaction:

$$C_6H_{12}O_6 + \text{Yeast enzyme} = 2C_2H_5OH + 2CO_2 \quad \ldots \quad \text{E14.3}$$

Brazilian facilities are generally batch-processes, but many are also continuous processes.

Yeast is generally recovered from the effluent, acid-washed to remove bacterial contamination and re-used.

From Corn Starch Crop

In US, ethanol is produced from cornstarch by dry or wet milling process (see Fig. F14.5). The corn generally contains followings-

Starch	70%
Crude Protein	10 to 11%
Oil	4.5 to 6%
Hemi cellulose	2 to 3%
Lignin	1%
Ash	1%

Process

In a typical dry mill, corn grain is milled to powder, heated with water to 85° Celsius mixed with amylase enzyme for an hour, then heated further to 110 to 150° Celsius to liquefy all starch and also to reduce the bacterial level. Then the content is cooled back to 85°C steadily and more alpha-amylase is added. The mixture is cooled till complete conversion to dextrose sugar. This overall saccharification operation is:

$$(C_6H_{10}O_5)n + nH_2O + \text{Amylase} = n.C_6H_{12}O_6 \quad \text{E14.4}$$

The sugar is then fermented in presence of the catalyst Yeast, similarly to the cane sugar fermentation. The fermented broth, known as beer, contains 12 to 14% ethanol and distilled to 95% purity. If necessary, the hydrous ethanol can be further dehydrated to anhydrous ethanol with molecular sieve.

The solid remnants containing 27% protein are dried as Distillers' Dried Grain (DDG) or with soluble (DDGS).

A portion of the corn protein is sold as corn gluten meal (CGM), a 60% protein product.

From Cellulose Biomass

Cellulose Material

Finally, Ethanol can also be made from the sugar contained in the structural part of the plant biomass.

Plants contain the followings by deadweight percentage –

- 40 to 50% cellulose, which is 10,000 glucose molecules joined by beta 1-4 linkages as long line molecules, aligned parallel to one anther joined by hydrogen bonding.
- 25 to 30 % hemi cellulose, which is amorphous cellulose made up of following five sugars.

  Arabinose, Galactose

  Glucose, Mannose

  Xylose,

  and hexuorine acid, methoxyl acetyl, free carboxyl group.
- 15 to 20%, Lignin, which is complex phenylpropanic acid polymers.

  Lignin and Hemi cellulose are joined by covalent bonds and cellulose fibres into microbial bonds.

  Plants also contain lesser amounts of minerals, free sugar, protein, and other compounds that vary with species.

  Together, this composite of cellulose, hemicelluloses and lignin and other lesser ingredients are called lignocelluloses or just cellulose biomass.

Cellulose biomass is available abundantly as

- Agriculture residues (*e.g.,* corn stover,, sugarcane baggasse),
- Forestry residues (*e.g.,* forest thinnings)
- Industrial wastes (*e.g.,* paper, sludge)
- Municipal bio-wastes
- Woods (*e.g.,* trees & plants)

Basic Process:

The method to convert cellulose to ethanol involves three basic processes.

(*i*) Sulphuric and other acids can convert both cellulose and hemicellulose into sugars, (*ii*) Alternatively, enzyme cellulase can hydrolyze cellulose into glucose. Similarly, a collection of enzyme loosely called hemicellulase can breakdown hemicelluloses into above five sugars. The balance lignin cannot be converted into sugar but can provide boiler fuel feed.

(*iii*) Sugars are then fermented into ethanol as given by equation E14.3.

Detailed Processes

(*i*) Cellulose is reacted in 0.5 to 1% dilute Sulphuric acid at 160º to 190º C for approximately 10 minutes, releasing most hemicellulosic sugar by a hydrogen reaction similar to equation E14.4.

(*ii*) The content mixture is then quickly cooled down to 100º C to stop the reaction and also to remove acetic acid, furfural and other fermentation inhibitors which are released from the biomass.

(*iii*) The solid and the liquids are separated to remove further the remaining acetic acid and much of the sulphuric acid. Sulphuric acid is further neutralized by lime treatment and the solution pH increased to 10 by excess liming.

(*iv*) A small portion of the solid alone or also including the treated liquid is fed with a microbial fungus trichoderma reesai to convert the feed into the enzyme cellulase. The effluent is now combined with the bulk of the pretreated solid to hydrolyze cellulose into glucose. Enzyme Yeast is added to ferment the glucose to ethanol soon after the release into a simultaneous saccharification and fermentation called SSF approach.

(*v*) The fermentation beer broth containing 5% ethanol is then distilled to 95% ethanol and then could be further enriched to 100% anhydrous ethanol passing through molecular sieve.

Process Improvement

Further process research was carried out to achieve both economic and high productivity in following two routes:

(*i*) SSF approach was refined to achieve improved economies and productivity.

(*ii*) Secondly, ordinary yeast could not ferment or the different sugars of the biomass, which was needed for high productivity. In 1980, bio-scientists were successful to genetically engineer new bacteria to make ethanol from all five sugars.

Energy Balance

It will be interesting to know how much energy input is required to produce a liter of the ethanol by any of the above three methods. From the detailed energy balance investigation, the following observations in Table T14.7 have been derived:

Table T14.7: Energy Balance

S.No.	Process Description	% Ethanol Fuel value (approx)
1.	To grow biomass	10%
2.	To transport to factory	5%
3.	To factory processing	5%
4.	To fuel distribution	5%
	Total Energy input	25%

14.6 Bio-diesel Fuel

What is bio-diesel?

Vegetable oil has properties strikingly similar to those of Petro-diesel and hence it may be considered as a potential fuel for Internal Combustion engine as auto-fuel or other transportation system like Railways. In fact, Professor Diesel invented the Diesel Engine to be run with vegetable oil. The major problem associated with direct use of vegetable oil in the engine with its high viscosity due to three way interlinking of fatty acid chain in triglyceride form of the raw oil. One possible way to break open the interlinking is by transesterification of the Triglyceride into mono-alkyl ester of single fatty acid chain.

The reaction is shown below:

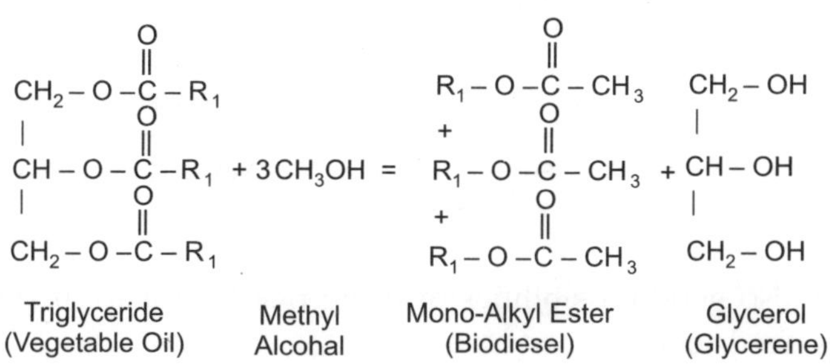

Feed Stock

In USA, refined soyabean oil and in the European Union refined Rapeseed and sunflower oil are used in large plants of capacity 3000 tonnes per day.

India, being an edible oil deficient country and net importer of such edible oil at high costs, it is uneconomical to use purified oil for bio-diesel production. Moreover, with a very large population to support, all irrigated and arable lands are engaged in food crop cultivation.

India therefore requires to use waste arid and semi-arid lands to grow non-edible tree-borne oil (TBO) bio-diesel of the following botanical families:

— Jatropha Curcas (Ratanjot / Verenda)
— Pongamia Pinnata (Karanj)
— Azadiracta Indica (Neem)
— Madhuca Indica (Mohua)

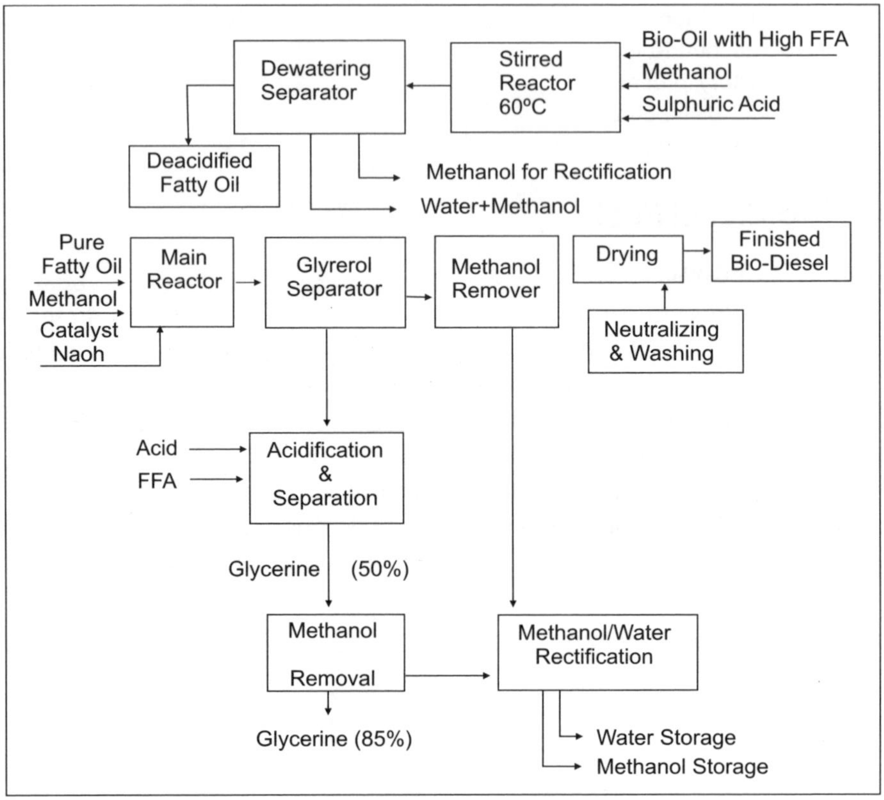

Fig. F14.6 : Process Flow For Production of Bio-Diesel (Klass 2004)
(Methyl Transesterification)

There are also good possibilities of using rice bran and sal fruits (Shorea Robusta) as feedstock.

Of all the feed stock plants, Jatropha Curcas , most favoured in India, is a drought-resistant perennial plant growing in marginal soil. It grows quickly, starts producing fruits in 2 years, maturing in six years and living thereafter for 50 years. The annual output is 1 to 5 tons per hectare. The oil content is 37%, the residue cake makes good fertilizer.

India has about 60 million hectares of waste lands, of which, if 30 million hectares are utilized for energy cultivation with preferably Jatropha plants producing 2 tons of oil per hectare per annum, amounting to a national output of 60 million tons of Bio-diesel in say, 10/15 years, being equal 50% of total Indian consumption today.

Bio-Diesel Production

The process of flow diagram of Bio-diesel Production is shown schematically in Fig. F14.6 below. If the Triglyceride oil contains less than 5% Free Fatty Acid (FFA), no pretreatment of the oil is necessary. In that case, the primary reactor is fed with the following proportions of inputs:

Triglyceride oil	60 kg
Methanol	13.6 kg
Catalyst	0.6 kg

The mixture is heated to 70ºC at 1.5 kg/cm^2 pressure for an hour to achieve transeterification of the glyceride to Bio-diesel (mono-methyl ester) and glycerol to about following proportion:

Bio-diesel (CH_3OOCR)	58.0 kg
Glycerol	6.5 kg

To obtain fuel grade bio-diesel to ASTM or Euro norm, proper quality control is necessary to remove / separate the following during the process shown diagrammatically in Fig.F14.6.

(*i*) Excess Methanol

(*ii*) Catalyst (NAOH)

(*iii*) Process Water

(*iv*) Soaps, if any,

(*v*) Unconverted triglyceride, if any

(*vi*) Unconverted trace of free fatty acid

(*vii*) Glycerol.

Pre-Treatment

As mentioned earlier that if the triglyceride oil is doped with small amount of the respective fatty acid up to 3%, the same can be neutralized by adding enough alkali catalyst, excess alkali still active to catalyze the reaction.

But if the FFA is more than 5%, FFA must be converted to methyl ester catalyzed by sulfuric acid with large excess of methanol (20:1) for just about an hour at 60ºC. The byproduct water must be decanted with methanol.

Thereafter, the transesterification process must start with NAOH catalyst as indicated in the process scheme.

Physical Properties

The Physical properties of bio-diesel (tranesterified from Jatropha Curcas oil) are now compared in Table T14.8 with the Jatropha oil itself and also with the European standard of Bio-diesel (ref. Raheman & Mondol).

Table T14.8 :Comparative Physical Properties of Bio-diesel & Petro-diesel

Fuel Properties	Jatropha Oil	Jatropha Methyl Ester	European Std. for Bio-diesel	Petro-Diesel
Density at 15 C (kg /m3)	920	880	860—900	820
Pour point (C)	5	4-	-	-23
Viscosity at 20C (m-Pa-s)	75.6	4.8	3.5 - 5	2.6 – 4.1
Calorific value (MJ/kg)	37	39	-	42
Flash point (C)	340	191	>120	74
Acid value (mg KOH /g)	8	0.41	< 0.50	0.35
Carbon residue %	4.2	0.05	<0.3	0.17
Ash content %	1.1	0.014	<0.02	0.17

Source: Raheman et al 2006

From the above Table T14.8, it would be seen that Bio-diesel has very similar properties to Petro-diesel. But keeping in view that bio-diesel has little higher viscosity than petro-diesel, somewhat higher pour point temperature and slightly lower calorific value, bio-diesel being widely used in diesel engine at 20% mixed with 80% petro-diesel (Mixture popularly known as B20 without any technical problem. So much so, a single cylinder diesel engine has been run on pure Jatropha bio-diesel (B100) and engine torque, power and specific fuel consumption has been found generally satisfactory.

We will now mention a few points of advantages and disadvantages of using bio-diesel in lieu of petro-diesel.

Advantages of Bio-Diesel

1. Bio-diesel has flash point of 191°C compared to 74°C of petro-diesel.
2. Bio-diesel has better lubricity.
3. Cetane number of bio-diesel is much higher.
4. Bio-diesel is the only alternative fuel that can be used in any conventional unmodified diesel engine.
5. Bio-diesel is non-toxic and biodegradable. It is a 100% domestic fuel.
6. Biodiesel produces significantly lower pollution carbon dioxide, CO, and hydrocarbons. It has no sulfur dioxide emission.

Disadvantages of Bio-diesel

- Bio-diesel may give rise to winter starting and operational problem because of high pouring point and somewhat higher viscosity.
- Biodiesel is highly solvent. Hence to avoid to filter deposits being dissolved and carried to the engine. The filter deposit is to be changed more often.

- Quality of bio-diesel can easily be tampered.
- Neat bio-diesel demands compatible elastomer (Hoses, brackets, etc.)

Economics of Bio-diesel

In the USA and European Union, the bio-diesel is produced from edible oil and hence the production cost of bio-diesel may still be higher than the petro-diesel, which is processed economically in large refineries working continuously.

However, in India, bio-diesel is proposed to be produced from non-edible oil (*e.g.,* Jatropha, Karaja, etc,) grown on fallow arid wasteland. Hence the seed price will be quite low and the final bio-diesel product will definitely be cheaper and conspicuously competitive to petro-diesel.

Further, cultivation of 30 million hectares of waste land producing about 2 tons of bio-diesel per hectare i.e. 60 million tons will immensely enrich the poverty-ridden Indian agriculture, creating nearly US 20 billion dollars per annum agro-based bio-diesel industries. Bio-diesel thus promises to bring about both economic prosperity and equity in 10 years time.

Finally, the availability of petro-diesel is concentrated in the politically unstable Middle East, however long it may last. But bio-diesel being produced in all parts of the world will ensure energy security and energy independence.

International Oil Energy Scenario

Following the oil crisis in the early seventies in the aftermath of the Gulf War, all nations undertook measures for energy independence. Brazil undertook to cultivate sugarcane extensively and ferment bio-ethanol so much so that by the year 2002, it was capable of producing 13.9 million tonnes and all cars were required to use E85 gasoline (15% gasoline + 85% bio-ethanol).

Next step in the global evolution of bio-energy was the production of bio-diesel, the global output of which, including USA and EU, reached 4 million tons, representing about 2% of global consumption of 16,500 million tonnes of petro-diesel as shown below:

Annual Global Pero-diesel Consumption (2002)

Global consumption	16,500 million Tons
USA	800 m. t.
China	210 m. t.
India	120 m. t.

Source: Waghray 2005

14.7 Development of Bio-Refinery

We are all acquainted with Petroleum Refineries which produce various categories of energy fuels, *e.g.,* Aviation , Petrol /gasoline, light diesel, heavy diesel, furnace fuels etc. besides a host of chemicals. Table T14.9 shows a schematics of a Bio-Refinery which also cover all the Processes of Fig. F14.5 and Fig. F14.6.

Table T14.9: Bio-Refinery Scheme

Refining Processes	Products / Byproducts
Fermentation of Sugars Sugarcane crop	Ethanol Lactic acid Acetone Butanol
Fermentation of Sugars – Dry Corn Mill – Wet Corn Mill	
Fermentation of Sugars Acidolysis of Cellulose, Hemicellulose & Lignin	
Extraction of Bio-oil – Triglyceride of Fatty Acid, – Triesterification (Monoalkylester)	Bio-Diesel
Wood Pulp Mill	Paper
Chemical Processes Hydrolysis Hydro-generation Oxidation	Furfural Levuline acid, Polyol
Gasification Steam Reforming Pyrolysis Combustion	Synthetic gas, Tar chemical, Methanol

Source: Elliot 2004

The Bio-refinery can now be defined as a Refinery Factory incorporating a number of processing steps, including pre-treatment, separations, catalytic and bio-chemical transformations, for the production of chemicals and the fuel products from biomass.

The future of bio-refinery will have a major impact on society as fossil-derived resources become scarce and more expensive. Indeed, use of biomass is the only option for maintaining the supply of new carbon-based chemical and fuel products to meet needs of tomorrow's society.

❑❑❑

15

CHAPTER

Solar Energy– A Promising Alternative Power Source

15.1 Overview of Solar Energy

The energy from the sun is virtually an inexhaustible source of energy, which promises an environment-friendly quite viable alternative power source to the energy-starved world. This is very important, considering that the non-renewable fossil fuels coals, oil, and gas, which have been sustaining the world industrial-urban-infrastructure intensive contemporary civilization so far, are on the wane. The alternative energy system must start contributing substantially in about 50 years' time

The huge energy reservoir of the sun releases about (3.8×10^{20}) MWe energy at 5500°C, of which the earth receives only a tiny fraction (1.6×10^{11}) MWe. But the total electricity generated globally is approximately ($1.5\text{x}10^{13}$) KWH (equivalent to 4.4×10^{6} Mwe). It is said that assuming full utilization of solar radiation, a radiation-collecting Solar Power Station measuring 30,000 km^2 in area could theoretically meet the global energy demand. (Please note that Taiwan measures about 30,000 km^2).

Now there are several technical limitations to be overcome in the fuller utilization of solar energy, which are enumerated below:

- Solar radiation is of very low intensity @1000 watt per sq. m. approximately.
- Sun shines intermittently in a daily cycle, which again changes by seasonal variation.
- The sun's position by azimuth and elevation in the sky and the angle of incidence on the earth vary continuously.
- Solar radiation cannot be stored directly and must be converted immediately into heat, current or chemical reactions.

However, modern engineering has been able to surmount some of these problems that have led to substantial utilization of solar energy into electrical power. In 2003, the largest sizes of solar thermal power plants are 80 Mwe parabolic trough systems that have been operating in California since 1980. There are nine such Power plants with variable sizes that have a combined capacity of 354 Mwe and generates approximately 800,000 MWH of electricity every year (running time 3000 hours annually). This is sufficient to service a medium size city of 200,000 people. This is a high achievement of the human endeavour.

There are two important ways to convert solar energy into electrical energy form to be of high use to the world at large.

- **Collection of Heat:** This is by far the largest and the most common practice. The solar radiation is captured into a collector, which heats a liquid (*e.g.,* water) or a solid (*e.g.,* packed bed of bricks, etc.) or gas (*e.g.,* air). The steam generated from water heating may run a turbine to generate electricity. The final uses could include, inter alia, space heating and also cooling.
- **Direct Generation of Electricity:** Another important use is to use solar photon energy to activate a series of photovoltaic cells to directly generate electricity.

In the next section, we will deal with the various solar energy collection systems. (See also Fig. F15.1).

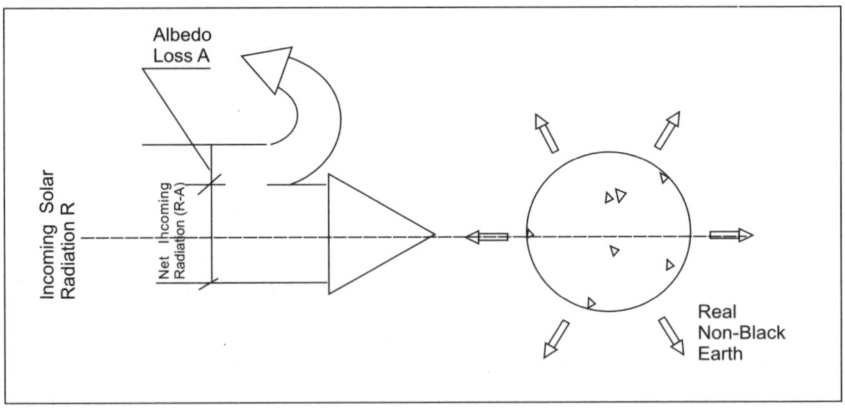

Fig. F15.1 Solar Radiation on Real Earth with Albedo Loss A

15.2 Types of Solar Energy Collection systems

There are three important elements in the heat collection system, namely the Heat Collector, the Concentrator and the Receiver. If the temperature desired is below 100°C, the heat concentrator can be eliminated. Sometime the heat collector and the receiver are common, as in a heat pond.

Flat Plate Heat Collector

The simplest of the heat collector is the Flat Plate Collector, as shown in Fig. F15.2A for liquid heating and Fig.F15.2B for air heating without any heat concentrator. The collector, a metal plate, painted black and placed in the direct sun, heats up and reaches a temperature at which it radiates and loses heat as much as it absorbs heat from the sun radiation. To minimize heat losses, a cover layer of transparent glass is placed above the collector to prevent heat losses by air convection.

Concentrator

The concentrator further increases the temperature at which energy can be usefully recovered. It requires some method of optimally concentrating the radiation, so that the energy absorption per unit area of the heat-collecting surface increases several times. There are several types of concentrators as described below:

(*i*) **Heliostat:** The conceptually simplest approach is to employ a series of flat mirrors, called Heliostats, that gets continually rotated to direct soar radiation onto an absorbing surface like the 'Central Receiver system' containing a high temperature working fluid such as molten salt mixture and which is mounted on a 'Power Tower' (see Fig. F15.3A).

(*ii*) **Parabolic Concentrator:** The mathematical equation of a parabola is

$$Y = (1/4f) \times (X^2),$$

where X and Y are the two orthogonal coordinates. (See Fig.F15.3B)

All rays of light parallel to the y-axis of a mirror trough parabola will be reflected and focused at the focal point at a distance *f* from the vertex, as shown in Fig. F15.3B and F15.4.

A heat absorbing metal tube, carrying liquid situated along z axis, receives the heat from its concentration system (See Fig. F15.3B). The largest solar thermal power plants in California works on this principle.

(*iii*) **Paraboloid Concentrator:** In Fig. F15.3C is shown three-dimensional Paraboloid concentrator, which has the highest concentration capability.

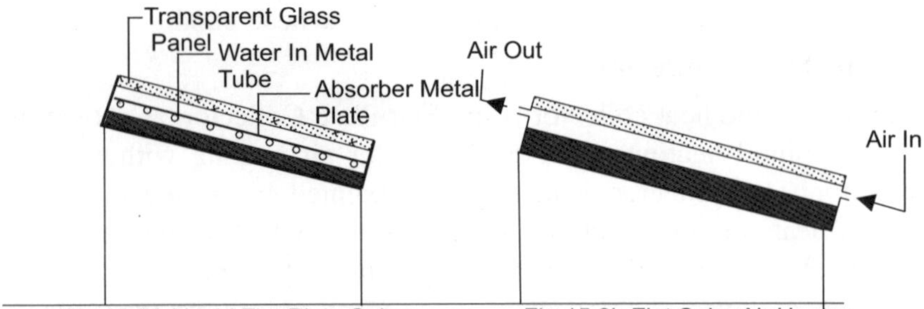

Fig.15.2A Liquid Flat Plate Solar Heat Collector

Fig.15.2b Flat Solar Air Heater

Fig. F15.2: Flat Solar Heaters

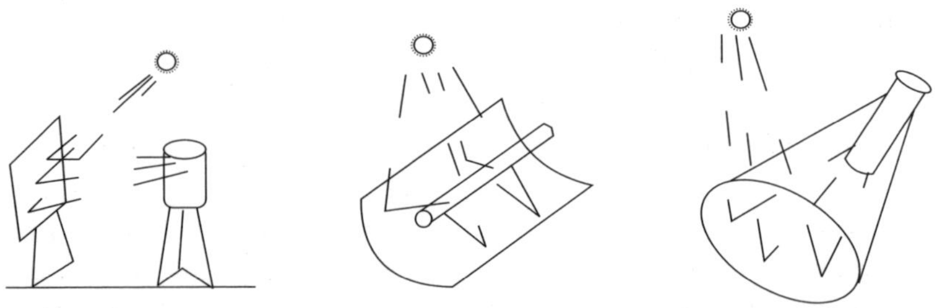

Central Tower With Helostat Mirrors
Fig. F15.3 A

Linear Parabolic Concentrator
Fig. F15.3 B

Paraboloid Concentrator With Central Recover
Fig. F15.3 C

Fig. F15.3: Different Types Of Solar Energy Concentrators

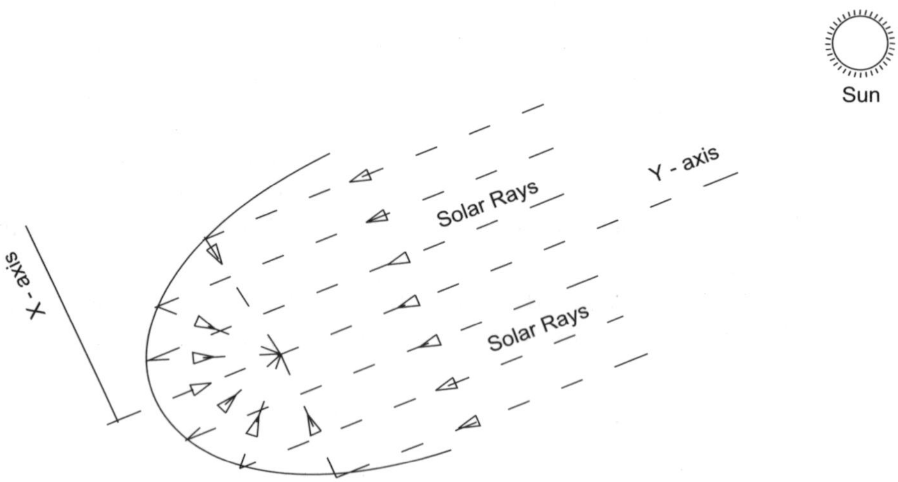

Fig. F15.4: Parabolic Mirror Heater

Each of these approaches to solar energy collection has a typical concentration ratio, which physically determines the ratio of the surface area of the concentrator to the surface area of the Receiver. The various concentration factors of different solar energy collector technologies are shown in Table T15.1.

The Energy Conversion Efficiency is given by the ratio–

(Energy converted into heat) / (Solar energy absorbed)

The limiting conversion efficiency of converting received heat to steam for electricity generation is given by the Carnot Cycle Efficiency as below,

Maximum Conversion Efficiency = 1 – (Tcold/Thot);

where Tcold and Thot are in absolute temperature Kelvin representing respectively incoming and outgoing temperatures of steam.

Table T5.1: Operating Characteristics of Various Solar Concentrators

Solar Concentrator Types	Typical Operating Temperatures C	Concentration Ratios	Tracking Axe	Maxm. Conversion Efficiencies %
Flat Plate Collector	30 – 100	1	-	21
Parabolic Trough Collector	260 - 400	8 - 80	One axis	56
Heliostat Field + Central Receiver	500 -800	600 - 1000	Two axes	73
Paraboloid Dish Concentrator	500 - 1200	800 - 8000	Two axes	80

Source: Tester et al (2005)

15.3 Storage of Heat Energy

Because of variable and uncertain nature of solar radiation, there is almost always a mismatch between the availability of solar energy and the user demand for the energy. That raises the necessity of storage of solar heat, which can then supply the energy to the user as and when required in the quantities demanded.(See Fig. F15.5A)

Fig. F15.5A shows a simple buffer storage where the demand of solar energy roughly coincides with the collection of solar radiation. Buffer storage requires minimal storage capacity.

Fig. F15.5B shows a storage system where solar heat is collected, say, during the noontime daily but the demand continues more or less through 24 hours during day and night. This is then an example of Diurnal storage.

Fig. F15.5C shows the energy storage system for use throughout the year with demands also spread throughout the year. This is the annual storage system, which is really an extension of the Diurnal storage system taking care of the seasonal variation of the sun.

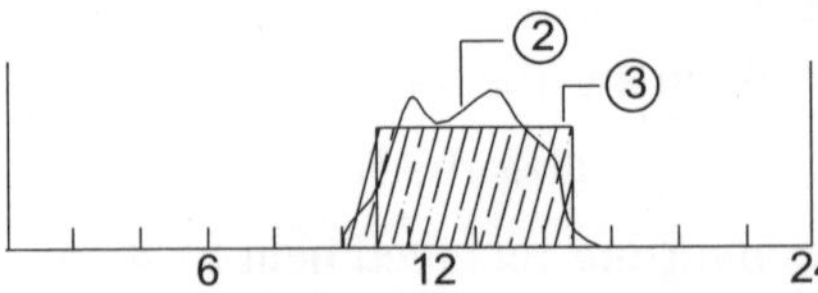

Fig. F 15.5 A : Buffer Storage

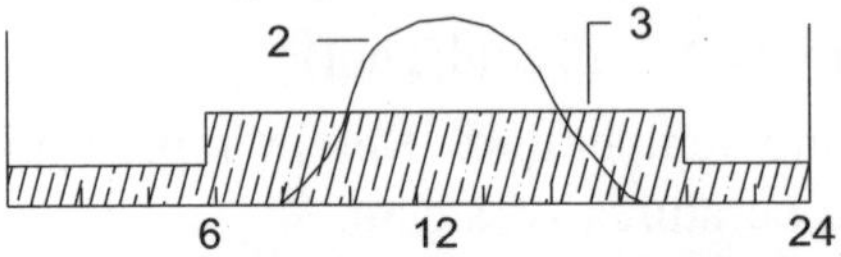

Fig. F 15.5 B: Diurnal Storage

② Energy Collected

③ Load Requirements

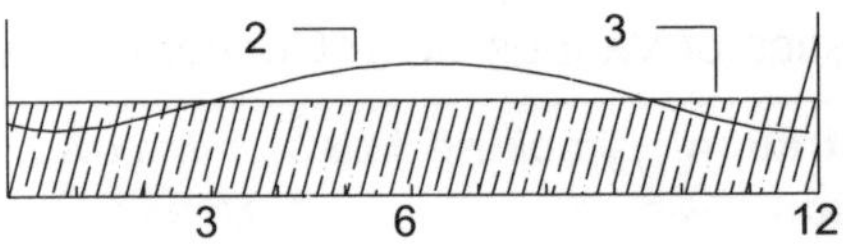

Fig. F 15.5 C: Annual Storage

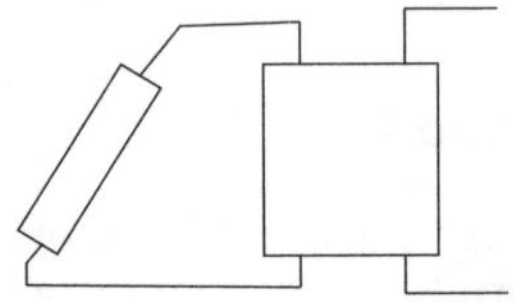

Fig. F15.5D: Different Types of Heat Storage

Principles of Heat storage

The heat transfer can take place in any of the two principal modes of Sensible heat and Latent heat.

The sensible heat transfer law is given by

$$\text{Esh} = m.\int_{t_1}^{t_2} Cp.dt$$

where ∫ is an Integral from temperature limits lower T_1 to T_2 higher of heat storage design, m the mass and Cp the specific heat of the heating medium.

The latent heat transfer is given by,

$$\text{Elh} = \text{m.L},$$

where L is the latent heat of the material undergoing phase change.

The heat transfer can theoretically be effected by reversible endothermic and exothermic chemical reactions on small scale perhaps.

The concept of an underground Aquifer sensible heat storage is shown in Fig. F15.6.

The Solar Pond, which is really an expanse of water of small depth for directly absorbing and storing solar radiation. Heat is captured and stored in the bottom, while the upper layer acts as an insulation shield. To prevent convection heat loss, salt is dissolved in the water to maintain a concentration gradient.

Table T15.2 shows the properties of some typical liquids, which can store sensible heat. Similarly, Table T15.3 shows the properties of some solid media, which are suitable for sensible heat storage.

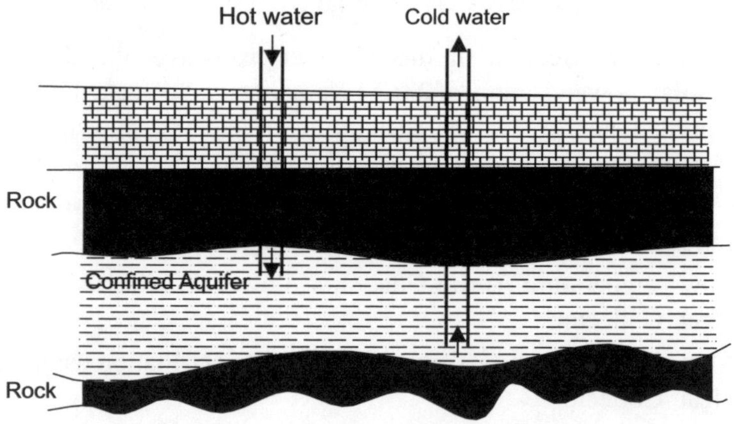

Fig. F15.6: Scheme Of Heat Storage In Underground Aquifer

Table T15.2: Properties of Some Sensible Heat Storage Liquid Media

Properties	Water (Properties at 100 C)	Servotherm (Properties at 100 C)		Hitec (Properties at 400 C)	Sodium (Properties at450 C)
		Light	Medium		
Highest temperature for use C	100	150	250	425	760
Density (kg,m3)	958	867	880	1760	847
Specific Heat (kJ/m3-K)	4.22	2.24	2.22	1.51	1.27
Thermal Conductivity (W/m-K)	0.683	0.126	0.123	0.54	68.3

Table T15.3: Typical properties of Some Sensible Heat Storage solid Media

Properties	Rocks	Pebbles	Magnesium oxide	Aluminium oxide	Silicon oxide
Density (kg/m3)	2254	1350	3575	4000	2600
Specific Heat (kJ/kg-K)	0.81	0.90	1.06	1.02	1.26
Thermal Conductivity (W/m-K)	0.13	0.85	10.5	6.3	2.3

Source: Sukhatme (1996)

15.4 Solar Thermal Electrical Power Generation

In the Introductory Section of this Chapter, we have already mentioned about the solar power station of 80 MWe capacity. In fact, the application of solar thermal energy for the generation of electric power is the most important economic recovery and use of solar energy.

Depending upon the type of solar thermal energy collection system, the Power station may be grouped into three categories:

(*i*) Low temperature Turbo-Generator Power Station using Flat-Plate Heat Collectors;

(*ii*) Medium Temperature Turbo-Generator Power Station using Parabolic Heat Collectors;

(*iii*) High Temperature Turbo-Generator Power Station using either Paraboloid Heat Concentrator-cum-Collectors or the Central Heat Receiver energized by Heliostat Heat-Concentrator.

Fig. F15.7 shows the schematic working of a Low-Temperature Power Station, which also incorporates a Heat storage System. While the primary solar heat collecting liquid media is water, the secondary medium normally recommended is methyl chloride, Tuoline or refrigerants R-11, R-113 or R-114, which vaporizes at low temperature. The water temperature could rise to 100°C and the secondary fluid vapour at 90°C at several atmosphere pressure. The turbo-generator system works on Rankine cycle at low efficiency. The hot and cold temperature difference could be as low as 55°C, the thermal efficiency as low as 7 to 8 % and overall Power generation efficiency approaching 2 % only.

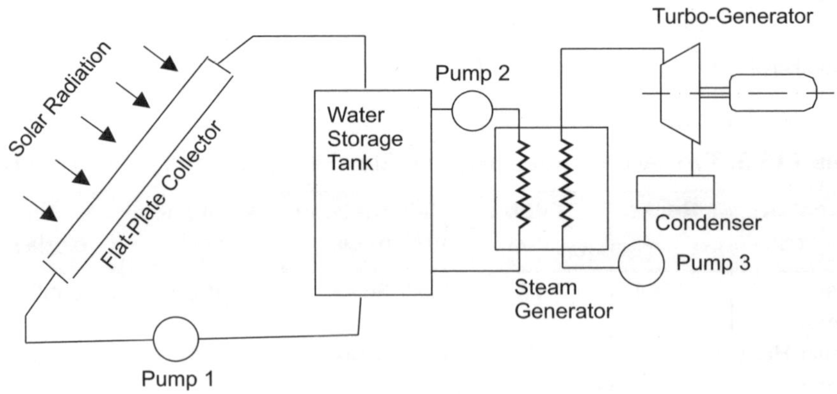

Fig. 15.7: Low Temperature Segs Plant

A low 10 KW plant was installed in IIT Madras, India using Flat-plate collectors, whereas an 150 KW plant in Israel employed Solar pond.

Fig. F15.8 shows the schematic working of a medium temperature Power generation system employing parabolic concentrator and linear pipe heat receiver. The primary fluid is oil, while the secondary fluid operating the turbine is water steam. The oil temperature could be raised to 400°C and the power generation efficiency will be higher compared to the low temperature system described earlier. Table T15.4 shows a series of Solar Electrical Generating station from SEGS I to SEGS IX with total generating capacity of 360 Mwe, and the largest generator having capacity of 80 Mwe. Some of the power stations have been working for several years since installation. The parabolic heat collection surface was as high 484,000 sq.m. for the 80 MWe plant.

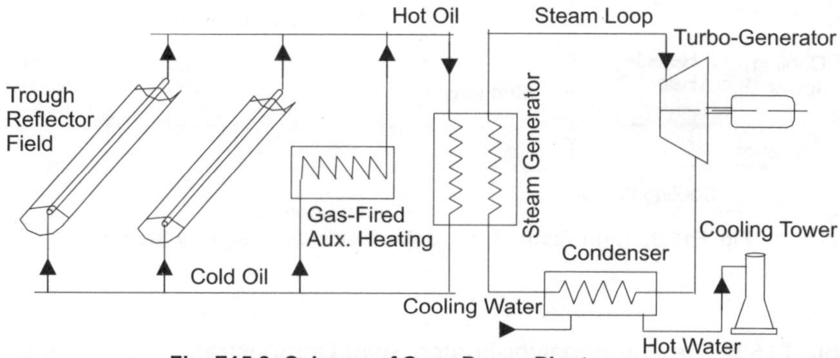

Fig. F15.8: Scheme of Segs Power Plant

The high temperature Solar Power Plant shown in Fig. F15.9 is still in the development stage. Some Demonstration Power stations using this technology and having understandably low power output are shown in Table T15.5. The primary fluid of molten salt reaches temperature as high as 566 C, so that if the solar radiation is low, an auxiliary electric heater keeps the molten salt in the liquid phase.

Table T15.4: parabolic Trough SEGS* Solar Thermal Power Plants in California

Power Plant	Net Electric Power (MWe)	Total Aperture Area (m2)	Year of Commissioning
SEGS I	13.8	83,000	1984
SEGS II	30	19,000	1985
SEGS III	30	230,000	1986
SEGS IV	30	230,000	1986
SEGS V	30	251,000	1987
SEGS VI	30	188,000	1988
SEGS VII	30	194,000	1988
SEGS VIII	80	464,000	1989
SEGS IX	80	484,000	1989

*SEGS (Solar Electric Generating System)

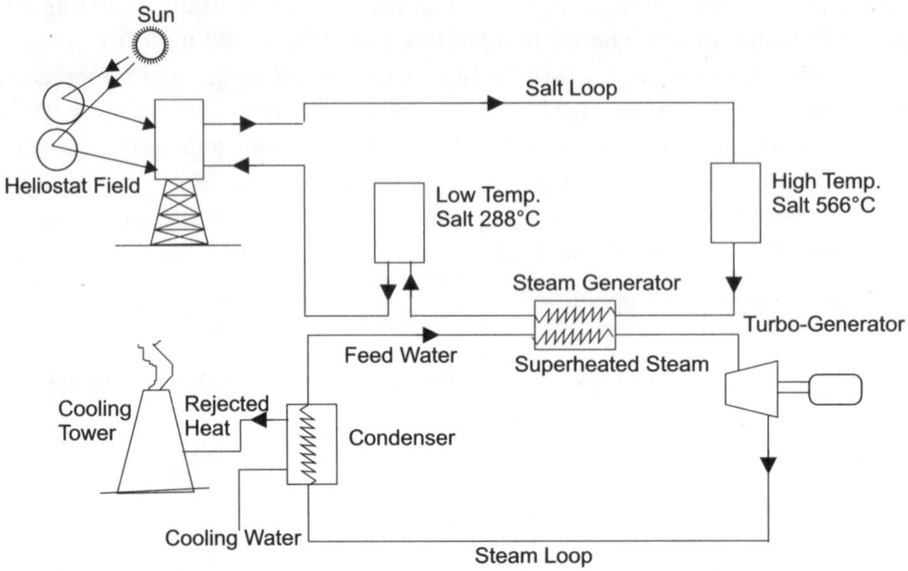

Fig. F15.9: High Temp. Sges (Solar Electric Generating Station)

Table T15.5: Central Receiver-heated Solar Demonstration-cum-Research Power Plants built over 20 Years

Plant Names	Net Electric Power (MW)	Total Reflector area (m2)	Heat Transfer Fluid	Effective Storage Heat Capacity (MWhe)	Period of Operation
Envelios (Italy)	1	6,216	Water/Steam	0.5	1980-84
Sunshine (Japan)	1	12,912	Water/Steam	3.0	1981-84
IEA-CRS (Spain)	0.5	3,655	Sodium	1.0	1981-85
Solar One (U.S.)	10	71,477	Water/Steam	28.0	1982-88
Solar Two (U.S.)	20	81,344	Molten Salt	107.0	1996-99
CESA-1(Spain)	1.2	11,880	Water/Seam	3.5	1983-84
Themis (France)	2.5	10,794	Molten Salt	12.5	1983-85
MSEE (U.S.)	0.75	7,845	Molten Salt	2.5	1984-85
SES-5 (former USSR)	5	40,584	Water/Steam	1.5	1985-88
Weizman (Israel)	0.5	3,472		none	1988-2002

Source: Luzi (2004)

Table T15.6: Estimated Costs of Solar Energy by Different Technologies

CSP Technology	2002 Capital Costs ($/kw)	2010 Capital Costs ($/kw)	O & M Costs (c/kwh)	2002 Approx Electricity Costs (c/kwh)	2010 Approx Electricity Costs (c/kwh)
Power Towers	4,000	2,500	0.7	5 – 9	3.5 – 5
Parabolic Troughs	4,000	2,500	1.0	7 –11	6 – 9
Dish - Engines	1,500	1,500	2.0	9 - 13	4 – 6

Sources: SunLab (1998), US DOE (1996) (from Tester et al 2005)

15.5 Economics of Solar Thermal Power

Solar thermal power generation technologies have achieved impressive developments over the past two decades. Some of these progresses, most notably parabolic troughing, have resulted in technology commercialization, delivering the lowest cost solar power yet achieved. We now give an outline of the status of the emerging Capital and Operational and Maintenance costs of solar power.

Capital Cost

Capital cost includes land, site preparation, storage of energy, plants etc.

In the past, the solar thermal power capital costs have varied between USD 3,000 – 9,000 per kw.

In 2003, 10 to 50 Mwe power plants were offered for approximately USD 2,500 to 3,000 per kw, whereas 0.5 to 10 MW power plants required investment of USD 3,000 to 5,000 per kw nominal plant capacity.

Operation and Maintenance Costs

O & M costs include labour and maintenance costs for operation and maintenance.

For large plants, the ratio labour : materials = 2 : 1 ;

For small plants, the ratio labour : materials =1 : 2 ;

Whereas current commercial solar Thermal power plants have typical O & M costs in the range of USD 0.025-0.035 per net kWh of continuous power generation, the trend of optimization is towards USD 0.010-0.015/ kwh.

Other Cost Data

Similar cost trends are also reported by Tester et al (2005), both in respect to capital costs and O & M costs, as reflected in Table T15.6.

In Fig F15.10 from Tester et al (2005) give the economic trend of the solar power over the period till 2010.

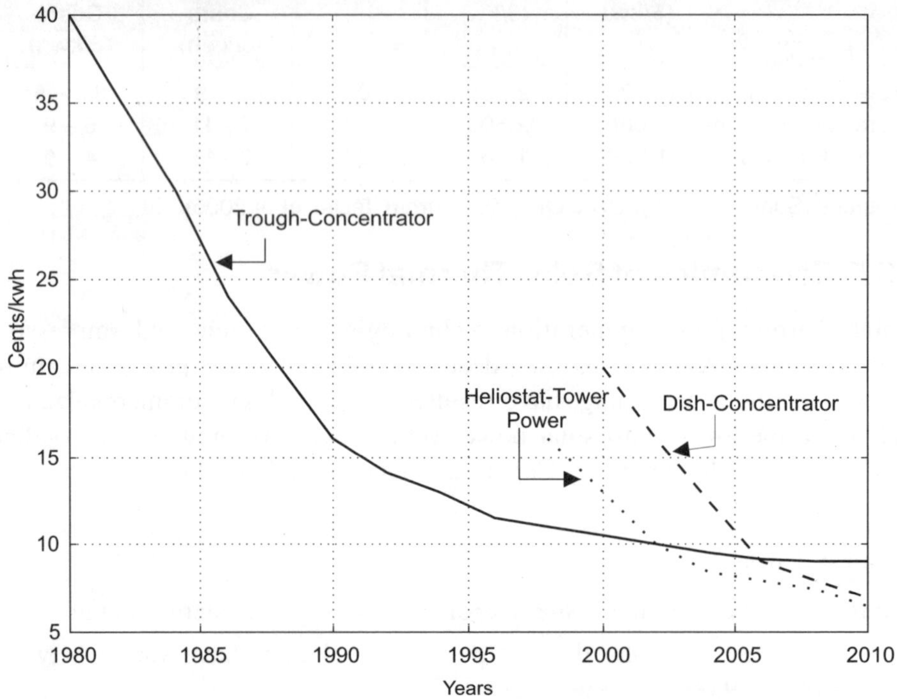

Fig. F15.10: Solar Power - Costing Trends Till 2010

15.6: Photovoltaic Cell – Direct Electric Power

So far we were considering extracting solar thermal energy and converting the same into electric power customizing the conventional Turbo-Generator. Photovoltaic cell (PV) is a device, which converts solar radiation directly into Direct current voltage. One of the main advantage is that these cells have no moving parts, require no maintenance and the energy capacity can be increased or decreased by adding cells in series and in parallel as required. Power conversion efficiency could today reach about 25 %. The PV cell systems have naturally limited power output but they are very portable and can be used in the remote jungle or far desert.

Construction of a PV Cell

Fig. F15.11A shows schematically the general construction of a silicon cell. P-typed single crystal silicon wafer, 300 micron thick, has on top a very thin negatively doped N-typed impurity. The electrode on top is made of twin-fingered Ti-Ag metal solder allowing the sun radiation to excite the exposed silicon crystal surface. The other electrode is of the same Ti-Ag metal solder sheet covering the entire back surface of the cell. Other details of the construction is the application of a very thin anti-reflection SiO coating on the sunlight surface and a general transparent encapsulation of the cell for protection and longevity.

Fig. 15.11B: shows the schematic electrical connectively

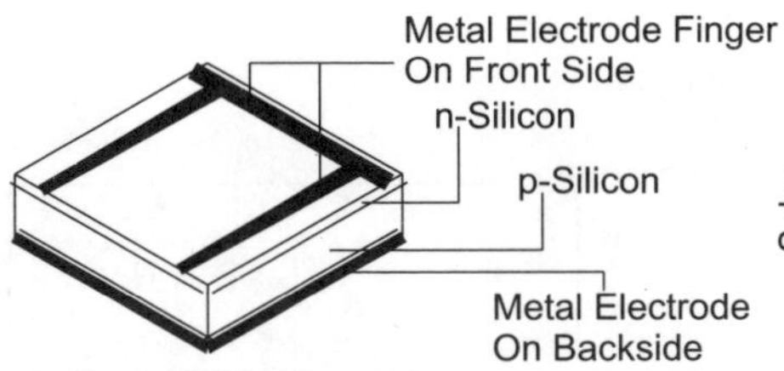

Fig. F15.11A: General Construction of A Silicon Cell.

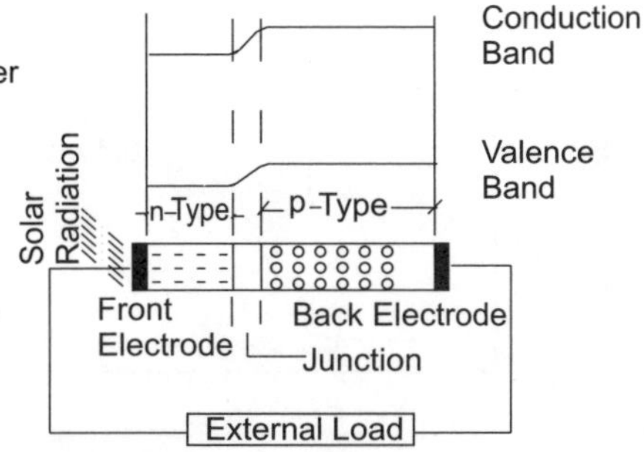

Fig. F15.11B: Principle of Working A Solar Cell

Fig. F15.11: Silicon Solar Cell

Operation of a Solar Cell

The materials, which can absorb the energy of photons of sunlight, are known as semiconductor, *e.g.,* Silicon, gallion, arsenide, and cadmium sulphide. In a semiconductor, the electrons occupy one of the two energy bands – the valence band and the conduction band. The electrons in the valence band are at lower energy level and are more strongly bonded with the atom, while the electrons in the conduction band are at higher energy and comparatively less bonded. The difference between the energy levels of the electrons in the two bands is called the band gap energy, represented by Eg in Fig. F15.11B.

The Solar Cell is really a P-N type junction diode. Fig. F15.11B shows conceptually the electrical connectivity. The electrons in the N-type band and the free holes in the P-type semiconductor are mobile. They can be made to flow through an external circuit by applying a driving potential provided by the solar photon energy, overcoming the bias gradient in the junction of the silicon diode cell. The P-type sandwiched silicon wafer is doped with same atoms of Boron, while silicon of N-type is doped with atoms of phosphorous. The depletion region adjoining the N-type absorbs some holes acquiring a higher positive voltage and repels further holes approaching the diode junction. The depletion region adjoining P-type layer absorbs some electrons and develops a negative voltage repelling further flow of electrons. The reverse voltage bias Vb at the junction is shown in Figs. F15.12A and F15.12B.

The current flow on applied photo-voltage is shown in Figs. F15.12C and F15.12D. The total current flow It is sum of electron flow Ie and hole flow Ih and is given by,

$$It = Ie + Ih$$

Where Ie and Ih are respectively the majority and the minority carriers or *vice versa* depending upon the voltage direction or bias.

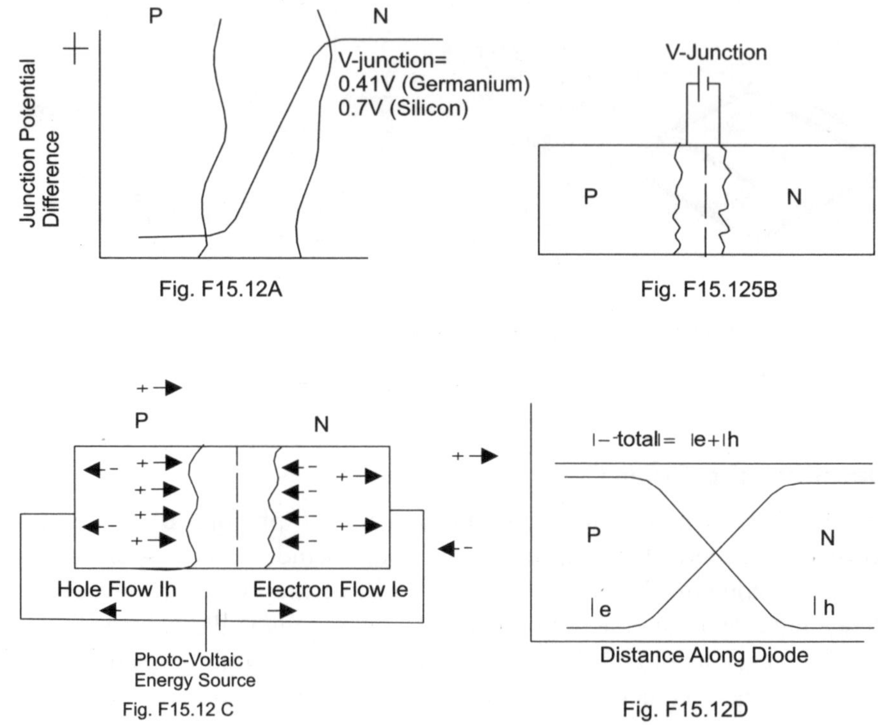

Fig. F15.12: Properties of a Photovolatic Cell (Mottershed 1988).

Specification, Application and Cost of Solar Cells

The specification of a single commercial module of 36 is shown in Table T15.7. In example X15.1, we show how 9 nos. of the above module produce about 425.8 watt energizing a pump delivering water at 10 tonne per day to head of 50 metres. This demonstrates the viability of using solar cell in a remote agricultural land.

Then there are esoteric applications, where portability of photovoltaic power wins over its natural high cost. It is believed that solar cells were extensively used in the Mariner and Skylab Satellite Missions of USA. The cost in the early Mariner mission was several thousand US dollars per peak watt, while in the later Skylab Mission, the cost came down to USD 100.00 per watt. For terrestrial application like the above pumping system where the demand for quality and reliability are not as stringent, the cost may today range around USD 1 to 2 per watt.

From Tester et al (2005), we get the cost trend of PV cells over the years 1990 to 2004 (see Fig. F15.13)

Table T15.7: Typical Specification of a Silicon cell Module (Sukhatme)

Module Size	100 x 50 cm
Cell size	10.4 x 10.4 cm
Number of cells	36
Maximum output	50 W
Maximum Voltage	17.1
Maximum Current	2.92 A
Open Circuit Voltage	21.6 V
Short Circuit Current	3.22 A
Conversion Efficiency	12.8 %

Example X 15.1

Example of Use of PV Energy for Rural Water Pumping.

Problem: Design a Photovoltaic cell to raise underground water at 10 tonne per day with a head of 50 metres. Pump works 8 hours per day. Use PV of Table T15.7.
Assume Pump efficiency Ep = 0.4; Water density d = 995 kg/ m3;
Pump capacity Q = [10,000 / (995x8x3600)] = 0.349 x 10**(-3) cu. m. per second.
Power demand of the Pump P net = [(Q.d.g.H) / Ep] W = [0.349 × 10**(-3)] × 995 × 9.81× 50/0.4 = 425.8 watt.
To use the PV module of Table T15.7, assume,
Inverter efficiency Ei =0.8;
Solar cell efficiency Esc = 12.8 % = 0.128 (given);
Solar radiation = 1000 w /m2;
A.C.power output of a single module of PV cells = 1000 x (0.104 x 0.104) x 36 x 0.128 = 49.84 watt.
Number of such T15.7 modules required = 425.8 / 49.84 = 8.55 units.
Therefore an array of 9 modules, each module containing 36 cells will be required to supply approximately 450 watts.

15.7 Future Outlook of Solar Energy

Along with the wind energy, the solar energy offered a very viable alternative source for the expanding energy demand. The search for the alternative energy has become more intensive in the light of the alarming effect of global warming caused by the GHG emission from the combustion of the fossil fuel. In any case, the global stock of fossil fuel is also fast dwindling and the economy is in frantic search of one or more alternative energy fuel. In this scenario, both the thermal power and the photovoltaic power will play an over increasing role.

Larger power plants will preferably employ thermal energy using parabolic heat concentrator. Table T15.4 shows nine power stations generating about 350 MWe using parabolic trough heat concentrator technology. Table T15.5 shows some grid-connected smaller power plants employing central heat receiver with heliostat concentrator.

On the other hand, photovoltaic cells-supplied power is eminently suitable for isolated smaller consumers some distance away from the main grid. For example, it is more economical to install a stand-alone PV system instead of a transmission line to village having a load of say, 10 or 20 kw, and which is situated some 20 or 40 km away from a main high tension grid line.

PV system is also eminently suitable for special application for satellite powering for its portability irrespective of its cost. Solar cells have already been used on the Mariner Spacecraft and Skylab. PV system is also useful for traffic warning light on high voltage power transmission towers, communication towers etc.

There is increasing production of photovoltaic cells over the last two decades as shown in Table T15.8 Fig. F15.13 shows both the lowering of costs per watt and increasing manufacturing capacities in mwe.

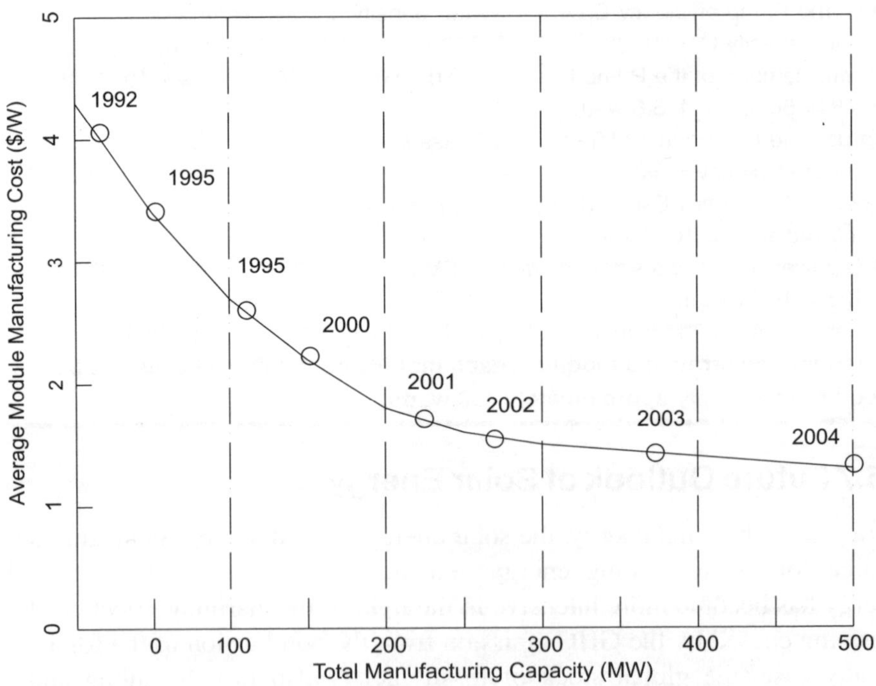

Fig. F15.13 Cost Reduction Trends Of PV Cells.

Table T15.8 Increasing PV Module Production (million) Worldwide (1988-1999)

	1988	1989	1990	1991	1992	1993	1994	1995	1996	1997	1998	1999
Rest of world	3	4	4.7	5	4.6	4.4	5.6	6.35	9.75	9.4	18.7	20.5
Europe	6.7	7.9	10.2	13.4	16.4	16.35	21.7	20.1	18.8	30.4	31.8	36.4
Japan	12.8	14.2	16.8	19.9	18.8	16.7	16.5	16.4	21.2	35	49	80
USA	11.1	14.1	14.8	17.1	18.1	22.44	25.64	34.75	38.85	51	53.7	60.8
Total	33.6	40.2	46.5	55.4	57.9	60.09	69.44	77.6	88.6	125.8	153.2	197.7

Source: NCPV (2002)/ Tester(2005)

❑❑❑

16

CHAPTER

Wind Energy & Power

16.1 Introductory Perspective

Sail Ship – Ancient Use of Wind Power

Wind, storm, tornado are carriers of tremendous energy, which the modern world aspires to capture, to control and to put to useful work. From the time immemorial, winds have been waving the seas carrying the silt, buffeting the continental shelf, but also rustling the leaves and filling the sails.

In the ancient time, sailing vessels ferried the Indian spices, goods and culture to countries of South, Southeast and East Asian. In old time, Egyptians also plied the Nile River as early as 3100 B.C. with craft equipped with sails. For many centuries till the 19th Century, nations depended on sailing vessel for all ocean transportations.

In this Chapter, we are proposing to present the application of the perennial wind energy to generate mechanical and mostly electrical power.

Some Early Wind Mills

Like the sail for water transportation, windmills capture wind energy on the blades of the wheels to pump water, grind grains, and now may as early as the 11th century. Among others, we can mention two types of windmills in old Europe (Paasqualite et al 2004);

Post Mills: In 1937, William of Almour designed the Post Mill. The whole of the Post mill including blades, axles, milling cabin rests on a massive vertical post. The miller orients the post mill into the wind swinging windward the entire windmill assembly around the post, the so-called Post Mill. The use of the post mill spread throughout Europe including England.

Tower Mill: During the golden age of the European windmill, the Dutch alone operated 10,000 windmills. They did more than pump the water, and grind the grain. It is understood that the mills provided more than 25 per cent of Europe's industrial energy from 1300 to the coming of the steam engine and cheap coals during the nineteenth century.

Some Early Wind Turbines

The first to build a practical large-scale wind turbine was Charles Brush in 1886. The Brush dynamics worked well for 15 years but did not become popular because of high capital cost involved.

Experimentation continued between 1890 and 1920, but the most successful was Jacob's wind electric and wind charger. By 1930, wind turbines had found niche in the rural landscape in both Europe and the America. However, as noted in Oxford Report, their use was recommended only where there is no general supply of electricity.

Wind Energy Initiatives across countries

We now present some accounts of developments of wind power in countries US, Denmark, Germany and India.

U.S.A.

The American dream of abundant nuclear power was very much shaken with the accident at Three Mile Island nuclear power plant in April 1979. This trauma followed by the 1986 Chernobyl catastrophe in Russia led to the stoppage and slow-down of all U S development projects in nuclear power. In the midst came the Middle East oil crisis in 1973.

The power development vacuum saw very fast growth in the wind Turbine Power initiatives in California State between Los Angeles and San Francisco, the most economically growing state but having no coal nor oil.

Denmark

It had always played a leadership role in the development of wind power. Unlike US wind turbine design, which was concentrated in the hands of the aerospace industry, Danish wind technology grew out of the agricultural sector, a natural byproduct of Danish economy and the long history Danes have with wind power. They used it to power the ships of early conquests and to carry Norsemen as far the new world.

Germany

Wind developments in Germany followed a Government Law in 1991 requiring all utilities to pay 90 percent of their average retail rates for purchases of electricity from renewable energy source such as wind turbines. This led to very fast development wind turbine energy all over Germany from the polders in the North Sea to the hilltops of Central highlands and also to some extent in other parts of Germany.

India

India has a long coastline of about 4000 km with quite high winds. During the eighties and nineties, significant progress has been made in the design of small capacity standard multi-blade wind mills for pumping water and hundreds have been since installed across the country. During the nineties, a demonstration wind farms of total installed capacity 3MW have been installed at five coastal locations. The capacity at each location being at 0.5 MW, consisting of 10 nos, propeller type turbo-generators of individual capacity of 55 kw. Later bigger wind Turbines Farms of 200mw have been installed .The Indian wind power scenario is given in Table T16.1.

Table T16.1: Indian Wind Power(MW)

Cumulative Installed Capacity (MW) Dec.2005	Annual Energy Additions (MW) (MW) 2005-2009				Cumulative Installed capacity (MW) by 2009
	2006	2007	2008	2009	
4300	1000	1000	1000	1000	8300

Source: Kar (2007)

Global Wind Power Distribution

In Table T16.2 are shown the main wind power generating and user countries in 1999. Germany occupies the top position producing 4072 Mwe power in 1999.

Table T16.2: Global Generation of Wind Power

Country	*2008 (MW)	1999 (MW)
Germany	23,903	4072
United States	25,107	2502
Denmark		1733
Spain	16,754	1722
India	8,896	1077
United Kingdom		534
Netherlands		428
China	12,210	300
Italy		249
Sweden		216

Source: Chambers (2000)/* Wikipedia website 2009

16.2 Wind Farms Worldwide

Wind farms (Wind Park) is defined as a cluster of wind-turbo-generators connected to a single Power Substation. It is postulated that a wind farm should have at least several wind turbines connecting in parallel. Many independent studies conclude that the increasing power demand of today's world can only be met by the renewable energy sources. In fact, it is believed that in 2050, more than 60 % of energy could be available from renewable source, about 20 % of which should be wind energy and a major part of power supply will come from wind farms on land or offshore at sea.

From early 1980, development has moved towards larger wind farms; for example in 1985, only 20% wind farms were larger than 5 mw, whereas in 2001 nearly 70 % exceed 5 mw. Today large wind farms have capacities from 10 mw to 100 mw and the individual wind turbines are larger than the 1.5 mw (See Table T16.3). Because wind turbines are clustered into bigger units, system integration requires special attention. For example,

(*i*) It is estimated that a penetration (supply fraction) of wind energy to a large grid can be a much as 20 % without special precaution to secure grid stability.

(*ii*) For a larger penetration beyond 20 %, attention needs to be focused on control system of the wind farm, improvement of power quality, system and grid stability. Turbine equipped with state-of-art power electronics may even add stability to the grid.

Good wind conditions are essential for the economics for any wind project. In south Germany, a wind turbine produces 1000 hours/annum; in North Germany 2000 hours/annum; in China/ Kazakhstan 400 hours/annum; in Gulf of Suez 6000 hours/annum.

The Horns Rev offshore wind farm, 20km in sea from the Danish coast of Jutland in North Sea, was constructed in 2002 and consists of 80 turbines of 2 mw each. The area of the farm is spread over 20 km squares.

During 2002, new wind farms reached 7227 mw, bringing the total global installed capacity to 32,000 mw supplying 0.4 % of world's electricity demands and with more than 60,000 wind turbines installed worldwide, 85 % Of the total capacity is installed in Europe led by Germany, Spain and Denmark.

Offshore wind farms are still a small fraction of the total global installations. By the end of 2002, a total of 279 mw wind farm capacities are installed offshore, 233mw of which are in Denmark.

Table T16.3: Size Distribution of Wind Farms Globally

Market Segment	Percentage of world capacity							
	1995	1996	1997	1998	1999	2000	2001	2002
1. Dispersed Development, single wind turbines and wind farm < 5 mw	75	70	60	45	42	46	35	32
2. Commercial Utility owned wind farms from 5 to 199 mw	24	28	40	54	57	53	64	65.5

Source: Petersen et al (2004.)

Wind farms: Issues of siting, sizing & layout

Having decided to use wind energy, one has to decide where to place the wind farms (siting), the size of the wind farms (sizing) and the optimum layout of the wind farm.

As examples, we cite a few wind farms. The Midelgrund Offshore farm outside the harbour of Copenhagen was first planned as a straight line, but on the aesthetics ground, it was later aligned to an arch. The Alta Mesa wind farm is located on top of a mountain ridge. In the United Kingdom, some of the wind farms are placed in an irregular pattern, required by legislation not to affect visual impact.

Technical Issues of wind farm

Turbine spacing & wake field

When the wind turbine extracts energy from the wind, the rotor creates a pressure field that reduces the speed of flow immediately behind the turbine disc and expands the flow stream tube around the wind turbine. The thumb rule distance between two consecutive rotors should be X = k.D, where k = is on the order of 10 times the rotor diameter (See Fig. F16.1).

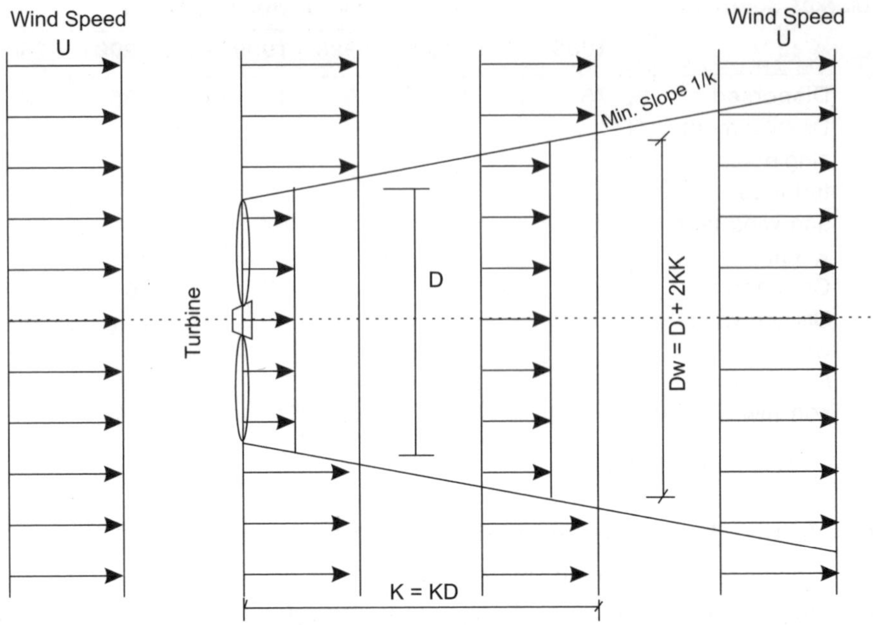

Fig. F16.1: A Simple Wake Model of a Winf Turbine

Wind Farm Efficiency

The reduced wind speed in the wake behind means wind turbine operating in the wake of an upstream turbine will produce less energy. Deep in a wind farm, a wind turbine may experience several wakes and a wind farm will produce less energy than the same number of turbines operating singly and independently under equal wind conditions. The ratio between the annual production of a wind farm and the production of corresponding isolated independent turbines is the wind farm efficiency.

Wind Farm Examples

The two wind farms in Table T16.4 have been built as large-scale demonstration projects according to an agreement between the Electricity Sector and the Danish Government. The combined electricity production corresponds to about 3 % of the electricity consumption in 2002 in Denmark.

While large, these wind farms are not the largest in the world. At present the record goes to King Mountain Wind Ranch in Texas with a capacity of 278 mw and consisting of 214 number 1.3 mw Bonus wind turbines.

Nor will the turbine size in future offshore wind farms will be limited to 2 mw size; most manufacturers today are developing wind turbines in the 3 to 5 mw range.

Table T16.4: Comparative Data of Two Large Offshore Wind Farms

Wind Farm Particulars	Horn Rev Farm	Nysted Wind Farm
Installed capacity	160 mw	160 mw
Number of turbines	80	72
Wind Turbine type	Vestas 2 mw	Bonus 2.2 mw
Expected annual production	600 Gwh	595 Gwh
Hub height	70m	70 m
Wind farm area	20 km2	24 km2
Water depth	6.5 – 13.5 m	6 – 9.5 m
Distance to shore	14 –20 km	10 km
Distance between rows	560 m	850m
Distance between turbines in a row	560m	480m
Internal grid voltage	34kv	33 kv
Transmission to shore voltage	150 kv	132 kv

Source: Petersen et al (2004)

16.3 Technology and Design Aspects of Wind Turbine Energy Systems

Wind Resource

Wind is a renewable resource that can generate significant amount of electricity for both grid-connected and stand-alone applications. However, a major barrier to the deployment of wind energy in many regions of the world is the lack of reliable, detailed wind data, which has recently been made good by the US Department of Energy and Meteorological Department. India is also developing its own wind data.

Wind Speed

Wind speed is the simplest representation of the wind at a given point. Anemometer or other calibrated instruments measure wind speed.

Wind Direction:

Wind direction is measured by a wind vane, usually located at the same height as the anemometer. Knowledge of prevailing wind direction is important in assessing the available resource for the wind turbine siting and spacing. The wind direction distribution is often presented as a wind rose.

Air Mass Density:

$d = [p / (R.T)]$

Where d is the density in kg/m^3;

p is the air pressure in Pascals *i.e.*, N/m^2;

R is the specific gas constant for air *i.e.,* 287 J/kg.K;

T is the air temperature in degrees Kelvin (K= C+273).

An approximate value of d based on altitude only is given by

d = 1.225 – [1.194 x 10**(-4)].z,

where z is the geodetic elevation in metres.

For design purposes, it is quite common to use d = 1.2 kg/m^3

Wind Power Density

Now V is the speed of wind over an area A and air density is d, wind power Pw is obtained from the Fluid Dynamics

Pw = 0.5 (d.V.A).V**2 = 0.5 d.A.V**3; E16.1

(Pw / A) = 0.5d.V**3; E16.2

where (Pw/A) is defined as the wind power density, *i.e.,* Power over a unit area.

Wind Energy Density

Energy is defined as (power x time). Hence, we can obtain wind energy density by integrating eqn. E16.8 over time 0 to t,

$$Ew/A = \int_o^t (Pn/A).dt = 0.5d \int_o^t V^3.dt \qquad \text{..... E16.3}$$

Normally, wind data are available as a series of wind speed ranges say 0-2, 2-4, 4-6 km. per hour, each range blowing in hours per month or per annum (See Tables T16.8). In actual calculations, the integrals are to be replaced by summations as follows:

$$Ew/A = 0.5d \sum V3.(dH) \qquad \text{...E16.4}$$

where dH indicates time in hours.

Wind Turbines – Developments

We have already spoken of two early turbine types - Post Mill and Tower Mill. There were also two cylindrical turbines (See Fig.F16.2) with vertical axis of rotation.

(*i*) Savonius types, where the blades are two opposing scoops, each scoop is actually a half hollow cylinder.

(*ii*) Darius types, whose blades are two or three solid aerofoils, fitting tangentially with turbine circle.

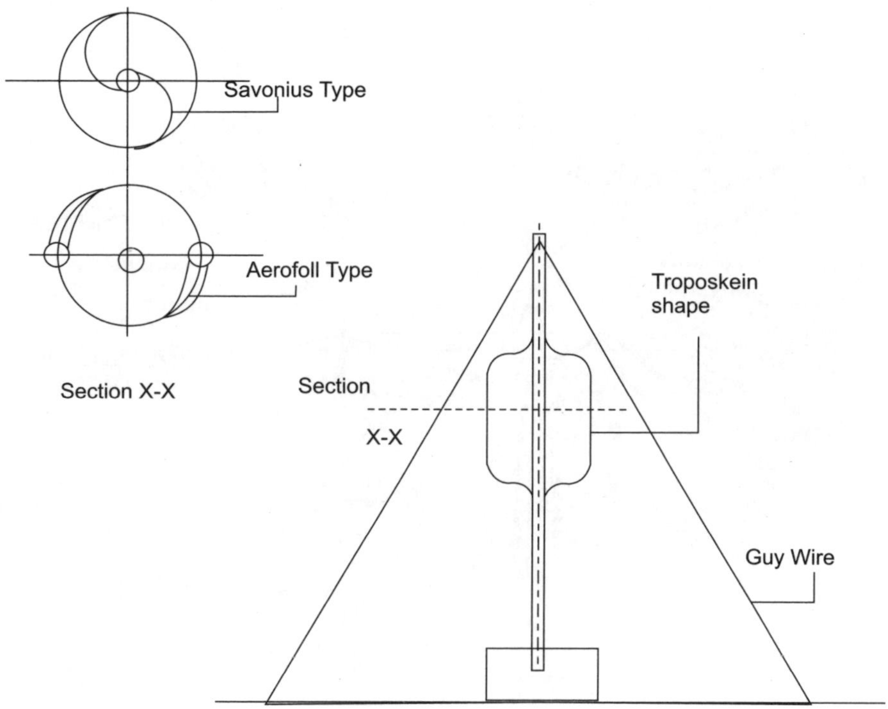

Fig. 16.2 Cylindrical Wind Turbines

Multi-bladed Horizontal-axis Wind Turbine:

More contemporary is the multi-bladed Wind turbine rotating about the horizontal axis. Each blade is made of curved sheet metal like air-fan blades facing incoming wind at right angle. The number of blades ranges from twelve to twenty. They are fixed to circular rims, one at the inner end and the other near the outer end for strength and rigidity. The multi-bladed horizontal-axis wind turbines have diameter varying from 2 to 5 m (See Fig. F16.3)

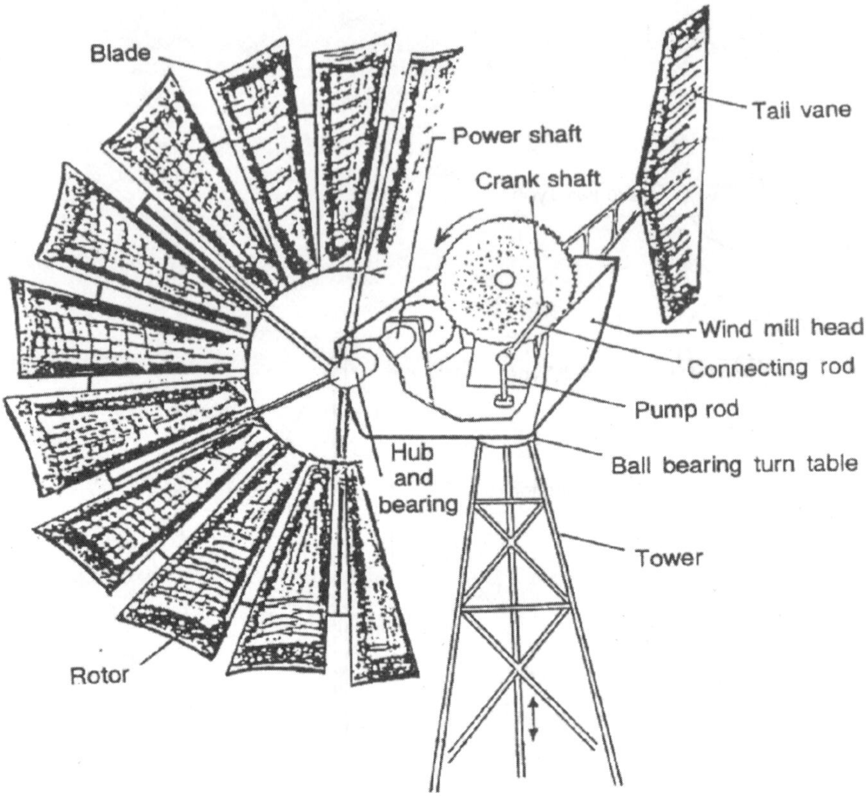

Fig. 16.3: Multi-bladed Turbine (Horizontal Axis)
(Source: Sukhatme)

Propeller Type Wind Turbine

The modern Propeller type wind turbine are the most technologically advanced machines, which have 2 or 3 blades, molded from rigid glass-reinforced plastic, which are quite strong. They also rotate about the horizontal axis. The turbine diameter can be quite high to 20 m to 50 m and the latest turbines in operation can generate 2 mw. Further, future turbines are being designed to generate 10 mw for each turbine. The hub height can be a high as 70 m above ground or offshore sea level. (See Figs. F16.4 & F16.5).

The Table T16.5 shows the principal specification of a modern Wind Turbine Electric Generator of rated power 500 kwe. The main physical components of such a Turbine are also illustrated in Fig16.5.

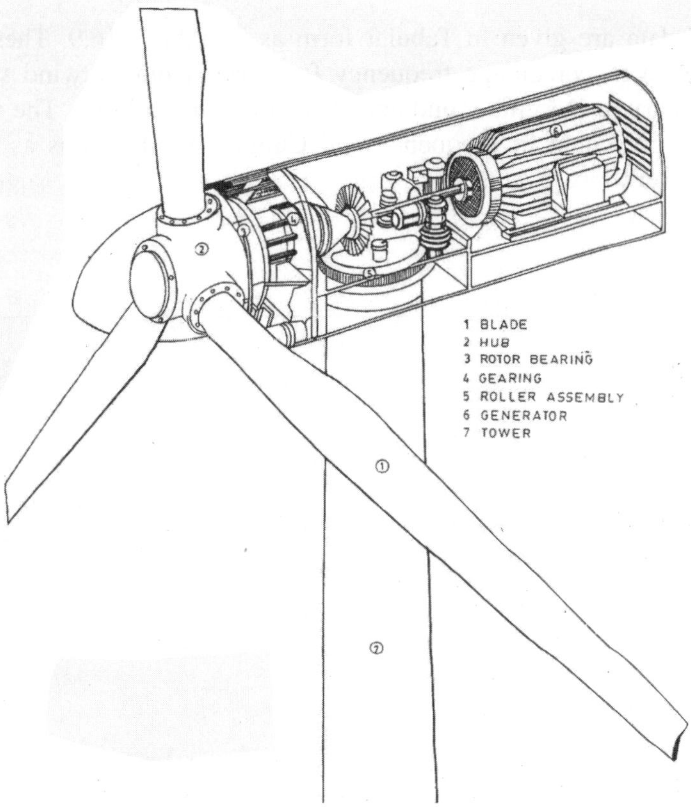

Fig. F16.4: Propeller Type wind Turbine (Source Suknatme)

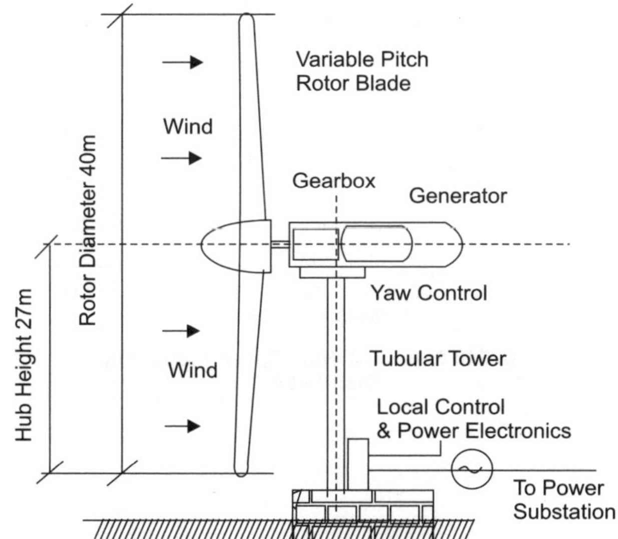

Fig. F16.5: Principal Elements of a Modern Turbine Generators

Frequency and Distribution of Wind Data

Raw Wind data are given in Tabular form as in Table T16.9. These tabular data are given in percentage frequency for varying hourly wind speeds for months January to December and annually in the last column. The wind data can also be presented in Frequency and Cumulative diagrams as shown in Figs. F16.6 and F16.7.

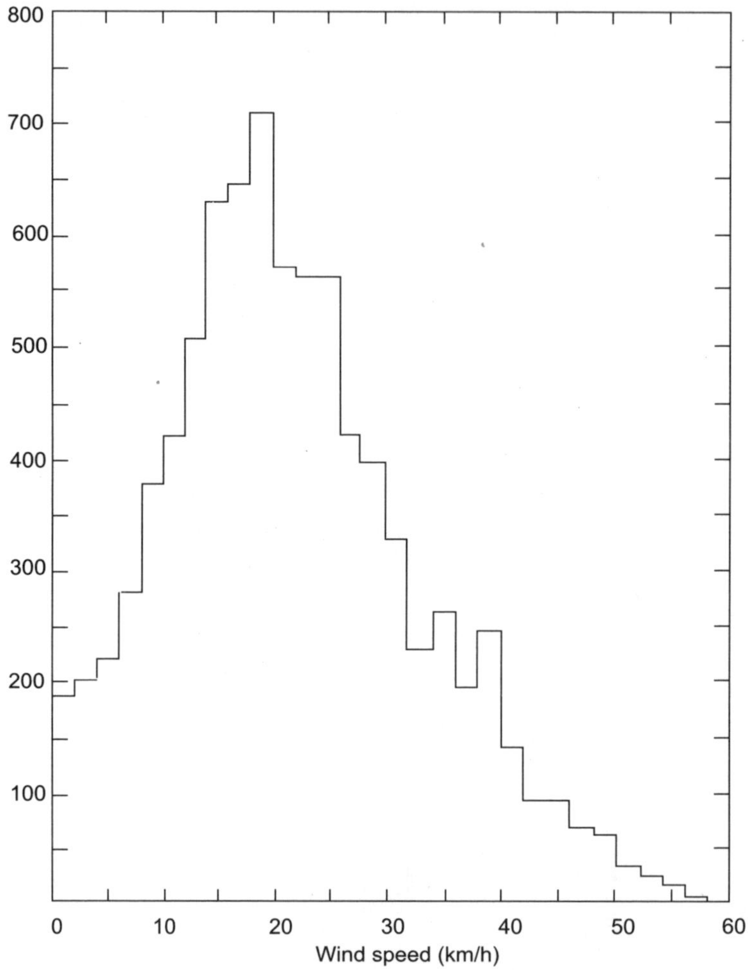

Fig. F16.6: Annual Wind Speed Frequency Distribution

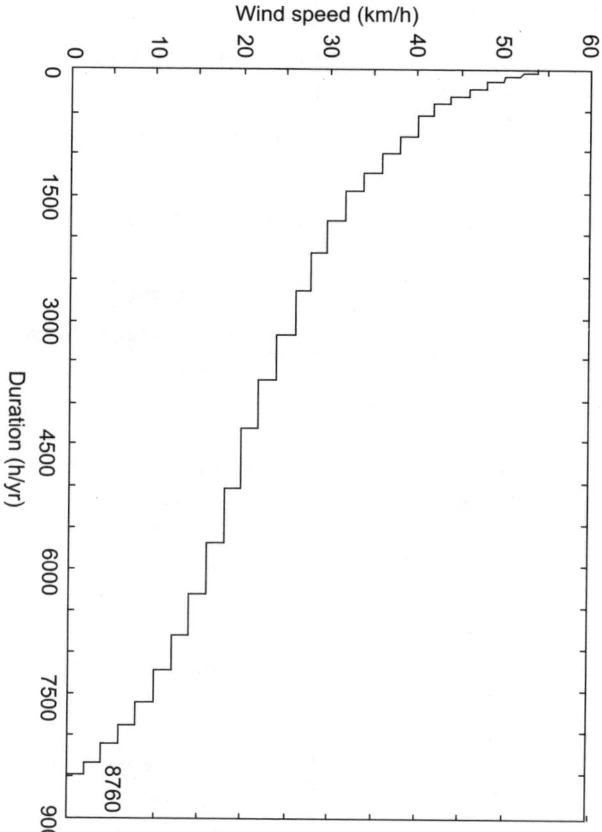

Fig. F16.7: Annual Wind Speed Duration

The Frequency distribution gives the hours in a year or month for wind of a particular speed. The Cumulative Distribution gives the total hours in a month or year or often percentage time for wind ranges from zero to a particular wind speed magnitude.

The wind speed frequency distribution characterizes the wind at a given location in two ways. *Firstly*, the frequency distribution determines how often a given wind speed is observed at a given location; *secondly,* it identifies the range of wind speeds observed at that location.

The wind speed frequency distribution in many areas can be closely approximated by the Weibull distribution function.

Wind Speed at Different Heights & Power Law Index

Friction effects cause the wind speed to slow down near the ground, so that the wind speed at the ground level is lower that at greater heights above the ground. In the near surface layer <300m above ground level (agl), the variation of wind speed (*i.e.*, wind shear) w.r.t. height is approximated by the power law, $V_2 = V_1[h_2 / h_1]^{**}I\delta$ E16.5

where I is the index , whose value is commonly taken at I = [1/7] = 0.14; according to Sukhatme (1995), I = 0.3.

Note that Sukhatme's value has been used by us for all our design exercises that follow:

Table T16.5: Principal Specifications of a Modern Wind Electric Turbo-Generator

Parameters	**Specifications**
Type	**Horizontal Axis, Upwind Rotor**
Rated power	500 kwe
Rotor diameter	40 m
Hence swept area	1257 m2
Hub centre above ground	50 m
Wind speed –related specifications:	
Cut-in	4 m/s
Rated	13 m/s
Cut-out	25 m/s
Survival	60 m/s
Rate of rotation	30 rpm (constant)
Tip sped 62.8 m/s	
Tip speed/rated wind speed	5.2
Installed cost	1000 $ / kwe rated
Hence specific cost	398 $ / m2
Rotor	3-bladed, Fibreglass, variable pitch
Generator	Fixed Speed
Type	4-Pole induction, 66o V, 60 Hz
Rate of Rotation	1800 rpm
Gear Box	3-Stage, planetary/parallel
Yaw control	Active, hydraulically driven.
Over –power protection: reduce	Reduce blade pitch Brake
blade Mechanical load	Disc & Caliper
Tower	50 m, steel, Tubular.
Site wind field-related:	K = 2.5
Weibull shape factor	
Average wind speed	7 m/s
Specific Power available	300 w/m2
Integrated performance:	
Overall Power coefficient (COP)	0.3
Power availability	95 %
Specific energy delivered	824 kwh/m2 per year
Total energy	1.036 x 10**6 kwh/year

Source: Adjusted from Tester et al (2005).

Example X16.1 : To Determine Wind Energy

Given:

Wind turbine site:	Kandla Port;
Time period:	Full month of April;
Hub Center above ground:	33.3 m;
Wind data given in	Table T16.9.

Problem: To determine

A. Total wind energy per sq. metre available during the month;

B. Wind energy per sq. m. available to the turbine during the month under the following conditions:
Turbine cut-in wind speed: 10 kmpH;
Turbine cut-out wind speed: 48 kmpH;
Turbine Design wind speed: 36 kmpH.

Solution A:

Using the equation E16.4 and assuming that the wind speed data given in the above Table are average values over the range of 2 kmpH,
Number of hours in April = 30 x 24 = 720 hours.
[Em/A] = [0.5x1.2x720/100]x[2.1x1**3 + 2.3x3**3 + 2.1x5**3 + ...
0.9x45**3 + 0.4x47**3 + 0.2x49**3]
(kg/m**3).(H).(km**3/H**3)
= 4.32 x [1591043.2] / [3.6**3 x 1000] kwH/m**2.

Note:

1 wH/m**2 = 1 [(kg.m/s**2)x(m/s).H] /[m**2] = 1 [kg.H/s**3];
1 [(kg/m**3).H.(km**3/H**3)] = 1 (kg/m**3).H. (1000**3).m**3/(3600)**3.s**3
= [1/(3.6**3)] kg.H/s**3 = [1/3.6**3] w.H/m**2
= [1/(3.6**3 x 1000)] kw.H /m**2.

Solution B:

Em/A = [0.5 x1.2 x 720/100] x [(4.41x11**3 + 5.0 x 13**3 + ...+3.2x33**3 + 3.5x35)
+(2.4+2.7+1.5+1.1+0.9+0.4+0.2) x 36**3]

Example X16.2 : Design of a Multi-Bladed Wind Turbine

Given:

Required Power output = 7.5 kw (10 hp);
Wind speed at hub centre Vw =27kmpH;
Height of hub centre = 33 m.

Solution:

We assume air density at d = 1.2 kg/cu.m.

To determine the output power, we have to also know the optimum Power Conversion Efficiency Epf of the Turbo-generator. From Fig. F16.8, we take,

Tip Speed Ratio St = 0.9, Epf = 0.3,
Power Output = Epf x [0.5x1.2xPi(R**2).(Vw**3)]; (Vide E16.1)
7500 w =0.30x0.5x1.2x3.1459x(R**2).(27x1000/3600)**3

From above Turbine Radius,

R = 5.5 m

The number of blades Nb actually affects the power conversion efficiency. Hence, experimental data are required to select the number of blades, which usually may vary between 12 to 20. We select rather arbitrarily

Nb = 15.

From Fig.F16.9, we select solidity ratio Sr = 0.4. Hence, Chord length of blade is determined a follow,

$15 \times 5.5 \times C = 0.4\pi \times 5.5^2$ or C = 0.46 m

The angular velocity of the rotor from the definition of the Tip Speed,

St.Vw/R = 0.9x 27 x 1000/3600/5.5 = 1.23 radian/second = 23.4 rpm.

Example X16.3 : Design of Propeller Type wind Turbine.

Problem:

A: To design a 3-bladed propeller type wind turbine to produce 0.5 mw power at the design wind speed at Kandla Port.

B: To determine the annual energy output at the Kandla Port.

Given:

Design wind speed Vw =42 kmpH (11.67 m/s);

Hub centre above ground = 40 m;

Power output Pw = 0.5 mw.

Solution A:

Assuming that the wind turbine works at the optimum power conversion efficiency, we get from Fig. F16.8

Epc = 0.45;

Tip speed ratio Tr = 5.6

The output power (vide E16.1)

Pw = Epc. 0.5d.[πl.(D**2)/4].Vw**3;

0.5x10**6 = 0.45x0.5x1.2x0.7854(D**2)x11.67**3;

Turbine diameter D = **38.5** m (take 39 m).

Turbine speed = Tr.Vw/R = (5.6x42x1000)/(19.5x3600)

= 3.35 radians/s = **32 rpm**

Therefore, the design particulars of wind turbine

Wind speed = 42 kmpH;

Diameter of the turbine = 39 m;

Speed of turbine = 32 rpm.

Grid Power available at various speeds:

Assume transmission efficiency to grid = 0.9;

Generator efficiency = 0.85;

Power at Grid Pge = 0.9x0.85x500x(Vw/42)**3 kw;

At different wind speeds, we get grid power Pge,

Vw	12	18	24	30	36	42	kmpH
Pge	9	30	71	140	240	382	kw

Solution B: Annual energy Generation:

Given:

Turbine diameter =39 m;

Hub height =40 m;

Design wind speed = 42 mpH;

Cut-out Speed = 72 kmpH; Cut-in Speed = 14 kmpH

Wind Data:

Table T16.9 gives the values of wind data at Kandla Port, India at base height of 33.3 m above ground, whereas our turbine hub height is 40 m. Hence, we need to convert all the wind data by using the Power law Index (See E16.5),

(V40) = [(40/33.3)**0.3]. (V33.3) = 1.056.(V33.3)

We also need to convert the two cut-in and cut-out speed to Tabular speed range,

Tabular cut-in speed = 14/1.056 = 13.25, say, 13 kmpH;

Tabular cut-out speed = 72/ 1.056 =68.15, say, 68 kmpH.

Contd...

Annual Energy output for Tabular wind Range 12-14 kmpH:
For the number of hours at speed range of 13 kpmH per annum (8760 hours);
= [(5.8%)/100] x 8760 hours;
Power at 42 kpmH at hub height 40 m = 500 kw;
Energy available at 14 kpmH at hub height 40 m
= 500x[13x(40/33.3)**0.3/42]**3 x [5.8x8760/100];
= 0.69617 x(13**3) x 5.8 kwH
The annual energy delivered by the wind turbine adding all speed intervals from 12 – 14 kpmH,
EkwH = 0.69617x [5.8x13**3 + 7.2x15**3 + 7.3x17**3 + ...3.0x35**3 + 2.2x37**3]
+ 500x[8760/100] x [2.8+1.6+1.1+1.1+0.8+0.7+0.4+0.3+0.2+0.1]
= **1102 mwH**

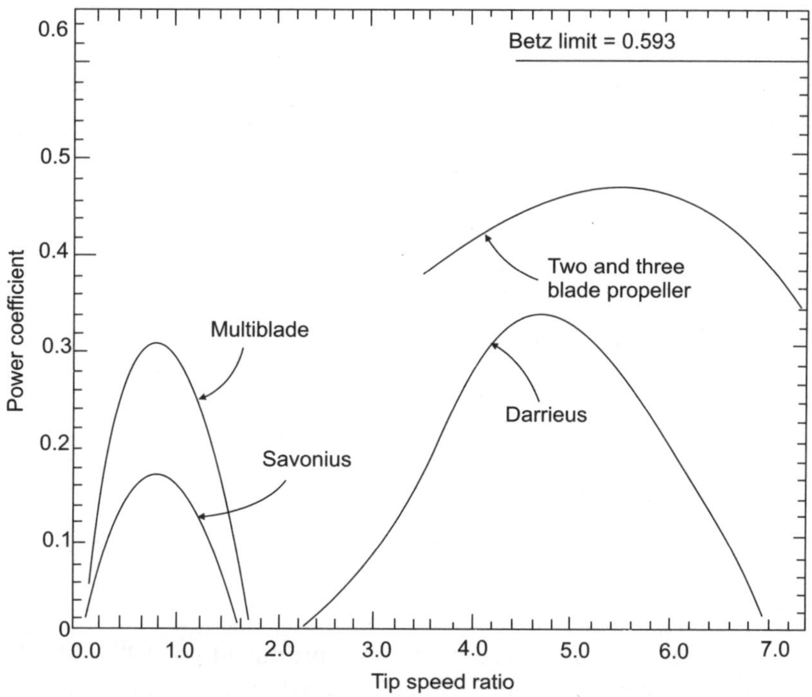

Fig. F16.8: Variation of Power Coefficient vs. Tip Speed (Sukhatme)

16.4 Global Wind Energy Outlook

Global Wind Resources

The fastest growing renewable power scenario is wind energy. Though presently, wind energy produces 2 % of world electricity, World Watch Institute estimates that wind energy could easily provide between 20 to 30 % of the electricity needed by many countries. In US, the American Wind energy Association (AWEA) estimates that wind power will provide more than 10% energy in the near future. The present global generation of electrical power from wind sources is presented in Table T16.6.

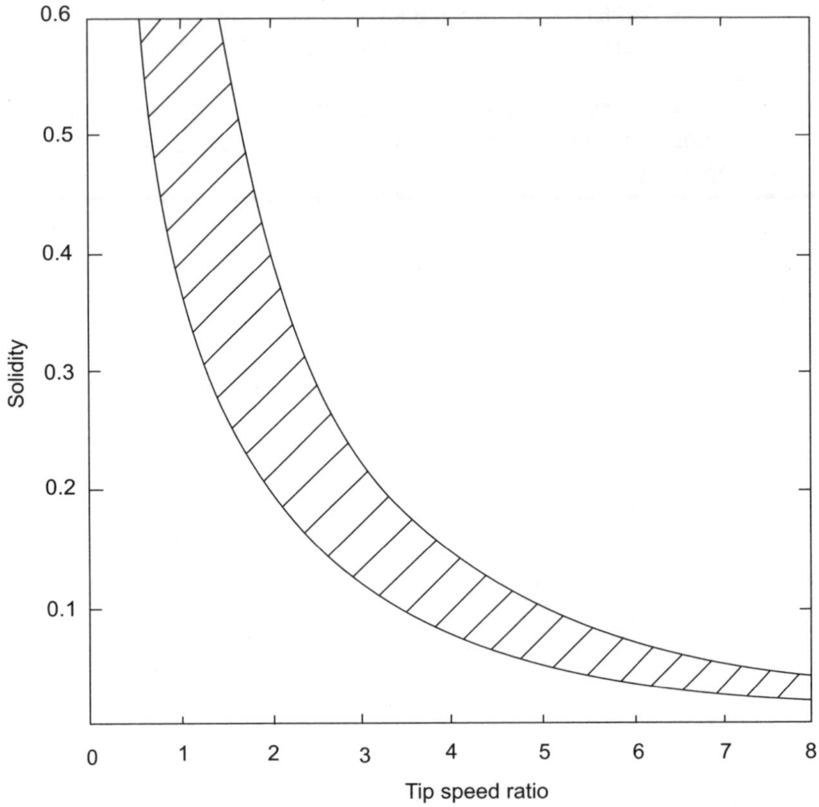

Fig. F16.9 : Wind Turbine Solidity vs. Speed ratio (Sukhatme)

A worldwide wind resource map was produced in 1981 by the US Department of Energy's (DOE) Pacific Northwest Laboratory and was published in a World Meteorological Organization document. The same map was first used in developing estimates of global wind energy potential in the same year 1991. A study by the University of Utrecht in the Netherlands estimated that 25 % of the world's land surface has at least moderate wind resource and 13% has at least good wind resource. In estimating the wind potential, the study assumed that wind turbines would be installed on one of the windy land area. The resultant global potential for onshore wind power production from the higher wind resource areas was estimated to be slightly higher than the world's electricity consumption (1 Terrawatt) in 1987.

Table T16.6: World Electricity Generating Capacities (MW)

Year	North America	Europe	Asia	Total
1980	250			250
1985	2,000			2,000
1990	2,500	1,000		3,500
1995	3,000	2,500	500	6,000
2000	3,000	10,000	2,000	15,000
2002	15,000	20,000	3,000	38,000

Source: adjusted from Pasqualite et al (2004).

US Wind Resource

In 1987, the DOE published the Wind Energy Resource Atlas of the United States. This assessment found that areas potentially suitable for wind energy applications are dispersed throughout much of the United States.

In a study completed for DOE in 1993, it has been observed that good wind resource areas, which cover 6% of the contiguous US land area have the potential to generate more electricity than the total US electricity consumptions in 1990. To provide 20% of the US electricity generation, the study estimated that only about 0.6% of contiguous US land area would have to be developed with wind turbines.

Table T16.7 shows the average wind intensity in several US cities of USA, which are all suitable for wind power generation.

Table T16.7: Wind Speed Values in USA.

Cities	Wind Speed (kmpH)	Cities	Wind Speed (kmpH)
Boston	21.4	Corpus Christi	19.1
New York	20.7	Milwaukee	19.5
Cleveland	18.7	Minneapolis	18.0
Chicago	18.0	Fort Worth	20.1
Oklahoma	22.5	San Francisco	17.0

Indian Wind Resource

India has made good strides in the utilization of wind energy claiming 5 th position in the wind power generation. The present generation capacity is over 1000 mw and the target for 2000 is 9000 mw (See Table T16.1). Valuable wind data have also been collected by Indian Meteorological department at its 40 meteorological stations spread all over India.

The data have been tabulated showing diurnal variation of the mean hourly wind speed (kilometer per hour) for each month and for the year, which are available (Sukhatme,1995). The typical wind data for the Indian Port Kandla in the Gujrat State are shown in Table T16.8.It is estimated from the wind speed data at 50 m agl that the wind energy potential will exceed 45,000 mw.

Wind Power Classification

The values on the Wind Power Resource Map produced by NREL,US (National Renewable Energy Laboratory of US) are based on wind power density, not wind speed. Wind power density is a better indicator of the available resource, because it incorporates the combined effects of wind speed frequency distribution, the dependence of the wind power on air density and the cube of wind speed. The US classification of wind resource based on wind power density is shown in Table T16.8.

Table T16.8: Classification of Wind Resources (USA)

Class	Resource Potential (Utility Scale)	Wind Power DensityW/sq.m. @50m agl	Wind Speed (m/s) a@ 50 m agl
1	Poor	0-200	0.0 - 5.6
2	Marginal	200-300	5.6 - 6.4
3	Fair	300-400	6.4 – 7.0
4	Good	400-500	7.0 – 7.5
5	Excellent	500-600	7.5 – 8.0
6	Excellent	600-800	8.0 – 8.8
7	Excellent	>800	>8.8

Source: Elliot et al (2004)

A Note: Mean wind speed is estimated assuming a sea-level elevation and a Weibull Distribution of wind speed with a shape factor (k) of 2.0.

Conclusion: In most ways, the wind power has come of age. The technology has vastly improved; the economics is fully viable today with the abnormal rise of fossil fuel.

Table T16.9: Wind Speed Data

Percentage Frequency Distribution of Hourly Wind Speed, Kandla Port, India.

KmpH	Jan	Feb	Mar	Apr	May	Jun	Jul	Aug	Sep	Oct	Nov	Dec	Annual
00	1.2	2.8	2.1	3.6	2.2	4.2	1.7	2.0	3.9	7.1	4.6	2.2	3.2
00-02	2.0	2.2	2.6	2.1	0.7	1.8	0.9	0.6	1.2	4.3	4.4	2.4	2.1
02-04	2.4	2.3	2.9	2.3	0.6	1.5	1.0	0.9	2.3	3.9	4.8	2.7	2.3
04-06	3.1	2.5	3.0	2.1	1.1	1.9	1.6	1.1	2.2	4.2	4.5	3.0	2.5
06-08	3.2	3.5	3.7	2.5	1.4	1.7	1.3	1.6	3.1	5.8	6.5	4.4	3.2
08-10	5.2	4.4	5.0	4.0	1.4	2.2	2.1	2.5	4.5	7.3	7.6	6.5	4.3
10-12	6.5	5.6	5.3	4.4	1.7	2.2	2.2	2.6	4.7	7.8	8.1	7.5	4.8
12-14	8.2	6.8	6.4	5.0	2.4	2.6	2.6	3.9	5.6	8.7	9.7	8.4	5.8
14-16	9.2	9.5	8.1	6.4	3.2	3.6	3.4	5.5	7.6	10.3	10.2	10.3	7.2
16-18	10.9	9.1	8.2	6.6	3.4	4.0	3.9	4.7	7.6	9.1	9.0	11.2	7.3
18-20	12.2	11.0	9.3	7.1	4.7	4.6	4.4	6.5	8.9	8.7	9.3	11.5	8.1
20-22	9.6	8.8	7.8	6.6	4.6	3.7	4.4	4.8	6.7	5.8	6.0	9.0	6.5
22-24	8.5	9.1	8.2	7.3	5.2	4.0	4.4	5.4	5.9	4.9	6.1	7.8	6.4
24-26	6.5	6.8	7.5	7.5	6.9	5.6	6.2	7.8	8.3	3.8	4.2	5.4	6.4
26-28	4.5	5.0	5.3	5.5	5.4	5.3	5.5	7.4	5.7	2.1	2.5	3.2	4.8
28-30	2.8	4.1	3.8	6.1	6.4	5.8	6.4	7.1	5.0	1.8	1.3	2.5	4.5
30-32	1.8	2.4	2.8	4.9	5.8	5.2	7.1	6.9	4.2	1.1	0.6	1.2	3.7
32-34	0.8	1.1	1.8	3.2	5.1	4.7	5.6	4.8	2.5	0.7	0.3	0.3	2.6
34-36	0.7	1.2	1.5	3.5	6.4	5.2	7.3	5.2	2.3	0.8	0.3	0.4	3.0
36-38	0.2	0.5	1.0	2.4	5.3	4.6	5.6	3.9	2.1	0.4	-	0.1	2.2
38-40	0.2	0.6	1.1	2.7	6.0	5.9	7.6	5.4	2.8	0.2	-	-	2.8
40-42	-	0.3	0.6	1.5	3.9	3.6	4.4	2.9	1.0	0.3	-	-	1.6
42-44	0.3	0.2	0.4	1.1	2.2	2.9	2.6	1.7	0.5	0.1	-	-	1.1
44-46	-	0.2	0.3	0.9	3.0	3.0	2.8	1.9	0.7	0.2	-	-	1.1
46-48	-	-	0.1	0.4	2.6	2.6	1.7	1.0	0.5	0.1	-	-	0.8
48-50	-	-	0.2	0.2	2.7	1.8	1.4	0.8	-	0.5	-	-	0.7
50-52	-	-	-	-	1.9	1.4	0.7	0.4					0.4
52-54					1.0	1.5	0.5	0.3					1.3
54-56					1.1	0.9	0.5	0.2					0.2
56-58					0.4	0.6	0.1	0.1					0.1
58-60				0.3	0.6	0.1	0.1						
60-62					0.3								
62-64					0.2								

Source: Mani et al

17
CHAPTER

Fuel Cells And Hydrogen Fuel Technology

17.1 Electricity from Hydrogen Fuel Cell

The recognition of Hydrogen as an energy fuel has a celebrated history. In Jules Verne's fictional masterpiece "The Mysterious Island", published in 1874, one of the character states –

"Yes, my friends, I believe that water will one day be employed as a fuel, that hydrogen and oxygen, which constitute it, either singly or together, will furnish an inexhaustible source of heat and light, of an intensity, of which coal is not capable."

A Fuel cell, as we understand today, is a device for directly converting the chemical energy of a fuel into electrical energy in a constant temperature process. Theoretically, a fuel cell can operate with a wide range of fuels, hydrogen is the principal fuel which chemically combining with the oxygen of the air generate electricity.

Operation of a Fuel Cell

A fuel cell is analogous to a storage battery, but one that is being charged constantly by the continuous supply of the reactants. In construction also, a fuel cell comprises of an electrolyte, which is an ionic conductor and two electrodes – a negative anode and a positive cathode.

Fig.F17.1 shows a typical fuel cell PEMFC (Polymer Electrolyte membrane Fuel Cell). Here hydrogen is supplied to the anode, where it ionizes into a proton and an electron, former passing through the electrolyte and the latter an electron is fed into the external circuit, driving a load. The H+ proton combines with the negatively charged oxygen ion of the air, producing water and a small amount of heat.

$2H^+ + O^- = H_2O$ E17.1

The anode is the negative electrode and the cathode the positive electrode of the fuel cell.

The efficiency of the Carnot cycle in an Internal combustion engine or a gas Turbine depends on the inlet high temperature Ti and the outlet low temperature To, as shown.

Carnot Efficiency= [Ti - To] / Ti,

Whereas in a fuel cell, the efficiency of a fuel cell is the ratio of two thermodynamic properties,

Fuel Cell Efficiency = dG / Dh;

where dG is the chemical energy or the Gibbs free energy and dH is the enthalpy change of the chemical reaction.

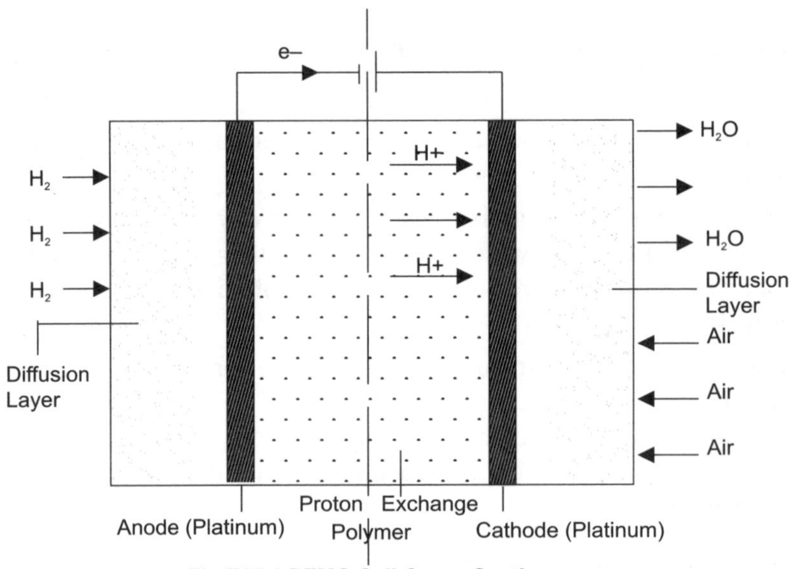

Fig.F17.1 PEMC Cell Cross-Section

17.2 Fuel Cell Types

Table T17.1: Five Main Fuel Cell Types

Fuel Cell Type	Alkaline	Polymer	Phosphoric Acid	Molten Carbonate	Solid Oxide
	Low Temperature Fuel Cells			High Temp.	Fuel Cells
Acronyms	AFC	PEMFC	PAFC	MCFC	SOFC
Operating TemperatureC	60 – 100	80 - 120	150 - 200	600 –700	600-1000

Source: Brandon 2004

Low Temperature Fuel Cells: Characteristics –

- They require relatively pure supply of Hydrogen as a fuel;
  (AFCs are sensitive to carbon dioxide, PEMFCs are to carbon monoxide.)
- They incorporate precious metal electro-catalyst to improve performance;
- They exhibit fast start up time;
- They are available commercially.

High Temperature Fuel Cells: Characteristics –

- They have fuel flexibility; they can operate on a range of hydrocarbons at high temperature.
- Their increased operating temperature reduces the need for expensive electro-catalyst.
- They exhibit slow start up time;
- They can require expensive construction materials to withstand high temperature;
- The reliability and durability are of concerns;
- They are suitable for integration with a gas turbine, offering very high efficiency combined circles.
- They are further away from commercialization, though there are a number of demonstration/pilot plants existing.

Alkaline Fuel Cells (AFC)

It was used in space shuttle producing power for onboard systems by combining the pure hydrogen and oxygen stored in the rocket fueling systems. The electrolyte is concentrated (85% by weight) potassium hydroxide (KOH) in AFC, operated at 250°C, or less concentration (35 to 50%) KOH for operation below 120°C. The electrolyte is retained in a porous matrix (typically asbestos), the electro-catalyst include nickel and other noble metals.

Polymer Electrolyte Membrane Fuel Cells (PEMC)

PEMFCs have high power density, rapid start up, and low temperature operation (around 80° to 120°C) and are therefore ideal for application as Transport batteries.

The electrolyte used is a proton -conducting polymer. This is typically a perfluorinated polymer, though other hydrocarbon-based membranes are under development enabling operation at 200°C approx. The catalytically active layer sits adjacent to the membrane, supported on a PTFE treated carbon paper, which acts as current collector and as gas diffusion layer. For operation on pure hydrogen, platinum and ruthenium are used, when higher level of carbon monoxide is present.

The present cost of PEMFC system is high, but in future the price will come down with mass manufacturing programme.

Demonstration plants of 250 kwe are being produced and used.

Phosphoric Acid Fuel Cells (PAFC)

PAFC fuel cells have been successful as stationary co-generation plants with more than 220 commercial plants already delivered.

PAFCs operate at 150-220°C, using 100% phosphoric acid electrolyte retained in silicon carbide matrix. Generally, platinum electro-catalysts are used in both anode and cathode.

Typical application is in hospital, where the waste heat can be used in the laundry etc.

Molten Carbonate Fuel Cells (MCFCs)

The electrolyte in the MCFCs is usually a combination of alkali carbonates, retained in a ceramics $LiAlO_2$ matrix. The operating temperature is 600 - 700°C, when the alkali carbonate forms a highly conducting molten salt with the carbonate ions providing the means for ionic conduction.

Solid Oxide Fuel Cell (SOFC)

SOFC operates at elevated temperatures, generally above 800°C for all ceramic high temperature variants and 600°C for all metallic-ceramic intermediate temperature FC. The electrolyte is a dense ceramic, usually yttria-stabililzed zirconium (YSZ), which is an oxide ion conductor at elevated temperature. The cathode is a perovskite material Strontium-doped Lanthanum magnetite, often mixed with YSZ in the form of a composite. The anode is a cernet of nickel and YSZ.

The main difference between HT-SOFC and IT-SOFC lies in (*i*) the thickness of the electrolyte, which is about 20 microns for IT-SOFC to reduce ionic resistance, (*ii*) the interconnect material; with stainless steel being used in lower temperature of IT-SOFC.

SOFCs lend themselves to application in which their high temperature can be used. Recently, advances in microelectronics have made possible a combined power plant of 250 kwe with an efficiency approaching 70%.

Fuels for FCs

All fuel cells can run on hydrogen, which combines with oxygen, supplied to fuel cells as air, to form water as in equation high temperature fuel cells can also work with hydrogen-rich methane. Low temperature fuel cells can also work with liquid fuel rich hydrogen.

The common fuel for the commercial stationary power systems with PAFC is natural gas, which is reformed by a separate steam reformer before the hydrogen is fed into the fuel cell stack. In contrast, high temperature FC can work directly on gas stream, though they may need to be cleaned by removal of sulfur and higher hydrocarbons. There are two methods of reforming:

Steam Reforming: $C_nH_m + n.\ H_2O = n.CO + (n+m/2)\ H_2$ ...E17.2

Partial Oxidation Reforming (Shaft Reactor):

$C_nH_m + (n/2).O_2 = n.CO + (m/2).H_2$;

$CO + H_2O = CO_2 + H_2$. (in shaft reactor) ...E17.3

Development of Fuel Cell Industry

In view of current concern for carbon pollution, there is spurt in growth of fuel cell industries. The actual FC industry is only 30% manufacturing, combined with the outsourcing of balance of plants (pumps, compressors, valves, etc) and 30 % integration work.

Fuel Cell cost

Being an emerging industry, there is no mass scale manufacturing and hence the general cost is at rather high level. An outline of costs can be given below:

(*i*) Some specialized FC may even cost $ 500 – 10,000 /kwe;

(*ii*) FC system for large scale power generation $ 1500/kwe;

(*iii*) After the mass production starts, the expected cost $ 300-400/kwe; whereas IC engine/ Gas turbine power costs range today $ 400/kwe.

Environmental Benefits

Emission of SOx and NOx will come down drastically, while using hydrogen group. With hydrogen as fuel, these emissions will be completely eliminated.

Fuel Cell Applications

(*i*) Stationary Power Generation: Capacity 500 kwe to 5 mwe High temperature Systems;

(*ii*) Transport (Motor Vehicles): Types PEMFC and ITSOFC;

(*iii*) Battery Replacement: PEMFC types Cells.

17.3 Hydrogen as an Alternative Fuel

(*i*) The hydrogen, as the lightest element, constitutes about 90 % of all matter in the universe.

(*ii*) Secondly, it is ecologically ideal, as it is carbon-free.

(*iii*) Thirdly hydrogen (H_2) can be produced from water (H_2O) using any energy source.

But there are also several technical problems in the widespread use of hydrogen, *e.g.,*

(*i*) Being lightest, it is difficult to compress, liquefy and store, ensuring adequate supply of energy like the fossil fuel.

(*ii*) Hence, even the highly pressed or liquefied Hydrogen has a low energy density relative to long molecule hydrocarbons, *e.g.,* gasoline (C_8H_{18}), methanol (CH_3OH), or ethanol ($CH_3.CH_2OH$), methane (CH_4).

(*iii*) Hydrogen as an energy carrier must be generated from other energy nuclear, solar or fossil fuel (possibly decomposing water) and hence it carries an economic premium on fossil fuels, unless fossil fuel becomes very cheap, which is now no longer the case.

(*iv*) The changeover to hydrogen economy has to be justified over the total ecological benefits of hydrogen.

Hydrogen Production from Fossil Fuel

Natural gas is now more plentifully available and priced economically vis-à-vis gasoline. It has therefore been envisaged to produce hydrogen from methane gas by process known as steam-methane reforming, essentially harvesting hydrogen (H_2) from water (H_2O) using natural gas by the following overall (endothermic) reactions:

$$2H_2O(l) + CH_4(g) + 245\ J = CO_2\ (g) + 4H_2(g)$$

Steam methane reforming can be 70 % efficient based on LHV of hydrogen (120 MJ) with substantially lower capital investment than other H2 production methods.

Multiple-step Thermo-chemical Cycle for H2O Decomposition

At the sufficiently high temperature (>3000 K), the thermodynamic work of steam decomposition (dG) will shrink to zero and H_2O will decompose spontaneously, theoretically permitting H2 production using only thermal energy. Unfortunately, at such high temperature, H_2 and O_2 molecules themselves become unstable and decompose, forming (OH) molecules and even individual atoms (O and H). This problem precludes generation of H_2 by single stage steam decomposition.

The simple illustrative example is a notional two-step thermo-chemical decomposition cycle in which a conceptual oxide material AO absorbs heat, decomposing at high temperature (eg. 1173 K), releasing O_2 and later reacts with H_2O to liberate H_2, releasing heat near room temperature (373 K);

$$\text{Heat} + 2AO = 2A + O_2\ (1173\ K);$$

$$2H_2O + 2A = 2AO + 2H_2 + \text{Heat}\ (373\ K).$$

Dividing steam decomposition into more than two steps relaxes the thermodynamic requirements on each reaction steps, permitting lower peak reaction temperature. The SI cycle is a three-step thermochemical water decomposition cycle conducted entirely during the fluid phase (Fig. F17.2). The SI cycle relies principally on the high temperature (and high entropy) decomposition of the complex molecules of Hydrogen Sulfate H_2SO_4 into many simpler ones (H_2O, SO_2, O_2), from which O_2 is removed. A second lower temperature reaction step regenerates H_2SO_4 by adding H_2O and captures as Hydrogen Iodide (HI), which is decomposed in a third step, producing H_2 at pressures of 20 atmospheres. The Iodine (I_2) is recycled for use in the second reaction, choosing the thermochemical cycle conceptually expressed by the following reaction cycles:

$$H_2SO_4 = SO_2 + H_2O + (1/2)O_2 \qquad \text{(at 1100 K);}$$

$$SO_2 + 2H_2O + I2 = 2HI + H_2SO_4 \qquad \text{(at 400 K);}$$

$$2HI = I2 + H_2 \qquad \text{(at 625 K)}$$

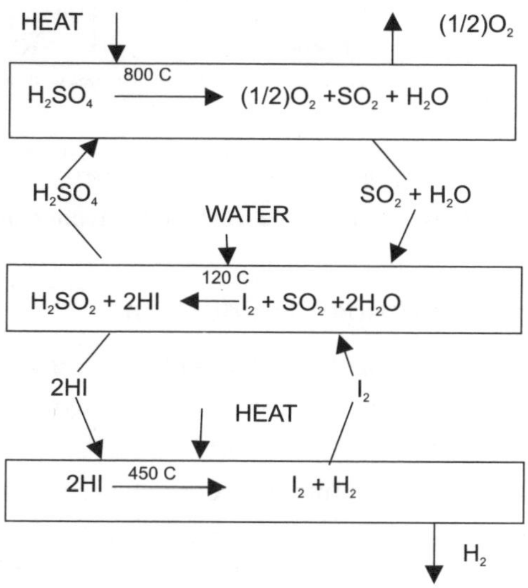

FIG.F17.2 Thermochemical Decomposition of Water

Electrolysis Process of Generating Hydrogen

The electrolysis is a conceptually simple and elegant method of decomposing water (and other exceptionally stable chemical compounds). The electrolyser is an electrochemical cell consisting of two electrodes – Anode and Cathode and an intervening electrolyte. Electrons flow from the anode (oxidation occurs) to the cathode (where reduction occurs). The intervening electrolyte conducts ions between the anode and the cathode, maintaining overall charge neutrality of an electrochemical reaction.

Water electrolysis technology is easily classified into three reaction types with each type conducting one of three ionic species across the electrolyte:

(*i*) Liquid alkaline electrolyte, which transports hydroxide ions (OH)-;

(*ii*) Proton exchange membrane, typically polymer medium, which transports protons (H) +;

(*iii*) Ceramic solid oxide membranes, which conduct oxygen (O)--

In Fig.F17.3A, (OH-) ions, in KOH electrolyte, travel from cathode to anode and oxygen and water are generated at anode. In Fig.F17.3B, the electrolyte is Proton Exchange membrane (PEM) instead of the corrosive KOH, when hydrated protons (H+) travel from anode to cathode. In the third system, in Fig.F17.3C, steam electrolyte in solid oxide ceramics (usually heavily doped ZrO_2), generates oxygen ions at cathode, which ions then travels to the anode and get released.

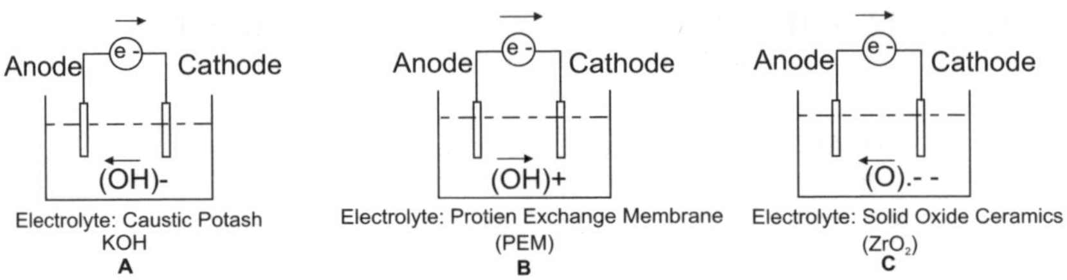

Note: Oxygen & water is generated at Anode

PEM, when hydrated conducts proton (H +) from Anode to Cathode.

Steam electrolysis acts on SOC to conduct oxygen ion from Cathode to Anode.

Fig. F 17.3 Working Principles of Three FC Technologies

Economic Status for Hydrogen Production

Currently, the most economic method of hydrogen production is \$1.0/kg by steam reforming of natural gas, but its dependence on fossil fuel and consequent GHG emission are unsupportable.

The next economic thermochemical energy source from nuclear fission may cost \$1.5 per kg of H_2 at moderate pressure, provided large scale plant size is adopted.

Electrolysis as a third alternative may cost \$2.5 to 3.0 per kg hydrogen is costlier but the system is environmentally clean.

17.4 Hydrogen Fuelled Automotive

A fuel cell engine couples a fuel cell into an electric motor to provide mechanical power. A fuel cell vehicle is an electric motor-driven vehicle having a fuel cell engine as its prime mover. A fuel cell vehicle has the advantage of electrical propulsion, which is quieter and offers high efficiency, through avoidance of mechanical friction. Another advantage is its high controllability and responsiveness providing smooth drive. If the fuel is not pure hydrogen but liquid hydrocarbon or alcohol, fuel processor is required. Fig. F17.4 (Decicco 2004) shows the schematic layout of the various subsystems of a fuel-celled Automotive.

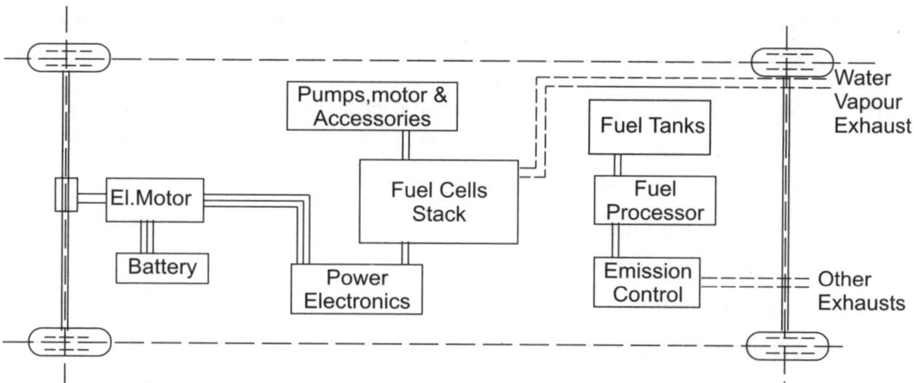

Fig. F17.4: Systems Layout of Fuel Cells-driven Motor Vehicle

It is the progress in the development of hydrogen-air PEM stack that has made fuel cells a contender of powering automobiles of the future. For many years, the energy and density of PEM cells were so low and the amount of platinum catalyst required was so high that mot of commercial applications seemed out of the question. For example, the amount of platinum required for PEM cells used in Gemini Space mission of the 1960s were on the order of 100g/kw, for a cost of \$1500/kw (assuming a platinum cost of \$15/gm). A typical automotive fuel cell stack would be 80 kw, implying cost of \$120,000.00 for the catalyst materials only. By comparison, current automotive catalyst converter requires 0.05 g / kw for an average car. Taking care of the stringent environmental emission requirements, the future gasoline car requires 0.1 to 0.2 gm per kw of platinum group materials.

Using current designs, a typical fuel cell stack of 80 kw of peak power, occupies a volume of about 80 litres and weighs roughly 90 kg. But more lighter designs are evolving. More auxiliaries are needed to complete the fuel cell stack, e.g. pumps, heat exchanger, sensors and controls, etc. A blower or compressor is needed to supply air to the fuel cell stack at positive pressure. Moisture control and thermal management are required. Heat removal presents some challenges, but PEM cells operate at low temperature.

The hydrocarbon fuel supply has better well-to-tank efficiency of 80 % for gasoline and 90 % for diesel. The major problem for use of hydrocarbon in fuel is the presence of carbon and difficulty of fuel processing.

Hydrogen is the ideal fuel for PEM cells, more than any other low temperature cell technology. Recently, the helium gas appears the commercial source of hydrogen, as it can be produced by steam reforming at a central plant and distributed to various filling stations.

The most difficult problem with the hydrogen fuel is its low specific energy density at 7 MJ per kg compared to 36 MJ/kg of gasoline, which makes it very challengingly difficult to arrange its storage on an automotive. Table T17.2 shows the energy storage options of prospective auto-fuels.

Table T17.2: Storage Options of Various Prospective Auto-Fuels

Fuel Storage Options	Specific Energy (MJ/kg)	Energy Density (MJ/litre)
Gasoline (Liquid in steel tank)	36	32
Methanol (liquid in steel tank)	18	18
Hydrogen (compressed at 5000 psi or 340 atmospheres)	7	2
Cryogenic liquid	15	5
Metal Hydride	3	12
Organic Hydride	9	7
Sponge Iron	5	6
Nanotubes	50	100

Note: Comparisons are made at higher heating values (HHV).
Gasoline 46MJ/kg; Methanol 23 MJ/kg; Hydrogen 142 MJ/kg

❑❑❑

18

CHAPTER

Hydro-Energy

18.1 Overview

As on now, the hydropower constitutes about 20 % of the global electrical energy supply. *Secondly*, hydropower is a renewable energy resource, which is the cleanest, not emitting any pollution to the environment. *Thirdly,* the hydropower occupies a very important space in the long term planning, because of its high contribution to irrigation, inland water transport, fishing and other economic areas.

All world rivers originate in the high plateau or mountains at high altitude and finally flow into the oceans at the lowest mean altitude. This water head difference represents huge potential energy and by utilizing this geodetic water head, that hydropower is generated. The technology of hydropower is already well established. The potential energy in the hydrostatic head is converted into kinetic energy, which rotates any of the three kinds of water turbines – impact-flow Pelton Wheel, mixed-flow Francis Turbine, and axial-flow Propeller Turbine, which in turn rotate electrical generator producing low cost electrical energy.

Although the technology of hydropower is well developed, it has not yet been possible to fully utilize the available hydro-energy; because it requires higher capital investment and long-term planning and implementation time. But the operation and maintenance of hydropower require low costs. The extent of non-utilization of hydropower is highlighted from the following data of World Energy Council.

Table T18.1: Global Utilization of Hydropower 2008

Various Hydropower Potential	Amount of EnergyTwh/year
Estimated Global Hydropower Potential	46,170
Technically Feasible Potential	14,650
Economically Feasible Potential	7,300
Utilized Hydropower	2,917

From the Table T18.1 is seen that utilization at 2917 Twh/Y is only 40 percent of the techno-economically feasible hydro-energy resource of 7300 Twh/Y. Further development of the hydro-energy is therefore a challenge to the world.

In Table T18.2A, are given a breakup of hydropower utilization continentwise. The study of such breakdowns allows regionwide planning of power generation without endangering GHG pollution. We see that even highly developed regions like North America, the utilization is about 44 %, in developed Europe, it is about 69 per cent but in the developing Asia it is only 37%.

Table T18.2A: Available Hydropower Continentwise 2008

Region	Installed Hydropower Capacities (Gwe)	Hydropower Generation Twh/y	Percentage of total power Generation	Potential Hydropower Twh/y
Asia (Including Middle Cast)	267	914	24	2500
North America	167	660	23	1500
South America	137	650	–	2000
Europe	225	693	20	1000
World	796	2917	21	7300

World Energy Council 2008.

Table T18.2B: Countrywise Generation of Hydropower 2008

Country	Installed Hydropower (Gwe)	Hydropower Generation Twh/y	Power Generation as Percentage of Potential
USA	99	255	
China	83	309	
Canada	67	315	
Brazil	64	282	
India	36	144	25%
Mexico	10	35	
Russia	–	180	
Norway	–	125	100%

Further from the data of the World Energy Conference under UN at MIT (See Table T18.3), we see that Norway uses hydropower as the total source of electrical power, but has further 80 per cent unutilized hydropower capacity. India seems to have made good strides having its 25 % electrical power supply from hydropower but it can still quadruple its utilization to meet the much needed electrical power.

18.2: Indian Hydropower Development Programme

The first Indian Hydropower station was commissioned in 1887 at the hills of Darjeeling town in the state of West Bengal. From the data of the Central Electricty Authority, shown in Table T18.3, we see India has considerable hydropower potential of 148.9 Gwe, of which only a small part 36 Gwe has been utilized. The hydropower capacity is now being further increased by 16.5 Gwe during the 11th five year period 2007-12.

Table T18.3: India's Hydropower Potential 2008

Name of River Basin	Probable Capacity Mwe at 100% power factor
Indus basin	33,822
Bramaputra basin	66,065
Ganga basin	20,710
Central India basin system	9,430
East-flowing River system	14,511
Others	4,162
Total	148,700

18.3 Generation of Hydropower

We have already indicated that it is first necessary to create the geodetic height or hydrostatic head, which is then to be converted to hydro-energy with the help of water turbines. There are principally two methods of creating the water head –

Run of River: In this method, the natural gradient of the river flow is used. A tunnel or penstock is used to divert or pat of the river flow in way of river loop at the lower level where a turbine is situated (See Fig. F18.1). After the flow through the turbine, the water is released back to the river channel through the Tail. Thus, down stream, there is no change in the flow rate. However, there might be instances, where the Tailrace diverts water to another river basin or lake, where there will be reduction in the river flow downstream (See Fig.F18.1 and F18.3).

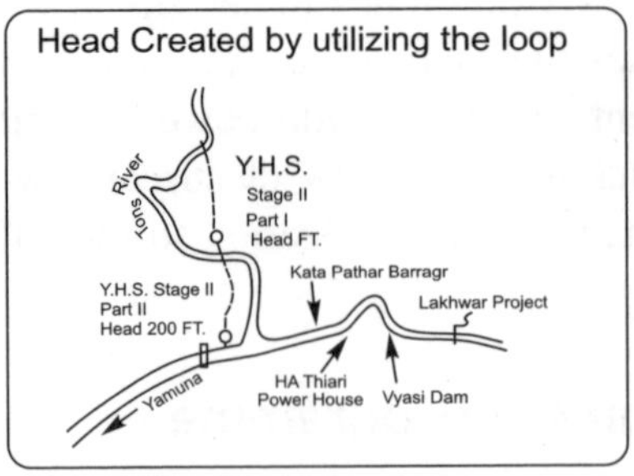

Fig. F. 18.1: Yamuna Hydel Scheme

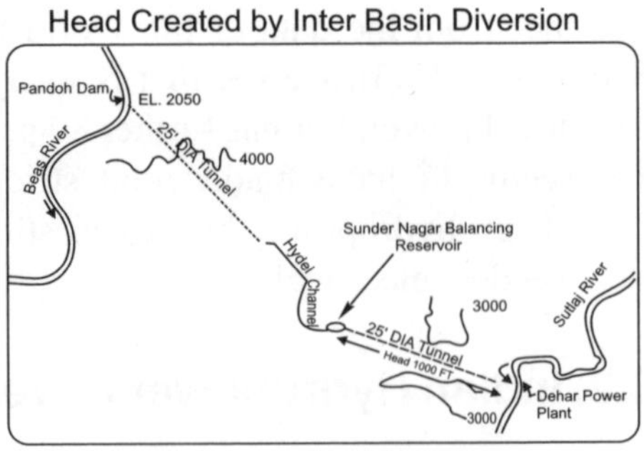

Fig. F. 18.2: Beas Sutlej Link Project

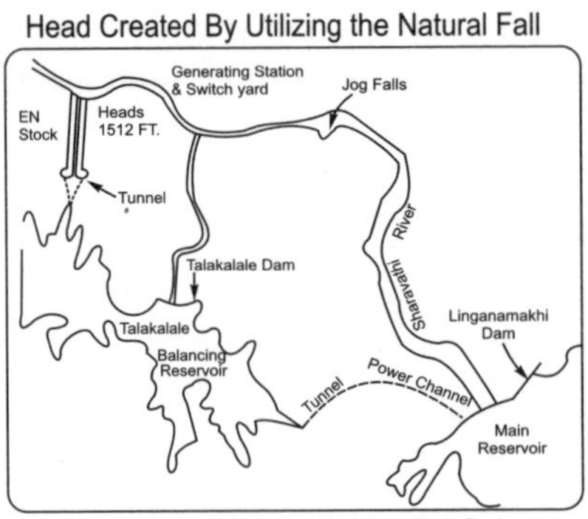

Fig. F. 18.3: Sharavathi Hydel Scheme

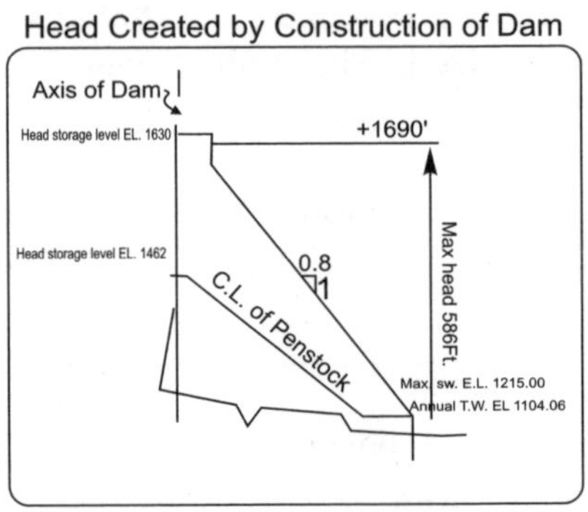

Fig. F. 18.4: Bhakra Dam

Fig. F18.1: Yamuna Hydel – Run-of River Scheme (Using River Loop)
Fig. F18.2: Beas-Sutlez – Run-of River Scheme (Inter-River Transfer)
Fig. F18.3: Sharavati Hydel – Run-of River Scheme (Transfer to Lake)
Fig. F18.4: Bhakra Impounded Dam. Source: Rao, 1975

Impoundment Dam: In this method a river is impounded in a valley by a large highrise concrete dam creating a great depth of water, which then is released through a large diameter Penstock to a turbine for generating immense hydropower (see Fig. F18.4). These types of high and heavy concrete dams are very capital intensive, but as they generally store large volume of water, they also serve multiple other functions, including agricultural irrigation, supplying water year round through a network of irrigation canals. The mammoth Bakra-Nangal Dam on the Beas-Sutlez Rivers in Punjab of India generates 2500 Mwe power at 60% load factor.

Pumped Storage systems: This is really not a direct system of power generation but really a storage of energy hydraulically. When the electrical energy supply is in excess in the night, the water is pumped up and stored at a sufficiently high head. Now the energy demand is peaked at daytime, hydropower turbo-generator is activated by the release of water stored at high head during the off-peak time.

18.4 Turbo-Generators

The created overhead is converted into mechanical energy by rotating water turbines, which in turn rotates electrical generators. There are three types of water turbines depending upon size of the water head available:

(*i*) **Pelton Impulse Wheel working with high Water Head:** High headwater is released as high-speed jet hitting the buckets of the Pelton Wheel tangentially (see Fig.F18.5) and rotating it at high speed.

(*ii*) **Francis Reaction Turbine working at medium Water Head:** This is a mixed flow turbine, the water enters the turbine radially and emerges axially, and thus the discharge is parallel to the axis of the turbine. In this reaction turbine only a small part of the available head is converted into kinetic energy and a major part remains in the form of the pressure head (see Fig. F18.6). A difference of pressure is established between the entrance and the exit of the runner. This difference of pressure is termed as "Reaction Pressure" (Vaishwanar 1997).

(*iii*) **Kaplan Propeller Turbine working at low Water Head:** This is axial flow turbine, water enters and exits the runner axially and almost all the water head remains as pressure head, used up in rotating the runner and the exit pressure is low (see Fig. F18.7).

The characteristics of the Hydropower systems are summarized in Table T18.4.

Table T18.4: Major Attributes of Hydropower

Positive Characteristics	Negative Characteristics
1. Hydro-energy is renewable and is thus most welcome with dwindling stocks of fossil fuels. 2. It is free of environmental pollution, not emitting GHG. 3. It is highly efficient, hydraulic conversion efficiency being on the order of 80% or above. 4. It has low operation and maintenance cost. 5. But it can have high initial cost, which can offset its multitude beneficent uses in Domestic and Agricultural irrigation water supplies.Long active life of the Hydro-plants.	1. Though the energy is renewable, supply may vary year to year due to vagaries of rainfall. 2. However, large projects lead to impoundment and inundation of large surface areas, which may lead to displacement of human habitation and fishing migration. 3. The capital cost is high. 4. Generally very long lead time in construction is required.

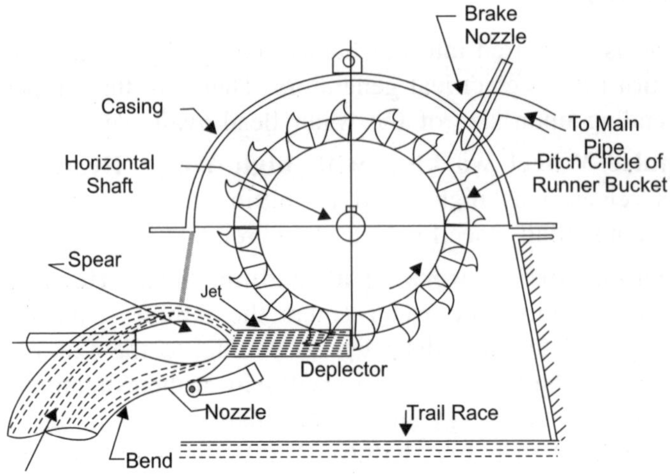

F 18.5 Pelton Turbine

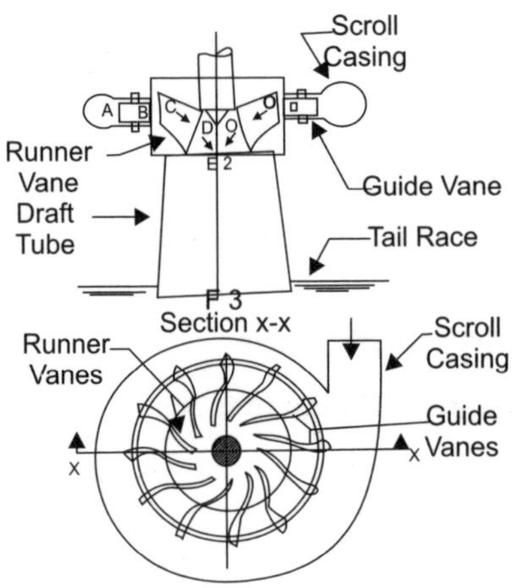

F 18.6 Franis Turbine

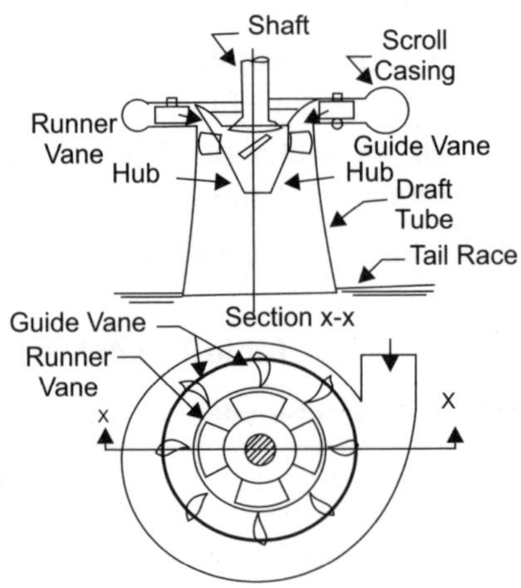

F 18.7 Kaplan Turbine

Fig. F18.5: Single Jet, Horizontal Shaft Pelton Turbine.

Fg.F18.6: Mixed Flow Francis Turbine.

Fig. F18.7: Axial Flow Kaplan Propeller Turbine.

❑❑❑

19

CHAPTER

Other Energy Challenges

In this Chapter, we will describe three more sources of renewable energy, of which, the Ocean Thermal Energy and Ocean Wave Energy are future energy systems. But the third system geothermal energy system is already generating electrical power all over the world in small measures.

19.1 Geothermal Energy

The geothermal energy is that obtained from the thermal energy entrapped in the earth. In 1902, electricity was first generated from the steam retrieved from the earth's crust at the Larderello, Italy at the boric acid plant there. Since then, the geothermal steam and hot water have been identified at the Geysers, California, Reykjavik, Iceland, Weirken, New Zealand, Cerro Pristo, and in Mexico, Philipine, Indonesia etc. The geothermal energy now contributes 10,000 Mwe of electrical energy and 100,000 Mwe of thermal energy.

Known Sources of Geothermal Energy

We now describe three types of geothermal sources:

(*i*) **Natural Hydrothermal Sources:** There are water aquifers, which get heated up from the heat source of the earth's core of molten rock or otherwise. The general arrangement is is shown in Fig. F19.1. If the water system is below the vapour pressure, the hot water is mixed with vapour as found in Geyser field at California, Lardello field in Italy and Matsukawa field in Japan (Tester et al 2005). There are also high-pressure hydrothermal fields, where the hot water is mixed with steam is obtained. Temperature may range up to 350°C.

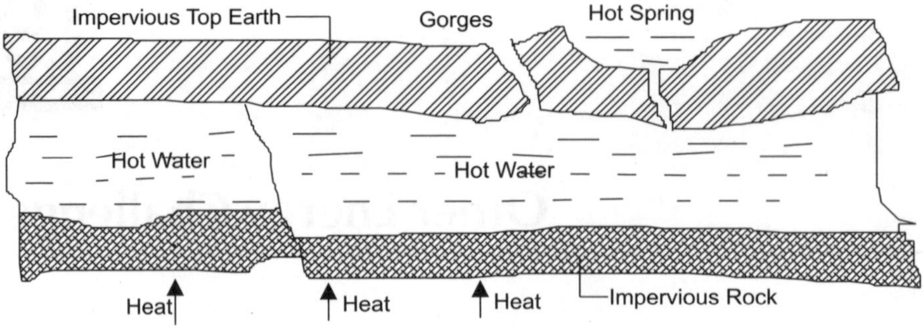

Fig.F19.1: Hydrothermal Source Aquifer

(*ii*) **Geo-pressurized Geothermal Sources:** In the Geo-pressurized basins, water, contained in large sediment-filled porosities, gets pressurized to a much higher level than the normal hydrostatic head. The fluid contained in the sandstone-shale and other shale structure may get over-pressurized upto 60 Mpa (600 bars) at temperature of 150-180°C (Tester et al 2005).

The geo-pressurized systems, like in Louisiana, US, may also have high salinity and cause serious scaling inside the heat transfer tubes.

(*iii*) **Hot Dry Rock Sources:** Because of the low permeability of the hot rocks, the quantitity of hot water available in the source may be limited, though the temperature of the rocks may range up to 200°C or more. To utilize the heat, the rock is drilled to the necessary depth of 4 to 6 km with two holes, one to ingress cold water into the rocks and the second to return hot water at about 150°C or more (See Fig. F19.2).

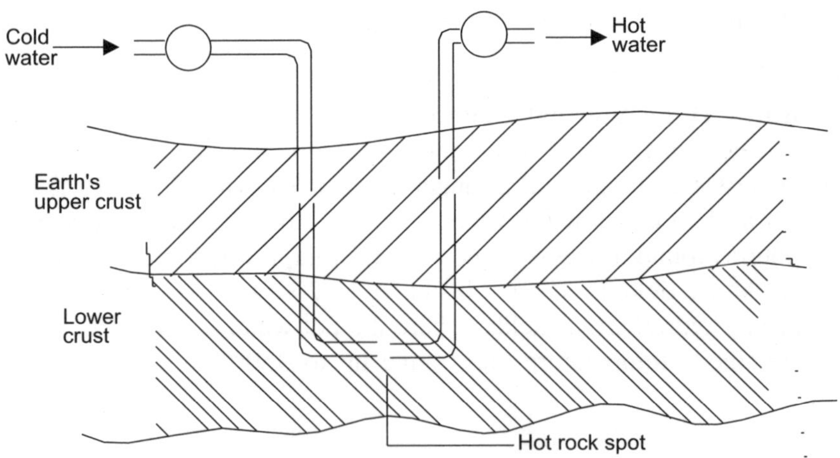

Fig. F19.2: Hot Dry Rock Heat Source

Geothermal Electric Power Generation

A simplified scheme of Power Generation from sourcing geothermally heated water under the earth crust is shown in Fig. F19.3. Hot Water is retrieved, passed through the heat exchanger and thereafter returned to the ground source. In the middle, we have the usual steam Turbine Generation System and on the very right side, we have the Cooling Tower to cool the condenser water.

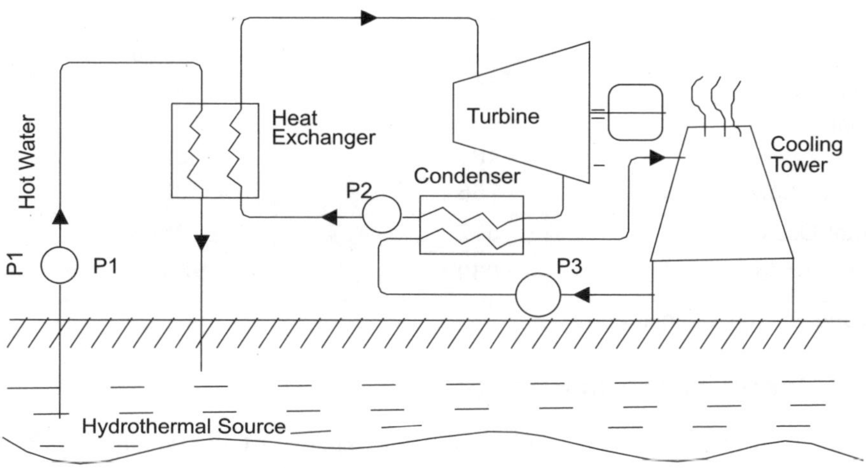

Fig. F19.3: Scheme of Geothermal Power Plant

Present Status of Geothermal Energy

Geothermal Energy is believed to be growing at 8.5 % globally. The global status of electric and non-electric capacities in 1979 is shown in Table T19.1.

Table T19.1: Global Geothermal Energy States, 2007

Countries/ Region	Installed Capacity (Mwe)	Generation capacity (Gwe)
Ethopia	7	–
Kenya	129	911
Total Africa	136	911
Costa Rica	166	1239
El Salvador	204	1293
Guadeloupe	15	95
Mexico	960	7404
Nicaragua	87	243
United States	2937	14,839
Total North America	4422	25,376

Contd...

China	28	96
Indonesia	992	5000
Japan	535	3120
Philippines	1970	10,215
Turkey	38	158
Total Asia	3563	18,578
Austria	1	2
Germany	3	–
Iceland	485	3579
Italy	810	5233
Portugal	23	178
Total Europe	1401	9077
Australia	–	1
New Zeland	461	3272
Papua New Guinea	56	3272
Total Occania	517	3523
Total World	10,039	57,460

World Energy Council 2009

19.2 Ocean Thermal Energy

Ocean is an enormous heat sink. The vast expanse of the ocean absorbs the solar radiation. The ocean surface is therefore relatively warmer compared to the temperature a hundred meters down the ocean. The temperature difference of about 20 C or more is used to evaporate low boiling point fluid, *e.g.*, ammonia, propane, etc., which in turn is used to drive a turbine of a Turbo-generator (See Fig.F19.4).

Reference to Fig. F19.4, warm water at 20 – 24°C from the ocean surface is pumped (by Pump P1) into the evaporator, into which is pumped (by Pump2) a low Boiling Point fluid (say ammonia or propane) at a higher pressure (say 10 bars or below). Saturated vapour of the fluid at say temperature 20°C and 8 bar pressure then drives the turbine. The low-pressure vapour is then cooled and liquefied in the condenser, by cold water (say at temperature 8 – 10°C) pumped by P3 from the ocean deep (200 – 300 metres below the surface). The Temperature-Entropy T – S diagram of the system is shown along in Fig. F19.4, which is a closed Rankin Cycle producing some net energy output.

Such a Pilot plant has been tested by the US Navy at Hawaii around 1979. But no actual Power Plant based on ocean thermal energy operates in the world. Though the concept of this power source is well known as OTEC (Ocean Thermal Energy Conversion), the operational technology of the same is a thing of future.

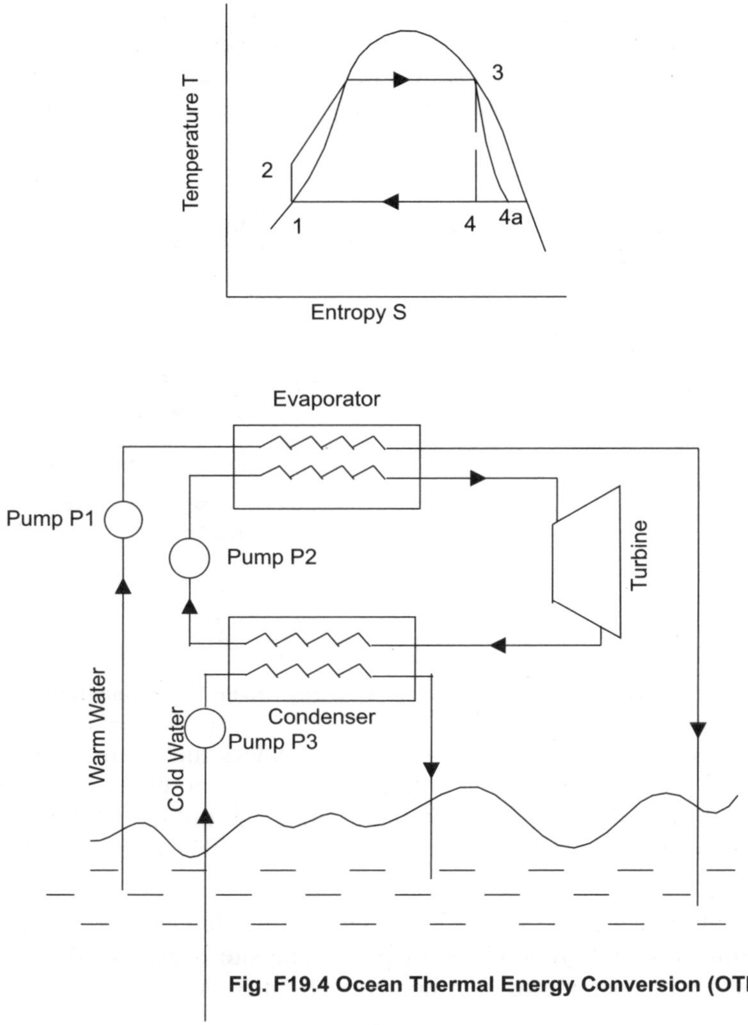

Fig. F19.4 Ocean Thermal Energy Conversion (OTEC) System

19.3 Ocean Wave Energy

The vast expanse of the ocean water surface, buffeted by the wind, gives rise to ocean waves. Considerable amount of kinetic energy is locked up in the ocean waves. But it is quite difficult to recover this large wave energy into effective energy applications because of wide fluctuation of wave amplitude and its random direction. Nevertheless attempts have been made to design several systems to convert the wave energy to generate power. It is estimated, on an average, there is 10 Kw power for every metre length of wave front.

We describe one common wave energy system – the Oscillating Water Column (see Fig. F19.5). The column is a heavy concrete column sitting on the ocean bed near the coast. An wave crest enters the column compressing the inside air upwards, causing the NR valve V1 to open and allowing the compressed air to enter the turbine from the left and making it to rotate and finally making the air to exit from the column through NR valve V4.

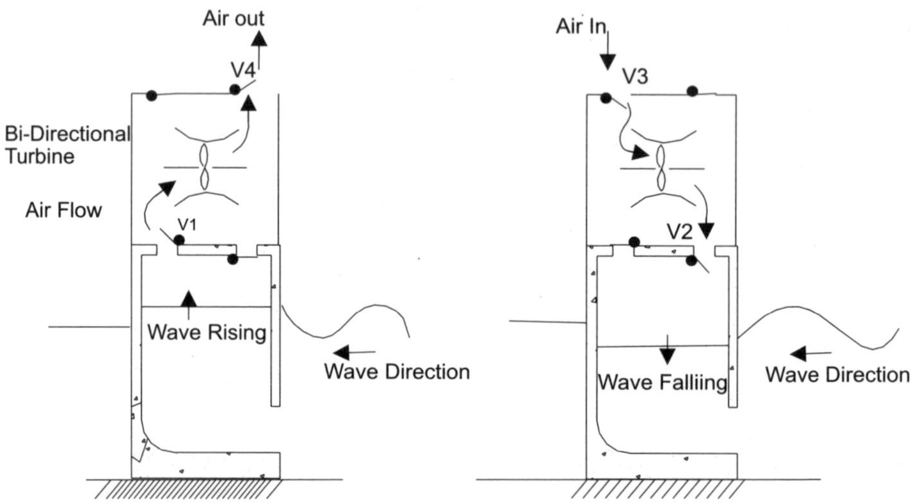

Fig. F19.5: Oscillating Water Column wave Energy Driven Turbo-Generator

Again it is shown that when an ocean trough enters the column causing the suction (low pressure) inside the column, causing air to enter the turbine chamber through the NR valve V3, and the to enter the turbine from the left and making it to rotate and air finally leaving the chamber into the column through the NR valve V2.

A pilot plant of capacity 150 KW is being installed at sea coast off Thiruvantapuram in South West Coast of India. The site is estimated to have a potential of 13 kw per metre at a depth of 10 metres.

Again this a futuristic energy source for which there is no production plant in operation.

❑❑❑

Part III

References

Chapter 12 Global Energy Challenges

Banerjee B P (2005), *Handbook of Energy & Environment in India*, Oxford University Press, YMCA Library Building, Jaisingh Road, New Delhi 110001.

Baumert K A *et al* (2005), *World Resources Institute,* Washington D C.

Dunn S (2004), *History of Hydrogen,* world Watch Institute New Haven, Encyclopedia of Energy, ed. C J Cleveland, Elsevier Press.

World Resource Institute, Washington D C..

Chapter 13 Nuclear Energy

Anderson P K *et al* (2004), *Nuclear Waste,* New Mexico Institute of Mining, Encyclopedia of Energy, ed .C J Cleveland, Elsevier Press.

Banerjee B P (2005), *Handbook of Energy & Environment in India*, Oxford University Press, YMCA Library Building, Jaisingh Road, New Delhi 110001.

Harper Collins India 1992, *"How Things Work – The Universal Encyclopedia of Machines"*, Indus Publishers.

Jagger J (1994), *The Natural World of Radiation*, Environmental Ethics – Readings in Theory & Application, ed. L J Pojman, Jones and Barlet, Boston.

Jagger J (1994), *Nuclear Power is Safe and we need it*, Environmental Ethics, ed. L Pojman, John & Barlet, Boston.

Lenssen N (1994), *Nuclear Waste: The Problem wont go away*, Environmental Ethics, ed.L Pojman, Jones & Barlet, Boston.

Mcfarlane H F (2004), *Nuclear Processing, Argonne Lab.*, Illinois, Encyclo. of Energy, ed. C J Cleveland, Vol. 4, Elsevier Press.

Magaud P & Marback G *et al*, *Nuclear Fusion Reactors*, Encyclo. of Energy, ed. C J Cleveland, Elsvier.

Masters G M 1991, *Introduction to Environmental Engineering & Science*, Prentice Hall, N J, USA.

Reyce Jr J N & King Jr B 2004, Nuclear Energy, Encyclo. of Energy, ed. Cleveland, Elsevier.

Subramanium T S 2007, *"Special Feature: BARC at 50"*, Front Line Magazine, vol 24, number 8, April 21-May 4, 2007, IISN 09770-1710, India.

Tester G M< Drake E M &Driscoll M J et al 2005, *"Sustainable Energy – Choosing among Options"*, Massachuchusetts Institute of Technology, Prentice Hall India, New Delhi.

Vogt D K 2004, *Nuclear Fission Reactor*, Lawrence Lab., California, Encyclo. of Energy, ed. Cleveland, Vol.4, Elsvier.

Walters L 2004, *Nuclear Fuel – Design and Fabrication*, Argonne Lab., Idaho, Encyclo. of Energy, ed. Cleveland, Elsevier.

Chapter 14 Biomass Energy

Elliot D C 2004, Biomass Chemicals, Encyclopedia of Energy, Ed. C J Cleveland, Elsevier Press

Faiji Andre P C 2004, Biomass Combustion, Ency. of Energy, ed. Cleveland, Elsevier

Green C & Byrne K A 2004,Biomass: Impact on Carbon Cycle and GHG Emissions, Ency. of Energy, ed. Cleveland, Elsevier

Klass D L 2004, Biomass for Renewable Energy and Fuels,

Schumacher L G 2004, Bio-Diesel Fuel , Ency of Energy, ed. Cleveland, Elsevier

Waghray A P, Appropriate Technology for Economic Production of Bio-Diesel from Vegetable Oil Seeds, Dhopeswar Engg. Pvt. Ltd., (dhopeswar@sancharnet.in), Hyderabad, India

Wyman C E 2004, Ethanol Fuel, Dartmouth College, Ency. of Energy, ed. Cleveland, Elsevier

Chapter 15 Solar Energy

Luzi A & Grove K L 2004, Solar Thermal Power Generation, Ency of Energy, ed Cleveland, Elsevier

Mottershed A 1988, Electronic Devices, Prentice Hall India, New Delhi.

Partin J 2004, History of Solar Energy, Encyclopedia of Energy, Ed. C J Cleveland, Elsevier Press.

Sukhatme S P 1996, Solar Energy 2nd ed., Tata-Mcgrawhill, New Delhi.

Tester J W, Drake E M, Driscoll M J et al 2005, Sustainable Energy- Choosing among Options, MIT Press/PHI

Chapter 16 Wind Energy

Chambers A 2000, "Wind Power Spins into Contentions, Power Engineering 104(2.)

Elliot D, Swartz M & Scott G 2004, Wind Resource Basics, Encyclo. of Energy, ed. Cleveland, Elsvier.

Kar M 2007, "Friendly Mining through Technology Integration" Seminar on Environmental Aspects in Mining Operations in India, Millennium Institute of Energy & Environment, Kolkata, India, March 2007.

Mani A & Mooley D A 1983, Wind Energy Data for India, Allied Publishers, India.

Pasquate M, Righter R & Gipe P 2004, History of Wind Energy, Encyclo. of Energy, Ed. Cleveland.Vol.6,Elsevier.

Petersen E L & Madsen P H 2004, Wind Farms, Encyclo. of Energy, vol.6, Ed. Cleveland, Elsevier.

Sukhatme S P 1996, Solar Energy- Principles of Thermal Collection and Storage, 2nd ed, Tata – Mcgrawhill.

Tester J W et al 2005, Sustainable Energy, MIT Press/ PH India.

Chapter 17 Fuel Cells & Hybrid Car

Basu R N 2007,

Brandon N 2004,Imperial College, "Fuel Cells", Encyclo. of Energy, ed. Cleveland, Elsevier.

Diccoco JM 2004, Fuel Cell Vehicles, Encyclo. of Energy, ed. Cleveland, Elsevier.

John M G 2004, American Honda Motor Co, Hybrid Electric Car, Encyclo. of Energy, ed. Cleveland, Elsevier.

Chapter 18 Hydropower

Rao Dr.K L 1979, India's Water Wealth, Orient Longman, New Delhi, India

Tester et al 2005, Sustainable Energy, MIT Press/ PH India.

Vaishwanar R S 1997, "Fluid Power Engineering", Jain Brothers, New Delhi.

Chapter 19 Other Energy Challenges

Sukhatme S P 1996, Solar Energy, Tata-Mcgrawhill, Inda

Tester et al 2005, Sustainable Energy, MIT press/ PH India.

Part IV
Challenges of Water Resources

20
CHAPTER

Overview Of Global Water Resources

20.1 Challenges of World Water Resources

Water is a precious resource

In the age of the Greek civilization, Thales Miletus (640-546 BC) is believed to have philosophized that " the best of everything is water". Down the centuries, Leonardo da Vinci (1452-1592) reaffirmed water as the "driver of nature". It is difficult to conceive of any programme of human development that does not presuppose a ready availability of water.

Because of its easy availability, water has traditionally been regarded as an inexhaustible free gift of nature; this has also led to extravagant and wasteful use, thus life-saving resource may become scarce and threaten human life, livelihood, and development and progress.

The demands of water will significantly and continually increase because of growing world population, accelerating industrialization and expanding agriculture and increasing economic well being of the people. It may consequently be inferred that there is likely to be critical shortage of water of required quality to sustain future growth and higher standard of living unless the water resource is adequately upgraded, scientifically managed, treated and vended as a valuable economic goods.

Dimensions of water problems observed by U N

Globally there may potentially be enough water to meet the forthcoming human needs but unfortunately much of water resource is not available at the right place, right time and right quality to serve human needs. Thus, in one way or the other, all societies, rich or poor, are generally affected. A UN Document (Mordel Plata 1977) summarises the worldwide water scenario as follows:

(*a*) Reasonably safe supplies of drinking water are unavailable for at least one fifth of the world's city dwellers and three-quarters of its rural people; in many countries, less than one half of the urban population and less than one tenth of the rural population are served with adequate and safe water supply.

(*b*) Increasing and unplanned concentrations of population and industry in large urban areas have strained water supply; this leads to problems of waste disposals, which in turn, degrade the environmental health and the quality of life.

(*c*) Proliferation of industrial processes, greater use of energy and increased use of agricultural activities are causing progressive and chronic degradation of the quality of available water by the increase of toxic compounds and other pollutants; the mutagenic and carcinogenic effect of these substances posses a potential threat to human life;

(*d*) Backwardness and relative isolation of rural areas where the great majority of the world population now lives aggravate the difficulty of providing adequate and safe supplies of drinking water, improved sanitation and waste disposal;

(*e*) Expansion of food production in water-short areas and in marginal lands has necessitated rapid development of irrigation and reclamation of lands to a degree that water and land resources have been exploited to their limit in many areas.

(*f*) Ever-growing land degradation from such causes as water logging, salinization and erosion is leading to losses in production potential, investment and employment;

(*g*) Ground water supplies are being exhausted, while both surface and ground water sources are deteriorating in many areas;

(*h*) Water is often needlessly used inefficiently and wastefully, calling for the application of scientific knowledge and setting of appropriate service level;

(*i*) Application of expensive technology to water development projects in some places is eating away the development fund earmarked for some other projects.

(*j*) Conflict about rights and priorities among different users intensifies as the increasing demands for available water grows.

These problems affect different countries in different ways over the local, regional and national levels. The problems of semi-arid and arid regions vary from the tropical and temperate climate regions.

Fresh water is not plentiful

In any case, the total amount of world water is theoretically constant and can neither be increased nor diminished. The ocean water constitutes almost the entire quantity of global water, but for a small portion of less than 3%, which is fresh water. Of these, the most of the fresh water is locked frozen in the Artic ice-caps, glaciers etc.; a further very small part of less than 1% in lakes, swamps, rivers and steams is available for human consumption. Hence, fresh water is not ,in reality, plentiful.

Estimates of the volumes of fresh supply can be derived from observation and measurement of precipitation, evaporation and ground water. The shortcomings in the appraisal methods and general lack of basic hydrological data do affect the accuracy of the estimate but the overall estimates for individual countries, continents and the world do serve very useful services. Such data reveal tremendous variations in local supplies among rocky terrain, semi-arid tracts, tropical rivers with surplus perennial water and temperate developed countries. Water quality by minerals, chemical or bacteriological contaminants also vary equally.

Water is vital for life

In 1930, Gartner (Furan1930) stated that a jellyfish weighing 500 g reduces to 0.45 g after desiccation. But Hill puts the dry residual to 3.2 percent. Be that as it may, Gartner proposed that all the constituents of the livings jellyfish are of equal importance and the water is as much a living substance as protein or carbohydrates, all being constituents of protoplasm. Referring to humans, 70 per cent of an able-bodied man weighing 70 kg is made of water weighing 50 kg.

Secondly living creatures cannot survive without water. The dehydration tolerance may vary with species and individuals, but essentiality of the water rule remains absolute.

Water is capable of dissolving many substances; it also produces decomposition of macromolecules offered to organisms as food.

Animals obtain the water they need from drink and from food of animal or vegetable origin. Vegetations obtain water either from the atmosphere or the soil. Living beings lose water by perspiration, transpiration, respiration and excretion.

Water is indeed at the source of all lives.

20.2 Global Water Stock

Hydrologic Cycle

The origin of water on our planet is not clearly known.. The current thinking is that the primordial earth had no oceans and perhaps very little atmosphere. It is surmised that volatile constituents, trapped inside the earth, leaked to the surface through various natural processes including volcanic eruptions, rock movements and hot springs and condensed to form oceans and the atmosphere.

The seawater, which constitutes 97 per cent of all water, contains high concentration of salts making them unusable for human, agricultural and industrial needs. It is possible to separate water from salts by desalination distillation technologies, but that is, by and large, highly energy-intensive and generally uneconomic. Fortunately, the sun performs that service for us when it evaporates seawater and leaves the salt behind. Sea surface constitutes 70 percents of earth's surface and the major part of the solar heat energy reaching the earth's surface gets converted into latent heat, removing seawater and water from the wet soil by evaporation and the water from the plant leaves by transpiration.

The resulting water vapour moves to the upper atmosphere and eventually condenses into water and returns to the earth as precipitation. Over the oceans, there is more evaporation than precipitation and over the land, it is the reverse, that is more precipitation than evapotranspiration. The difference between the precipitation and the evapotranspiration on land is the surplus water available for feeding the surface water consisting the rivers and the streams, lakes, springs and the ground water and the balance is the run-off back to the sea.

According to Budyko (Masters 200°), various quantities of evaporation and precipitation on land and sea are given in Table T20.1

Table T20.1: Annual Evaporation and Precipitation in thousand cubic kilometer

	Sea Area	Land Area	Total
Evaporation/Evapotranspiration	456	62	518
Precipitation	410	108	518
Surplus water on land	-	46	-
Run-off to sea	46	-	-

The above phenomenon of Evaporation/Evapotranspiration and Precipitation as a source of potable water is illustrated in Fig.F20.1 and the same is known as the Hydrologic Cycle.

Through the hydrologic cycle, water circulates from the earth to atmosphere to earth, driven by the sun's energy. Water moves endlessly from oceans to the atmosphere to the continents. But the precipitation that falls on the oceans does not contribute to the water resource. Similarly, of the precipitation on land, the amount that is evpotranspirated to the atmosphere is again lost to the water resource. As per the Table T20.1, only 46 thousand cubic kilometers of precipitation on land form the renewable water resource, which is renewed annually feeding the mother earth.

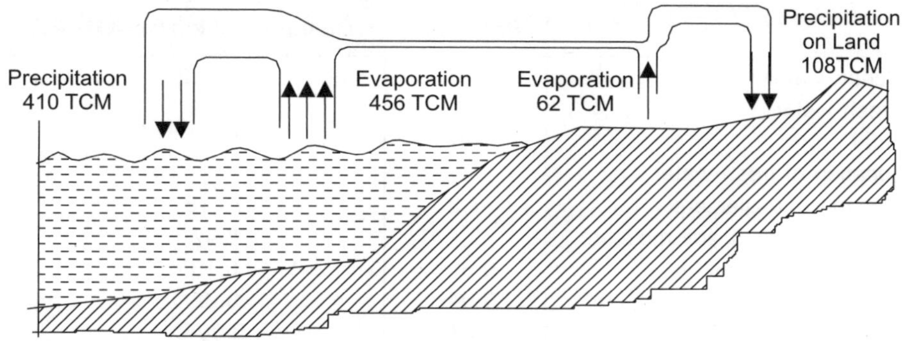

Fig. F20.1: Global Hydrological Cycle

Global Measures of Water

Recent estimates show that the total volume of water on earth is 1400 million cubic kilometers. The Table T20.2 shows that 97.3 per cent is ocean water. Of the 2.7 per cent fresh water, 77.2 per cent is frozen and stored in the icecaps and glaciers, 22.4 per cent is ground and soil moisture and 0.04 per cent in the atmospheric vapour. The fresh water that is readily available for the human and allied consumption consist of 0.35 per cent in lakes and swamps and 0.01 per-cent in streams.

While icecaps and glaciers and ground water account for most fresh water resources, for all practical purposes, it is the surface water in lakes, swamps and streams that constitute the basic available supply for the humans.

Another important feature to note that contrary to public perception, the stock of water in lakes and swamps is 35 times the total water in all rivers and teams of the world.

Although surface water provides most of man's supplies, groundwater use has been heavily developed in many parts of the world, especially for agricultural irrigation. This has resulted the overexploitation and depletion of the water bearing structure lowering the water level. There is inadequate knowledge about the groundwater and its recharging and even about the quantum of water in a specific aquifer. With more studies, it should be possible to plan higher utilization of groundwater in the overall management and development.

Table T20.2 : Water Stock of the World

	Per cent	Quantity (Million Cu. KM)
Ocean water	97.3	1362.2
Fresh water	2.7	37.8
Total	100	1400
Location of fresh water:		
Icecaps and glaciers	77.2	29.18
Ground water & soil moisture	22.4	8.47
Lakes & swamps	0.35	0.13
Atmosphere	0.04	0.02
Streams	0.01	0.004
Total	100	37.804

Seawater

Seawater amounts to 1360 million cu. km. At 97.3 per cent, as shown in Table T20.2, Sea covers about 72 per cent of earth's surface. But the average depth of oceans is 3800 m, which is only 1/1600 of the earth's radius and the mass of the seawater amounts to only 1/4500 of the mass of the earth. Though to a human observer the oceans appear immense and vast, put to an appropriate scale, it is merely a shallow lake.

The origin of seawater has long been contested. Some maintain that seawater was originally fresh water to which salt were carried by rivers. This view is contested by the fact that seawater contains 89% chloride, 10% sulphates and 0.3% carbonates, whereas fresh water contain 80% carbonates, 13% sulphates and 7% chlorides. The water of the great lakes inland sea (like Caspian sea) is not seawater; it is much less salty and contains three times more calcium carbonates. It seems reasonable to infer that seawater was salty from the beginning (Furan 1967).

Box B20.1 shows the broad compositions of the seawater.

20.3 Methods of Assessing Water Resources

Appraising Surface Water

Within a river basin, all water draining into the streams constitute the potential surface stock within the basin. The extent of availability varies with the flow of the stream at the specific location, its flow velocity and the river channel section.

Whether estimated or directly measured, the mean annual flow of a stream over a period of many years will thus represent the gross surface water supply. Full use of this supply would require that demand coincide with the flow at any time. But the actual supply that is used will depend upon the technology of water collection and utilization, actual distribution and flow available at the particular time and place. Consequently, any appraisal of water supply must take economic and environmental as well as hydrologic data in order to harmonize plans of demand and supply within acceptable technological and economic limits.

Hydrologists and other water engineers have many scientific tools and methods for assessing water supply in absence of direct measurements. The improvement of these methods and the dissemination were the main objectives of the International Hydrological Decade (1965-1974) and of the International Hydrological and Operational Hydrology Programme of UNESCO and WMO, including the water balance.

Fortunately, the water balance is derived from the equation of continuity. That is, for any hydrological unit and specific time period,

Outflow = Inflow +/- Storage E20.1

This principle is applicable to a small catchment, a major river basin, a reservoir, a continent or the world as a whole. Since changes in storage become negligible when projected over a long period of time, the mean water balance for a river basin can be stated as,

Run-off = Precipitation – Evapotranspiration E20.2

(Evapotranspiration, defined earlier, as transpiration from plant leaves and evaporation from soil, water and all other surfaces).

Maps of average annual precipitation, evaporation and run-off, are best derived from systematic water balance analysis. The isoclines can be constructed so that a balance of three terms of the equation E20.2 are available at all location, thereby enabling quick, approximate estimate of the run-off on basin-wide geographical basis.

The water balance concept is basic to water resource analysis. It provides the necessary hydrological data for all developmental plans. The water balance equation (E20.1) is applicable for any period of time. For short period of time, changes in storage become a predominant consideration and cannot be neglected. For short period, modeling technique can be applied with computer-generated data for the daily values of components in the cycle.

Appraising Precipitation and Evapotranspiration

Precipitation is the primary source of water supply for world food production, though irrigation for agriculture does still make additional demand on stream flow and ground water. Irrigation has to supplement precipitation in basic requirements of crop, where evapotranspiration is high. Precipitation data therefore play a vital role in water resource development and management.

Precipitation data, though are more numerous than stream flow; because, precipitation can be easily measured with simple instruments. What is perhaps more important is to draw precipitation isolines, correlate them to physiological factors and derive long-term trends, if possible, taking into consideration of any significant climate change. For example, it is important to know if widespread and prolonged draught can be expected within the natural year-to-year cycle, of the Sahelian zone of West Africa type.

Evapotranspiration is the most difficult term to evaluate in the water balance equation. Evaporation can be obtained by measuring free-water evaporation from water, knowing its weight loss or its draft in a graduated tank.

Box No. B20.1 SEAWATER COMPOSITION

A. Salt Content (approximate to 35 g per litre)

	gm/litre	gm/litre
Sodium Chloride (NaCl)	27.21	77.56
Magnesium Chloride ($MgCl_2$)	3.81	10.88
Magnesium Sulphate ($MgSO_4$)	1.66	4.74
Calcium Sulphate ($CaSO_4$)	1.26	3.60
Potassium Sulphate ($K2SO_4$)	0.86	2.46
Calcium Carbonate ($CaCO_3$)	0.12	0.34
Magnesium Bromide ($MgBr_2$)	0.08	0.22
Total	35.00	99.80

B. Ionic Dissociation: Seawater is a solution of electrolytes dissociateable into ions

Anions (Negative ions)	g/kg water	Percentage	Cations (positive ions)	G/kg water	Percentage
Chloride (Cl-)	18.98	55.04	Sodium (Na+)	10.56	30.61
Sulphate (SO4-)	2.65	7.68	Magnesium (Mg+)	1.27	3.69
Bromide (Br-)	0.06	0.19	Calcium (Ca+)	0.40	1.16
Bicarbonate (HCO3-)	0.14	0.001	Potassium (K+)	0.38	1.10
Fluoride (F-	0.001	0	Strontium (Sr+)	0.01	0.04
Boric Acid	0.03	0.07	Total	12.62	36.60
Total carried fwd.	21.86		B.F. last column	21.86	62.98
			Total ions	34.48	99.58

C. Trace Elements

Elements	mg/cu.m.	Elements	mg/cu.m.
Nitrogen	7000	Thorium	0.5
Silicon	1250	Molybdenum	0.5
Carbon	1500	Nickel	0.5
Lithium	100	Gallium	0.5
Phosphorous	100	Scandium	0.4
Iron	60	Cerium	0.4
Iodine	50	Vanadium	0.3
Barium	50	Yttrium/Lanthanum	0.3
Rubidium	20		
Arsenic	20	Silver/Mercury	0.3
Copper	10	Gold	0.6x10**-3
Zinc	10	Radium	0.6x10**-7
Manganese	10		
Lead	5	Total elements	20 mg/kg
Selenium	4		(20 ppm)
Tin	3		
Uranium	1		

Evapotranspiration can be measured by knowing weight loss from a potted plant and a similar pot with soil but without a plant. But it is important to study the influence of the methodological conditions, *e.g.,* radiation, wind, humidity and ambient temperature by employing suitable techniques.

It is interesting to note that evaporation from lake Nasser represents about quarter of the normal flow of Nile at Aswan.

Appraising Ground Water

Ground water is stored in aquifers (rock or soil containing and transmitting water). Groundwater and surface water generally form a somewhat linked water system. Flow can be in either direction, and the rate of flow varies geographically and chronologically.

The interchange is significant for some aquifers. But it has been estimated about 30 percent of total flow in a surface stream is supplied by from the groundwater, and seepage from streams is known to be principal source of flow to some aquifers. For example, water withdrawn from riparian wells along an alluvial stream can cause an appreciable reduction in surface flow and on the other hand, the diversion of surface flow can reduce ground water recharge. In other instance, withdrawal from either source has no apparent effect on supply from the other.

Accordingly to UN Document (Mordel Plata 1977), there are two ways to the exploitation of water. In the first instance, the withdrawal should match the average rate of recharge. In the second approach, the underground water is mined for a specific task for a specific period. For example, the aquifer, which supports 1.5 million hectares of irrigated land in Texas Panhandle, will be exhausted in a few decades at the current pumping rate.

There is again another instance; the rate of recharge is very insignificant like the artesian basin of the intercalary of the northern Sahara, which have immense stock of fresh water built over 40,000 years. The withdrawal of water from this type of aquifer to any significant degree will lead to gradual depletion of the aquifer.

It will be the concern of the development and management of the groundwater supplies on the regional or national basins to take an overview over the strategy of groundwater exploitation. It could be on the safe yield basis or fixed-period basis with alternative provision for time beyond the period. It is again a fact that supply recharge in many cases tends to accelerate with draw-down.

The appraisal of aquifer storage in particular requires extensive on-site exploration. Although the methods of appraisal are different for ground and surface water, supplies from either source cannot be evaluated separately unless it is established that the interflow and seepages are minimal. Flow rates inside the aquifers may be extremely slow. For example, a recharge at one area may take many decades to reach the other area of the extensive artisan aquifer where the withdrawal is planned. This time lag is to be taken into consideration in the ground water management.

The quality of water available under specific conditions may be estimated from the water basin equation. The equation can be written down for the surface and sub-surface components of a basin as a whole or for the groundwater reservoir only. The later approach is required for adequate understanding of the groundwater regime and its management, while

the former allows a macro-level evaluation of the total surface and ground water supplies. Either way, a sound estimate, requiring reliable hydrological data and aquifer characteristics study, is quite complex and computer-aided modeling and solutions may be called for.

Assessing Water Quality

The physical, chemical and biological properties of water determine its suitability of its quality vis-à-vis various uses.

Water is considered polluted when it is altered in composition, directly or indirectly, to the degree that it is less suitable or even unsuitable for a particular use or function. Pollutants reach water bodies from 'point source' such as power station, factories, mines and city sewer outlets and from non-point source such eroding basins and fields.

Pollution of water results mainly from human activities but there are instances of natural degradation, e.g. underground water getting toxically contaminated with arsenic, fluoride or other heavy metal traces, water in arid and semi-arid zones getting saline contamination.

Table T20.3 Networks of Hydrological Stations

Regions	Non-recording rain-gauges		Self-recording rain-gauges		Evaporation stations		Discharge stations		Sediment station		Water-Quality stations	
	Nos. existg.	1000 sq.km./st.	Nos. existg.	1000 sq.km./st.	Nos. existg.	1000 sq.km./st.	Nos. existg.	1000 sq.km./st.	Nos. existg.	1000 sq.km/st	Nos. Existg	1000 sq.km./st
Africa	18520	1.48	1214	22.61	1043	26.32	4390	6.19	740	36.73	801	33.94
Asia	17815	1.69	2489	11.92	1590	18.00	6386	4.58	1601	17.30	3290	7.94
S-Ame -rica	12200	1.05	2106	5.92	1091	11.72	3709	3.45	701	18.24	374	34.20
N. & C. America	19499	1.11	4547	4.78	2538	8.56	14521	1.49	1851	11.70	3291	5.94
S-W. Pacific	16419	0.69	1795	6.13	569	19.32	4801	2.19	749	14.06	2399	4.39
Europe	44704	0.13	5356	1.09	1004	5.53	12926	0.82	2057	4.63	8284	0.53

Table T20.4 Average Annual Water Balances, 1970

Region	Global Average Norms		
	P	E	R
Europe	7.4	11.4	3.0
Asia	31.9	18.7	13.2
Africa	21.3	17.2	4.1
Australia	6.9	4.6	2.3
N. America	15.9	9.2	6.7
S. America	28.6	17.4	11.2
Antartica	2.4	0.3	2.1
Land Areas	114.4	71.7	42.7
Oceans	418.3	461.0	-42.7
World	532.7	532.7	0

P = Precipitation; E = Evoporation; R = Run=off; Volumes in 1000 cu.km.

Developed from USSR Water Balance Report (1974), Luvovich (1974) & Banmgartner (1975)

It is not possible to infer water quality of a long stream by local measurement at a single station. Accordingly, observations on water quality must be based by collecting and analyzing samples periodically at a network of stations along streams, lakes and at bore holes. Typically, for a long stream like Ohio and Rhine, measurements are taken for temperature, sediment load, dissolved solids, major nutrients such as phosphorous and nitrogen, organic material by biochemical oxygen demand, dissolved oxygen, salts, acidity (or pH values), bacteriological count of coliform, heavy metals etc.

Reliability of Appraisals *vis-à-vis* Quality and Quantity

The reliability of appraisals is per se a function of sample size and its accuracy. The organization of comprehensive national network and data-processing services remain one of the most urgent tasks in water appraisal, development and management and framing a national water policy. At the international level, this is the core of the WMO Operational Hydrological Programme. The operation of unified network is critical, whenever basins or aquifers are shared by two or more countries that require using commonly agreed data for all operational management and sharing of water resources.

Table T20.3 lists the existing water Monitoring stations in comparisons with the densities recommended in the World Meteorological Practices. Because, these required densities are for average conditions, and because more than twenty countries, including China and then USSR are omitted, the resulting data are just approximation. If we assume that the European values are the lower limits and the North American values are the upper limits of the WMO recommendations, the network densities are far deficient in Africa and Asia and less deficient in other regions.

20.4 Availability of Fresh Water

Surface Run-off Water

In Section 20.2, we presented global stock of water as estimated by UN. We indicated there that all of the fresh water stock at 37.8 million cu. kilometre is not available to human use. As a matter of fact, the medium quality of fresh water that can be converted to human use is the annual run-off water to the sea. According to Table T20.4, the total global run-off amounts to only about 42,700 cu.km., which is only about 0.1 per cent of the global stock of fresh water. The more dependable value of fresh water run-off is 46,000 cu. km. as given in Table T20.1

Precipitation on land represents the total removable water availability. The evaporation represents the loss of the major part of that precipitation, which is more or less beyond control. The Balance is the run-off. In Table T20.4, the data of Baumgartner (1975), USSR Report (1074), and Luvovich (1974) are all averaged, in the triple columns of P, E, R., giving us normative values for future use.

It may be seen that over the Asian land, there is larger proportion of run-off, that is 13,200 cu. km., out of total precipitation of 31,900 cu.km. The run-off over the African is only 4,100 cu.km. out of total precipitation of 21,300 cu.km., whereas over Antarctica there is hardly any evapotranspiration. Finally, from Global water balance, the run-off is zero over the combined land and ocean areas, precipitation being equal to evapotranspiration.

Lakes and Reservoir Water

The total volume of fresh water stored in lakes amount to 200,000 cu km, equal to about 4 times the global annual run-off from all land areas. In comparison, the amount of water in man-made reservoirs totals to only 5,000 cu km, that is about 11 per cent of the yearly run-off.

Lakes and reservoirs help to raise the usability of run-off. While the run-off through the streams is seasonal, lakes and reservoirs store the water for use while the run-off might have nearly dried up. But the lakes are not the primary source of fresh water resource, which are to be renewed every year from the run-off.

Ground Water Resource

For the most part of the world, the ground water resource is not scientifically mapped nor the process of recharge properly understood.

From UN sources, it is understood that the quantity of groundwater up to 4000 metres, is quite large at about 10 million cu km, but the amount of groundwater which forms a part of the hydrological cycle is only13,000 cu km, about 1/3 rd of the total run-off. If deeper groundwater is mined, it will bring down their level without recharge.

However, most countries have drawn up estimate of their shallow ground water reserves, which are mostly used for agricultural irrigation.

Water Quality

The availabilty of suitable water is influenced by its quality. Surface water may contain suspended soil and many other organic and inorganic contaminants. On the other hand ground water contains far more dissolved salts than the surface water.

Only water, that is non-toxic free of bacterial contamination, should be included in the availability of water. However much of this water can be made potable after due processing. We will deal further more comprehensively with the quality of water in later section

20.5 Global Water Demand

Categories of Water Demand

In Table T20.5, it has been proposed to classify water uses by withdrawal, on-site and in-stream and take into effect the nature of use *i.e.,* consumptive or non-consumptive, so that it will be possible to have a better understanding of the effect of water use on the human environment.

Usages, which require to actually withdraw water from waterbody – drinking water, irrigation, industry, for example and thus deplete the source, unless water is returned to the source by drainage or otherwise after treatment.

In-stream uses – navigation, hydropower, fisheries, for instance, generally leave the source volume of water unchanged.

On-site uses – for rain-fed crops or wildlife habitat including wetlands, for example – may or may not affect the volume of flow.

Eventually, all the above uses, whether consumptive or non-consumptive, will generally cause a change in quality. Besides, the above uses will also involve in the potential losses of supply through evaporation, transpiration or diversion, so that there may net depletion of a source stream or aquifer. Admittededly, while some uses, such as irrigation or wider withdrawal for potable water can be wholely deplete a stream, other uses like repeated use of thermal cooling will reduce flow only marginally.

Table T20.5: Characteristics of Different Water Uses

Categories of Demand	Substitutes	Withdrawal	In-Stream	On-Site	Quality Effect	% Consumed
Drinking	None	-				1-15
Other Domestic		-				1-15
Public Urban		-				1-15
Livestock	None	-	-		Organic	1-15
Soil Moisture	None			-	Sediment	0-100
Irrigation	None	-			Salt	10-80
Drainage					Salt	0
Wetland Habitat	None			-		0-10
Aquatic Habitat	None	-	-			0
Navigation	Land transport		-			0-10
Hydropower energy	Other		-			0
Mining		-	-			1-5
Manufacturing						
Cooling	Air	-			Thermal	0-3
Processing		-			Organic/toxic	0-10
Mechanical Waste Disposal	Air/ mechanical	-	-		Organic	0
Recreation			-			0
Aesthetic			-			0
Flood Loss	Land use		-	-		0

Source: Gilbert White 1976

Gross Demand Basis

The demand for water has also an economic underpinning. It may mean the amount of water supply related services that would be demanded and used at a given price. But , in contrast to most other economic commodity, water, on the concept of plentifulness, has been regarded as a free natural commodity. Thus, it s not easy to establish reasonable levels of water demand and supply through the pricing policies. However in the coming decades, water might have to be priced, preferably publicly in some measure, for its conservation and improvement of its quality and equitable availability to all.

In general, demand for water can be projected in aggregate terms. In Hungary, for example the range of demand was projected till the year 2050 on the basis of historical relationship between per capita aggregate use and gross national product.

More detailed studies require examination of each categories of demands outlined in Table T20.5, projected at the national level or even for smaller areas such as basins and sub-basins. To do this, the most rudimentary method is extrapolation of rate of change observed a past time period. This method is of course used very widely. Unfortunately, the influence of socio-economic factors on demand or even the influences of technologies and policy framework on demand are not adequately included. Hence the long-term projection on basis is unlikely to be reliable. But the short-time projection will remain valid.

In the followings, we propose to look into various major sectoral demands already identified in Table T20.5.

Domestic Demands

To a person living in a semi-arid land, domestic water use may mean a few litres a day, manly for drinking and cooking and another few litres for washing and bathing. In such small communities in non-industrialised countries, likewise the water demand is generally low, say 10 to 20 litres per day, namely because of distant access of water. In higher income societies, watering of lawns and kitchen gardens accounts for some considerable water. In middle and upper income sections of large cities, household appliances, swimming pool and the washing of automobiles considerably increase water demands. A farming household may also use its supply to water livestock and irrigate a small garden.

Therefore, the quantity of water required domestically will vary with local circumstances as above, the method of excreta disposals etc and it is finally income-elastic. It is common to design new community supplies for delivery of an average 100 to 350 litres per capita per day for house connections and 10 to 25 litres per capita daily for people using public stand pipes. In rural areas, the design figures used by government agencies differ accordingly to the livelihood and the availability of natural water sources. The above general consumption ranges are illustrated in Fig. F20.2.

According to the 29th World Health Assembly, May 1976, the minimum targets of water supply and the human excreta disposal in the developing countries were to be achieved by 1980 on the following order:

In urban areas:

Safe Water supply for 91% population, consisting of

68% population by house connections,

23% population by standpipes.

Safe sanitation for 94% population, consisting of

38% population by public sewers,

56% population by household systems.

In rural areas:

36% population to have reasonable access to safe water,

24% population to have excreta disposal facilities.

In the above targets, there are wider sections of rural population uncovered even by the minimum health provisions. In this perspectives, UN Millennium Development Goals adopted in2000 exhorted the earliest remedial measures for eradication of poverty and provision of water and sanitation to all.

Industrial Demands

The economic development of a country and its application of technology influence the use and demand of water for industry to even greater degree than they influence the use of water for domestic purpose.

The major uses of water in industry are for cooling, processing, boiler water and other general purposes including air conditioning, cleaning etc. Cooling is the principal purpose of withdrawal and commonly accounts for as much as 60 to 80 percent of industrial withdrawal. This is the reason for the heavy use of water by the thermal electric power generating plants.

Table T20.6 lists a few heavy water using industries like primary metals, chemicals, petroleum, Pulp and Paper and Food Processing.

The critical considerations in the industrial use are how much water is taken in, how much is consumed and how much and what kind of waste is discharged, since this can gravely affect human and environmental health.

Consumptive use rarely exceeds 20 percent of withdrawal. In case of steam power generation, consumptive use is less than 2 percent, which economy is still being improved in recent times.

The waste discharge affects the quality of receiving water, precise effect depending upon flow and quality of the receiving water, its temperature and its current characteristics pH BOD load etc.

Today, inside the industrial plant, it is necessary to audit the water engineering as to its use, processing, and economy. Some less engineered plants may use several times more volume of water than a more modern processing unit. The waste may be treated within the plant or disposed off directly without treatment at great social cost. Thus, it is imperative that the volume and the quality of industrial waste be improved through process design, in-plant recycling and improved through process design, in-plant recycling and improved home keeping procedures.

Table T20.6: Water Intake and Discharge Waste Load in Certain Industries.

Industry & Selected Products	Units	Litres of water withdrawn per unit (Upper unit)	Pounds of 5-day BOD per 1000 gallons Process Water discharge (USA)
Primary Metals: Steel	Tons	8000	-
Chemical: Soap	Tons	10,000	16.70
Petroleum: Gasoline	Kilolitres	7,000	2.50
Pulp and Paper: paperboard	Tons	62,000	2.21
Food Processing: Sugar beets	Tons	10,000	9.16

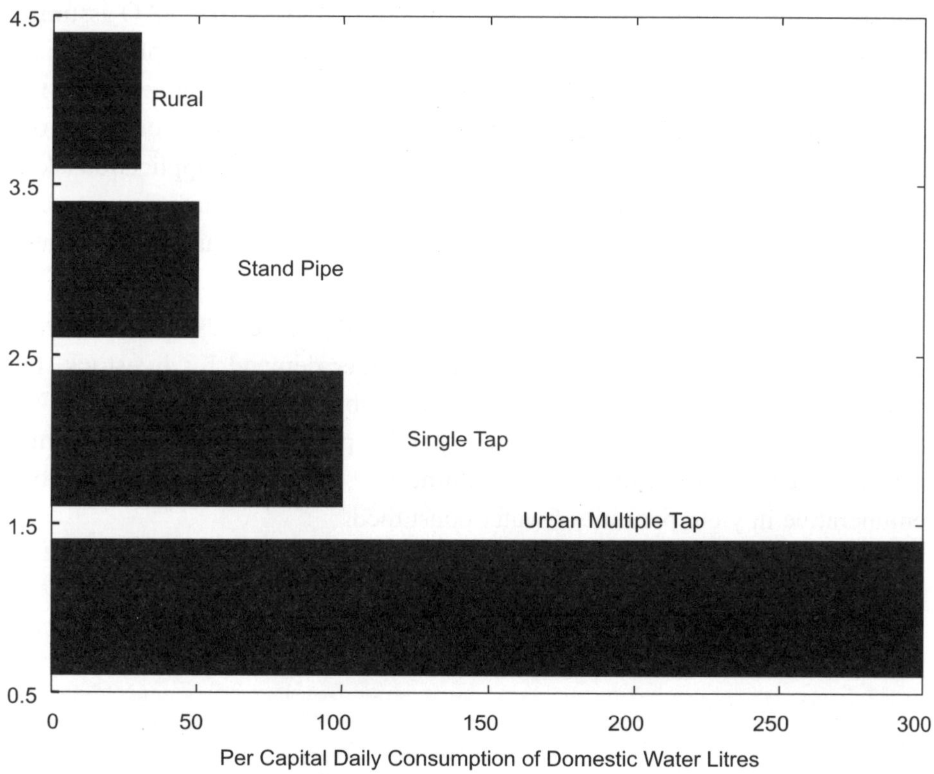

Fig. F20.2: Domestic Water Use

Agricultural Demands

As the world population grows, so does the need for increased food production. The augmented agriculture, in turn, demands greater supply of water. The supply of water will need more effective use of rainfall and greater conservation and management of water. To extend the agriculture to still higher level of productivity may call for irrigation by diverting and withdrawing lake, stream or ground water.

Agriculture is the largest user of water, accounting for some 80 percent world consumption, mainly through irrigation for higher production and also possibly for multi-cropping. In addition vast areas use rainfall as the principal source of water for crops. Approximately 200 million hectares of land were irrigated in 1970; almost half of these in the developing countries where food needs are especially acute.

While it is difficult to define irrigation per se, any planned agro-development still requires an ensured controlled water supply to the crop. The quantities and acceptable standards of reliability of supply are determined by the local climate, agronomic and economic conditions and water requirements are established at the local project level. In Table T20.7 are provided water estimates of gross water estimates for land irrigation and harvesting based on UN sources (1970). To arrive at the estimates of gross water demands of Table T20.7, FAO assumed the demand to 700 millimetres of water per season, when the staple food is a dry cereal and 1500 mm where it is rice. These values are presumed to be the total quantity needed by the crop, whether supplied from rainfall or irrigation, with the typical losses associated with storage, conveyance and application taken into account. Still, part of the water lost in application in one area may become a source of supply elsewhere through he seepage, flow back and ground water recharge.

Other essential agricultural needs, including rural domestic uses, make relatively small demands on overall water resources. Water demand for livestock, for example, is estimated at about 62000million cubic metres per year for 1700 million animals, and this allows 100 litres per day per animal to cover all animal needs, Fisheries represent a small volume use, which is of course highly remunerative in yield per unit of water consumed.

Table T20.7: Guidelines on Irrigation of Harvested Areas and Gross Water Demand (UN)

Regions Areas	Irrigated Harvested Demands (Billion Sq. Metres)	Gross Water (Billion Cu. Metres)	Depth of Water Per season (Metres)
Africa (excluding Egypt, Libya, Sudan)	20	14	0.700
Latin America and Caribbean	80	59	0.737
Near East (including Egypt, Libya, Sudan	160	109	0.681
Asia (excluding China and USSR)	750	899	1.199
Asia (China and other communist countries)	780	1167	1.496
USSR and Eastern Europe	140	95	0.678
North America, Australia and Oceania, Western Europe	330	228	0.691
Total	**2260**	**2571**	**0.690**

Hydroelectric Use

The turbo-plants for capturing hydro-energy are concentrated for the most part in Europe (notably in Russia), North America. Japan, China, India and Brazil. In some countries (Brazil, Canada, Morocco, Norway and Uganda for example), hydropower accounts for as much as 80 to 90 percents of electric power generated. Sometimes, Hydropower installations are connected through common grid with the thermal power generating Installations and share loads to meet peak demands Hydropower provides about half of electric power production in Japan and about 15 per cent in the United States. Table T20.8 shows the gross Hydropower potential of the world.

Often the hydroelectric water reservoirs are also used for irrigation programmes, navigation, domestic water supply or aquatic habitat. Thus, the development of hydroelectric power is often linked to multi-purpose programmes. Hydroelectricity plays a key role in the development of the Dez, San Francisco and Volga basins.

Apart from the fact that hydroelectric power is highly economic, it is one of the best source of environment-friendly energy, not emitting any polluting gases, GHG or otherwise.

Table T20.8: Hydropower Potential and Current Capacities

Region	Installed Hydropower MWe	Current Power Generation* Gwh	Polcutial Hydropower Capacity Gwe	Potential Hydropower Generation Twh
North America	167,082	660,761	682	2700
South America	137,908	650,643	636	3000
Asia (Incl. Middle East)	271,267	914,780	1545	5000
Europe	225,772	693,187	814	2500
Oceania	13,470	41,862	97	300
Total World	848,456	3,056,785		

*World Energy Council, 2007

Waterborne Transportation

Inland waterways make use of natural streams, lakes or artificial canals. Some of the great inland water systems are the Rhine, Seine, Danube, Elbe, Volga-Don, Ganges-Brahmaputra and the Mississippi. But the maintenance of the inland navigation ways involves quite elaborate management of flows, siltation problems and operation of various lockgates.

However, barges carrying as 3000 tons of cargoes with drafts of 2.5 metres are found on European waterways. Inland vessels in North America could have drafts of 3 to 4 metres.

Inland water transport faces stiff competition with the roadway and railway transport. But one thing is sure, given the infrastructure of inland port terminal with many road connections, maintenance of water navigation depths by dredging and water lockgates, water transport is the most energy economic for every ton-kilometre compared with the Road and – Rail. By this count, the inland transport will also be comparatively more environment friendly

Flooding

A natural function of all stream valleys is to carry the peak discharges that occur seasonally or at rear intervals. All forms of water channels must try to provide for all these flows as much as possible. But, as we know, it is not possible to prevent flooding of plains and valleys by peak discharges.

To mitigate the effect of flooding, there are several actions, which are advised as follow:

(*a*) Trying to control flood flows by engineering works as embankments etc.;

(*b*) Reducing flows by upstream land treatment and inter-basin water transfer;

(*c*) Modifying building and land uses to reduce to reduce vulnerability to loss;

(*d*) Sharing the burden of flooding through relief and insurance scheme;

(*e*) Finally accepting rare and really heavy peak flows leading to limited loss of property and lives, if any.

Unfortunately, on a world scale, the average annual per capita property damage from floods is estimated to be increasing, as is vulnerability to catastrophic damage caused by exceedingly rare floods with a probability of recurrence, say 100 years or more. Urbanization generally increases peak flows. And poor use of land has effects of even greater magnitude throughout the developing world.

Inland Fishing

It is understood that about 13 percent of the total world catch at 63, 000 million metric tons is obtained from inland fresh water sources. Unlike marine fisheries, which have shown marked decline in catch resulting from overfishing and inadequate harvesting since 1970,fresh water fish stocks are stable and increasing. In India, inland fish catch and matching harvesting have grown several times over the decades.

The demand for fish as a food supply is intensified by the shortages of other animal proteins, minerals and vitamins. Fresh water fish production thus plays a small, yet significant role in helping meet increasing food needs. In multiple purpose water planning, provision for fish production is of high priority.

20.6 Global Water Supply-Demand Management

In the world today, the essential problem in the management of water is how to reconcile increasing demand of the growing human society with the available fixed supply of fresh water. There are three typical scenarios in the global mismatch in the dynamics of demand and constant supply of water as obtained below:

(*i*) At the one extreme are the regions, countries or parts of countries with a large natural supply and a relatively low level of consumptive demand, which can be met without the requirement of an overall regulatory regime. Here water management implies management of various water projects on basically local considerations only.

(*ii*) At the other extreme are areas with scarce supply and relatively high levels of demand, which are satisfied only through overall regulation and rationing of water resources. The regulatory regime mainly concerns itself with the demand management including relocation of demands of humans and industries.

(*iii*) In the intermediate situations, where there is overall balance of water resource in supply and demand. Here the water management calls for both increased availability of supply at some locations and regulated demand by making more efficient use of water at some other locations.

Demand-Supply Management in Arid Areas

Demand-Supply management presents special difficulties in arid areas and semi-arid lands. The population in these regions is faced with overgrazed, perched or sterile pasture lands; a degraded or vanishing forest cover; very poor agriculture aggravated by uncertain rainfalls and river flooding. The surface water is very scanty or non-available. In many of these areas, there is considerable ground water potential, which needs to be developed as an integrated water work as in the artesian basins of sub-Sahara. Forestation, land development, dry farming and long-term integrated water sourcing planning are other measures. The last alternative is relocation of population to more benign neighbourhood

Long-term Water Development Planning

Solving problems of water supply and development need long years of infrastructure development works. From the UN report we know that Tisza River basin (a 156,000 sq. km. sub-basin of the Danube shared by five countries), water management began 150 years ago and likely to continue for another fifty years or so. The Salt River basin (32,000 sq. km. in the United States) developed complete regulation over 60 years period. Similarly, Netherlands undertook reclamation of sea over many decades to develop both new lands and water sources. In India too, river valley projects like Damodar Valley, Bakhra Nangal and Hirakud took more than one decade each for completing the core projects, the auxiliary developments having still continued.

To conclude our global overview of water resources, it should be emphasized that water development work is often undertaken on the consideration of its social benefits and economic costs. This is unavoidable, considering that there is always a demand for funds for other developments in the areas of alleviation of poverty, and improvement of education etc.

Table T20.9: Categorywise Water Visions – Withdrawal and Consumption 1995 and 2025

World Water Uses	World Water Use 1995 (Cu. km.)	World Water Vision 2025 (cu. km.)	Percentage Increase 1995-2025
Agriculture			
withdrawal	2500	2650	6
consumption	1750	1900	9
Industrial			
withdrawal	750	800	7
consumption	75	800	7
Municipal			
withdrawal	350	500	43
consumption	50	100	100
Reservoirs (Evaporation)	200	220	10
Total			
withdrawal	3800	4200	10
consumption	2100	2300	10
GroundwaterConsumption	200	0	-

Source: Cosgrove and Rijsberman(2000)

Table T20.10: Regional Water Withdrawal (cu. km./year)

Categories	Europe		Africa (Sub-Sahara)		Asia (East & South)		Total	
	1995	2025	1995	2025	1995	2025	1995	2025
Agriculture	140	130	42	40	1359	1266	1541	1436
Industry	130	130	6	20	153	263	289	413
Domestic	80	80	12	40	114	471	206	591

Table T20.11: Regional Water Stress: Global Perspectives

Parameters	Europe		Africa (Sub-Sahara)		Asia (East & South)		Total	
	1995	2025	1995	2025	1995	2025	1995	2025
Areas of water stress(million sq. km)	1.5	13.5	2.6	4	6.5	7.9	10.6	25.4
Population under water stress (million)	160	180	110	300	1.1	2.4	271.1	482.4

20.7 Some Critique of Global Water Scenario

An overview of global water resources has been outlined in the previous Sections. In this Setion, we propose to highlight some critical scenario of water resources, which beset the world today and likely to aggravate in the coming decades.

Freaks of Water Resources

We round up the total global annual available fresh water at 40,000 cu. km. But of this amount, only about 3,800 cu. km. (Cosgrove 1998), roughly 10 % were withdrawn (diverted or pumped) for human uses in 1995. Further, of the amount withdrawn, only 2100 cu. km., were consumed. The reminder was returned to the streams and aquifers, usually with significant reduction in quality. Now if we are withdrawing only 10 per cent of renewable water resources, and consuming only about 5 per cent, where is the water scarcity? This is the rub of the riddle.

The problem is that world's fresh water is very unevenly distributed in space and time. For example, in India's Brahmaputra-Barak basin, out of the total annual flow of 586 cu. km., only 24 cu. km., amounting to 4 per cent is utilizable, because of short duration of the rainy season with flood flow, geographical non-feasibility of large storage and thinner population. Similarly, human demands are very small in the Amazon basin, Northern Canada and Alaska. On the other hand, in the arid and semi-arid river basin, the withdrawal and use amount to 80 to 90 percent. In fact, the withdrawal of water is so high that some of the world's largest rivers do not reach the sea. In the wake of the multifarious developmental activities, total or near total of river run-off is either withdrawn or diverted. The Huang He (Yellow) River in China did not reach the sea on some days in 1972 and for seven months in 1997. The Colorado River in South-Western United States and the Indus River adjoining India and Pakistan are two of the many prominent rivers in the similar predicament.

Water Pollution

General pollution of rivers is a common phenomenon. Fast urbanization and growing industrialization along the river valley leads to high withdrawal of pristine water and return of polluted discharges to the river. Many rivers of newly developing countries are turning into large carriers of contamination like Thames was in the 19th Century England. The matter is further aggravated because of use of chemical fertilizer in modern agriculture, which seeps into the rivers and the ground water creating toxic contamination.

A new outlook in the overall water management is required to preserve purity and quality of water, which otherwise will affect human health very seriously.

Catastrophes of Floods and Droughts

Floods and droughts are principally due to extreme excess and absence of rainfall (or run-off) respectively. Both these extremes cause untold physical and economic hardship to the inhabitants of the basin and the destruction of the infrastructure. Death, devastation and damage follow from the floods and famine from draughts cause some scale of losses. It has been recorded (Cosgrove 2002) that between 1973 and 1997, an average of 66 million people suffered flood damage. The annual number of flood victims jumped from 19 million to 131 million in 1993-97. In 1998, almost 30,000 people died due to floods. Economic losses of floods in the 1990-decade amounted to 700 billion USD.

Flood Control measures, which today is technically feasible, need therefore be augmented considerably to mitigate human sufferings and property losses.

Unsustainable Scale of Water Use

The general pattern of sectoral consumption of fresh water is of the following order:

Agriculture / Irrigation	70 %
Industry	20 %
Domestic (Drinking/Cooking/Bathing)	10 %

As seen above, agriculture claims by far the largest share of water use. Much of that water withdrawn for agriculture is lost through evaporation from reservoirs, canals and soil and transpiration from crop plants, which also further absorb water.

A widespread revolution has come about in drawing groundwater for irrigation with hundreds of diesel engine driven pumps, privately owned, particularly in Asia. In Pakistan, it is believed there were 360,000 private pumps in 1993 alone drawing groundwater heralding new growth phenomena. Half of India's massive irrigated land measuring about50 million hectares sources water from the groundwater. In China more than 2million pumps irrigate 9 million hectares. In the United States, in the very large Ogallala aquifers, a privately funded groundwater well system feeds irrigation sprinklers is in operation. It is estimated that 150-200 cubic km more ground water is pumped out each year than is recharged in overexploited aquifers. As a result, ground water tables are falling significantly with risk of collapse of agriculture system in North China, US high plains and some major aquifers in India and Mexico. Naturally, a well-coordinated long-term water management plan is called for to regulate and to remedy this unsustainable and wasteful agricultural irrigation practice.

Industrial uses at 20 % withdrawal of water for power plants and processing appear quite profligate and wasteful. Some industries, for the same process, withdraws many times more water than a similar process in an efficient industry, practicing economy of water resources. Today, as the water is becoming scarce and more costly, some industries have introduced nearly 100 per cent recycling. Alternatively, reprocessed water is released for other multiple uses.

No doubt, municipal/domestic use accounts for only 10 per cent of world consumption of fresh water. Here again, there is much profligacy of water use for lawns, clothes wash, car wash etc. The following per capita consumption data speak for themselves:

	Litres per day
USA	400
EU	200
Asia/Latin America	200-600
Sub-Sahara	10-20

In Asian countries, of the water consumed, 70 per cent is lost through leaks. Improvement in maintenance and service will immensely increase the availability of the much-needed potable water. Similarly, OECD Countries can easily reduce their consumption substantially by redesigning and rationing their various washing, flashing, and lawn water processing.

Let us remember that Israel is efficiently managing their scarce water with annual consumption of water at 380 cu.m. per capita-year compared with the 1220 cu.m. per capita-year of the United Kingdom (UN Report).

Hydro-energy

Of the total global run-off of 40,000 cu.km., about 6,000 cu.km. is held dammed in reservoirs offering water for hydropower, drinking water, flood control and recreation, all highly benefiting the mankind. But large dams and hydropower projects lead to large scale displacements of human habitation and also cause many other adverse environmental impacts.

Majority of rivers of the developed countries have their discharges regulated by flow control, diversion for agricultural and even regulation of river channels by dams. Dams lead to accelerated evaporation also. Socioeconomic planning for hydro-energy projects is highly desirable, so that there is net socioeconomic gain and no specific section of the deprived and poor societies are made to suffer.

World Water Vision 2025

The World Water Council under the distinguished Co-Chairmanship of Ismail Segeldin and Frank Rijsberman, drew up a world water scenario for 2025 under business as usual (BAU) model, given the natural development of growing population, continuing economic progress and improving technology application. The gist of the findings of the model in respect of categorywise water withdrawal and consumption is presented in Table T20.9. It will be seen that the global withdrawal of water will increase from 3800 to 4200 cu.km. at 10 per cent growth annually. Similarly, water consumption will increase from 2100 to 2300 cu.km. again at 10 per cent growth annually

The scenario further highlights that the per capita availability of renewable water resources is likely to fall from 6,600 to 4800 cu.m. per capita-year. Given the highly uneven distribution of water resources, some 3 billion people will live in countries, wholly or partly arid or semi-arid and will have water less than 1700 cu.m. per capita-year, quantity below which a human is believed to suffer from water stress. Examined in an alternative perspective, 4 billion people will live in countries, where by 2025, about 40 per cent of renewable water resources will be withdrawn, which is another indicator of apparent water stress.

In further elaboration of Table T20.9, regionwise water withdrawals and consumptions are given for Africa, Asia and Europe in Table T20.10.

The portion of land areas and the people who will become conspicuously water stressed by 2025 is shown in Table T20.11. The Table T20.10 shows that in Europe, the total withdrawal of water in 2025 will, in fact, marginally decrease from 350 to 340 cu.km. mainly because of higher water efficiency in Agriculture.

The withdrawal levels in Domestic and industrial uses will remain unchanged, again because of technology and operational efficiency improving.

In Africa, the domestic water withdrawal/consumption will jump up conspicuously from 12 to 40 Cu.km. per annum because of expected substantial improvement in living standards and partly demographic increase. Consequently, overall withdrawal will increase from 60 to 100 cu.km.

The Asian scene is similar to the African scene, but increase in withdrawal is less pronounced. Domestic use will increase from 114 to 471 cu.km, but agriculture will withdraw less to 1266 from 1359 cu.km, because of superior agricultural practice and water management.

Towards the Water Crisis Resolution

The United Nation Millennium Development Goals (MDG) enunciated that in 15 years nearly all people in the world in the cities, towns and villages hall enjoy safe and adequate drinking water and sanitation and have enough food to meet their hunger and nutritional needs. The participants of the above 2025 vision exercises, having similar goals for water even before the MDG, had proposed five major actions to realize the goals:

(*i*) Involve all stakeholders in integrated water resources management;

(*ii*) Move towards full cost pricing of water resources of water services for all human uses;

(*iii*) Increase public funding for research and innovation in broad based public-private water resource management;

(*iv*) Seek international cooperation to improve integrated water resource development in multinational basins;

(*v*) Invest generously in all water-related infrastructure;

Among the specific measures, it must be recognized that agriculture is by far the largest consumer of water at the global average of about 65 per cent. Hence emphasis should shift to avoiding waste and saving as much water in agriculture. Dry farming and rain-fed farming including green farming should be increased. Even where irrigated farming is essential, excess water withdrawal shall be avoided and water should be utilized prudently as in sprinkler farming practiced in Israel. In any case, excess water deteriorates soil fertility by increase of salinity and possible soil erosion.

There is scope of economy in domestic use particularly in washing and gardening and flash-sanitation. New methods of eco-sanitation, not dependent on water as carrier, could be evolved, which may educe pollution and offer human waste as agriculture fertilizer.

Natural and artificial wetland might be used to improve pollution water and treat domestic effluents. There are manifold ways to improve water resource availability in the coming decades to raise quality of life, provided adequate effort and fund are directed to water resource development.

20.8 Ecology and Water Resource Development

Ecology perhaps originates from Greek word ‘oikos’ meaning home or place to live in. German Biologist Ernst Hackel is believed to have used this term in 1869 and defined this ecology as “the total relations of animal to both its organic and inorganic environment". The definition of Ecology was broadened by Charles Allee (1949) as "the science of interrelationships between living organism and the environment (both physical and the biota)". Environment with respect to a specie is the eco-system (*i.e.,* ecological system) of all living and non-living outside the particular specie. Each organism has its own environment. Here after our discourse is developed from the seminal work of Thanh and Tam (1990).

Now we have stated that water is the essential saver of life. With the growing world population, the rising economic well-being, the demand for potable and non-potable water is rising all the time. Since such water is not available from natural system, further development of water resources, its conservation and distribution have been taking place also at an accelerated pace.

Now it is seen that development and expansion of water resources on a large scale often have influences on the natural ecology and the environment, such harmful impacts can be avoided if definitive and appropriate corrective measures are analysed, considered and incorporated in the planning and management of the existing and new water resources. Box B 20.2 lists various components of ecosystems affected by water process and development. In this chapter we will study some dimensions of the ecological malfunctions that may impair a water project.

Ecology of Water Resources

We have water resources in two components:

(*i*) Storage or stock of water on earth amounting to 1400 million cu.km.;

(*ii*) Renewable component of 44,000 cu.km. per annum which sources from rain.

We see below that only about 0.2 percent only is available fresh water. This shows how precious is that very small amount of fresh water.

Sea Water			97.3 %
Fresh Water			
	Polar ice	2.1 %	
	Soil Moisture/Deep Ground Water	0.4 %	
	River, Lakes, Groundwater above 800m	0.2 %	2.7 %

Note: Water in Lakes 0.009 %

Water in Rivers 0.0003%

Even this small amount of fresh water is very unevenly distributed. For example, 50 % of Africa's total surface water resources are in Congo River basins. Lake Baikal in Serbia alone contains 10 percent of the total fresh water of the whole planet. Lake Ontario in North America can hold all the water of all the rivers of the world.

Renewable fresh water is the net balance water from precipitation less evapotranspiration from the land and the sea and roughly represents the total run-off to the sea as shown in Figs. F20.1 and F20.3 We should now infer that we understand the basic water ecosystem. Any of the human activities must not pollute the fresh water stocks. Similarly, any of the human activities and developmental work related to water should not disrupt the orderly functioning of the hydrological cycle. The UN agencies have already highlighted that human activities so far cause profound effect on the hydrological cycle arising out of atmospheric pollution, global warming, deforestation etc.

Quality of Surface Water

Water quality is very important consideration for all water development projects, as it affects all aspects of water use – for humans, for animals, for crops and even for industry. Box B20.3 shows some of the parameters, which affects the quality of water.

With respect to water quality in lakes and reservoirs, it is necessary to understand the problem of thermal stratification and its effect leading to the loss of dissolved oxygen. Further stratification also prevents the nutrients in the lower strata not getting distributed and mixed in the upper layers.

Among several other undesirable effects in quality, it may be mentioned that water under the spillway may become supersaturated with nitrogen due to high turbulence. The situation may also affect the mineral and organic content leading them to settle out. In the arid zone, the water often develops high salt concentration. With high evaporation and low precipitation regime, the impounded water gets high salinity.

Flooding of tropical forest may also cause hydrogen sulphide in impounded water, which causes aesthetic problem for human consumption, toxicity for fish and corrosion for metal parts. Decaying of plants in floodwater may cause high rise in nutrient contents, leading to eutriphication and loss of oxygen in water.

Box B20.2 Ecosystems affected by Water Developments

Climate Regimes:	**Vegetation Resources:**	**Wildlife resources:**
Rainfall	Virgin timber	Biological diversities
Temperature	Rare timber remnants	Animals
Wind	Reforestation activities	Fish
Evaporation	Natural forest	Birds
Humidity	Orchids	Breeding ground
Humidity	Nurseries	Sanctuaries
Solar irradiation	Games & inhabitant	Endangered species
Natural Disasters	Areas	**Recreational Facilities:**
(Flash Floods)	Important habitats	Hunting areas
	Endangered plant	Water-related activities
Water Resources:	Species	Campsites, picnic
Drainage pattern		Grounds
Significant streams	**Historical & Cultural**	Resorts & tourism space
Ground water	**Resources:**	
Lakes & ponds	Historical structure	
Beaches	Museum	
Wet lands	Temples & shrine	
Dams & reservoirs	Cottage industries	
Canals & locks	Cemeteries	

Box B20.3: Parameters of Water Quality

Aquifer safe yields	Heavy metals
Flow variations	Other toxic substances
Chlorides	Faecal coliforms
Oil & grease	Faecal sreptococci
Suspended solids	Temperature
Acidity, alkalinity & pH	Radioactivity
BOD and/or COD	Colour
Dissolved oxygen	Odour
Nutrients trace elements	Taste
(Mn, Mg, Bo, Se)	
Pesticides	

Impacts on Ground Water

Irrigation water may have impacted on ground water in various ways. In arid zone, if more irrigation water is supplied than required for evapotranspiration, it may lead to accumulation of salt in root zones. High irrigation water may also lead to seepage of salt, fertilizer and pesticides to the groundwater and cause harmful contamination.

Considering that there are 225 million hectares of irrigated land in the world, ground water contamination by irrigation projects could indeed be extensive.

Irrigation schemes in Punjab, Pakistan have in their first 10 years of operation has raised the ground tables by several metres.

On the other hand, excessive of pumping of ground water in India and USA has led to depletion of ground water levels conspicuously.

Water Logging and Salinity

Degradation of soil fertility by water logging and salinity development is common, but one of the most serious impact of irrigation. Increase in irrigation without proper drainage facilities usually leads to increase in soil salinization and alkalization as water evaporates, leaving behind salt particles on top soil. This occurred in Pakistan during the 1960s, where 24,300 hectares of fertile cropland in the country was lost every year (Biswas 1980). Soil salinization and soil erosion due to wrong irrigation practices are becoming serious problems throughout Canada's agriculture areas, particularly in the dry praire environment.

As much as half of world's irrigation schemes suffer in some degree from water logging, Salinization and alkalization affecting 1 to 1.5 million hectares annually.

Seepage from irrigation canals is a common problem along all unlined canals. This is evidenced in in Pakistan, where the ground water level, as mentioned earlier rose by 7 to 9 metres in last 10 years,

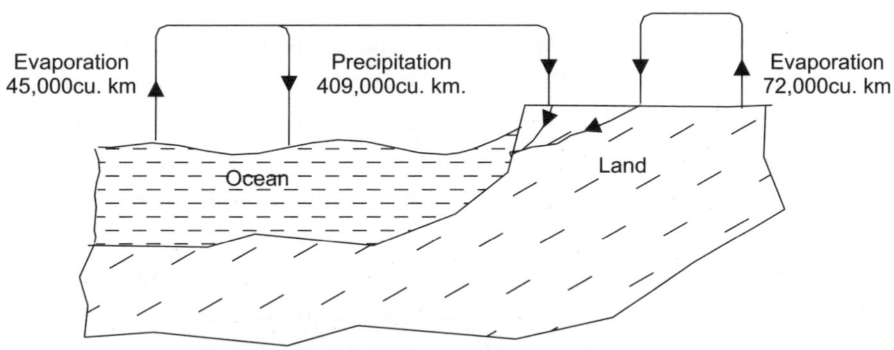

Fig. F20.3: Simplified Hydrological Cycle

Occurrence of Earthquakes

It has been proven that earthquake can be induced by large reservoirs. The main earthquake in Greece in 1966 had its epicenter under the reservoir of the Kremastha Dam. Likewise, the 1967 Koyna Dam disaster in India peninsula was caused by an earthquake whose epicenter was located at the dam itself (Biswas 1978).

Effects on Climate

Some small changes take place around a new water reservoir. For example, some change in rainfall pattern has been observed around Lake Volta in Ghana, the peak having shifted from October to July/August. Aswan High Dam, by more than doubling the surface area of Lake Nasser has increased the total evaporation from 6 cu.km. to 10 cu.km. The total loss by evaporation amounts to 11 percent. The point is made that such change will always have some effects on humans.

Biological Impacts

- — Direct killing of animals and plants in the area flooded by the dam;
- — Opening of new roads and navigational channel and migration of new inhabitants may affect the eco-systems and bio-diversity;
- — Development of fisheries in the new reservoir may be affected by water stratification, eutrification or other toxic attack;
- — Fisheries downstream might get adversely affected due to loss of nutrients;
- — Water resources development projects are often related to waterborne diseases.

Socioeconomic Impacts

- — Loss of timber forest and low-lying farmland due to flooding, clearings, construction or salinity;
- — Deposition of silt upstream reduces the agricultural and/or fishing yield downstream;
- — A Water development projects attract migration of people from other areas for better job or commercial opportunities;
- — A water development project may also cause considerable upheaval on the socio-cultural structure and some disruption of the local people's livelihood and resettlement in new areas.

Lessons Learnt

Generally all water resources processing and development projects have social and environmental implications, which are usually long-term, and widespread, but often reversible. Perhaps, the most well-known example is the Aswan High Dam, which created considerable adverse environmental impacts like excessive evaporation, decrease of fertile land downstream, and the killing of the sardine fishing industry in the Mediterranean and many other ill-effects.

But at the same time, Aswan Dam enables Egypt's electricity generation capacity to double, helped prevent disastrous floods, improve water navigation and created a vast potential fishery in the reservoir. Hence, while weighing the negative impacts, we should not ignore greater positive benefits, the real raison de etre for development project. This calls for a very balanced evaluation of the new water projects in respect of development *vis-à-vis* environment.

20.9 Management of Water Resources

Water resource is not so plentiful as could be splurged and wasted. Besides, water is very precious for life and living. From the UN Report, we already know that global annually renewable fresh water, available as run-off, is only 42,000 cu.km. and if all the water could be withdrawn and apportioned among the present 6 billion people, the share per person would have been 7000 cu.m., which is apparently sufficiently plentiful foe human needs. Unfortunately only 10 to 12 percent water can be harnessed today, such that only the actual

globally available water is 700 to 1000 cu.m. per person per year, which indicate high level water stress. Thus, there is great need to raise the recovery level to say15 to 25 percent of the global run-off depending upon the local water demand. Such upgradation of water resource could only be achieved through long-term planning and investment in waterworks and other related infrastructure.

Secondly, to make mater worse, the global run-off of 42,000 cu.km. of renewable fresh water is very unevenly distributed over the continents and regions. In the Amazon basins, Northern Canada and Alaska, vast amount of water flows away to the sea untapped and in India's Brahmaputra-Barak basins, only 24 cu.km. water is withdrawn of the gross flow of 1000 cu.km. annually. On the other hand, in the Indus river basin of India-Pakistan and Amu Darya and Syr Darya basin in the Central Asia, Huang He (Yellow) river basin in China, water is so heavily withdrawn away that sometimes no river water reaches the sea. This grave situation underlines the urgent needs of long-term planning for redistributing the water resources by possible inter-basin transfer of water, tapping ground water or corrective water shortage and conservation scheme.

The third dimension of the water problem is the local management of water resources, regulating supply, demand and consumption, u[grading water quality, minimizing leakage and wastage. Such measures will help to mitigate and improve the glaringly discriminatory annual water consumption levels — 2000 cu.m. in Europe, 1000 cu.m. in India, 1000 cu.m. in East Africa, 400 cu.m. in Israel, 100 cu.m. in Kuwait or less than 100 cu.m. in Sub-Sahara.

Let us now shift our attention to the developing countries. There is no doubt that for developing countries, which are all located in the tropical and sub-tropical regions, with erratic rainfall patterns, efficient control and management of water is an essential requirement for their continued development. Without proper water management, self-sufficiency in food and energy will suffer. Scarcity of water and reliability of its supply are major constraints for agricultural development of arid and semi-arid countries. Given the above scarcity-prone water scenario of developing countries, the demand for water is likely to increase continuously with the improvement of the standard of living of the people. No doubt, these countries need to undertake ambitious programmes of water resource development adopting principles of modern management and economic marketing with social forces for enduring viability of projects.

Thus, we are given to infer that the principles of modern management are needed to be applied to alleviate the water problem, adopting tools of long-range planning, modern engineering and project management for the fallowing broad objectives.

(*i*) To increase gross withdrawal of run-off water from 10 to 20 per cent by building dam and reservoir, distributional canals and any other means inside a basin;

(*ii*) To redistribute the increased available water to more water-stressed and arid regions through a network of storage facilities and distribution systems using measures of conservation and waste-minimization, soil conservation, forestry and if possible by inter-basin water transfer.

(*iii*) To modernize all local water distribution systems in the urban and rural habitation with the aim of high utilization, least wastage, recycle and reuse.

(*iv*) All programme of development and management of water resources should ensure as not to adversely impact the environment.

Outlining the above modern management of water resources, we enlist some of the prime considerations in the managerial assessments and executions enunciated by Prof. A.K. Biswas, a long-standing international water specialist of high repute (Biswas 1980):

(*i*) Comprehensive project Review including searches of all possible management options;

(*ii*) Recognition of water as an economic good;

(*iii*) Adoption of the river basin as a unit of water resource development and management;

(*iv*) The consideration of environmental impacts;

(*v*) The assessment of social impacts includes the involvement of public in policy formulation, planning, execution and financing.

We now elaborate these prime concerns.

Comprehensive Review of Projects

An overview of the tasks of water management is called for to enhance the benefits of the investment and also widen the scope of water utilization.

There are generally numerous potential uses of a given resource. For example, a stream can be used for hydro-energy generation, irrigation, municipal and domestic water uses, for fisheries development and also a somewhat negative use for disposal of toxic wastes. The last use may affect the earlier benign applications. Optimal uses of stream can only be attained if all potential uses are considered simultaneously and a combination is chosen that conforms to a selected criterion, such as the maximum net economic return or the conservation of particular ecosystem.

Secondly, such simultaneous preview of a water project allows its multiple applications. For example, Tennessee valley scheme comprises of hydroelectric power generation, flood control, irrigation development, and improved navigation. In India, we have similar multi-purpose river valley projects like Damodar Valley Corporation, Bakra Nangal and currently Normada Valley Project.

Thirdly, the interrelationship of water development with the development of other resources should be carefully assessed. Water schemes may result in the erosion and flooding of agriculture land, unforeseen inundation etc.

Fourthly, a comprehensive viewpoint may imply integration of water planning with overall economic and social planning. This is especially important when water is scarce and is vital for economy as in Egypt or Jordan.

Fig. F20.4 shows a scheme of gradually increasing scale of integration and sophistication in the planning and funding of a typical water project.

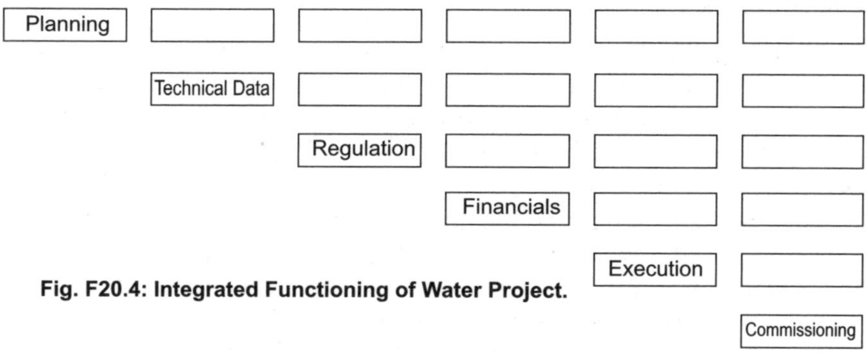

Fig. F20.4: Integrated Functioning of Water Project.

Water as an Economic Good

In recent time, there has been some loud thinking to levy a charge on the use of water with two principal objectives. *Firstly,* Charged water will be judiciously used avoiding wastages, preserving water for greater and wider use. *Secondly*, charges realized can be further deployed to enhance water infrastructure.

Though, other natural resources like forests, fisheries and minerals are chargeable natural resources, there is some basic objective to treat water similarly as economic goods as water is as essential for life as air and shall be equitably available to all, irrespective of their economic status.

Every society should therefore carry social cost-benefit analysis to determine if water should be chargeable even at subsidized rate.

River Basin as an Unit of Water Management

The River basin has increasingly become accepted as integral unit for water resource management. The rationale follows from the concept of a river as an organic whole, so that interference with one part of it will directly influence its operation at another point of the river system. Engineers and geographers dearly espoused the idea. Economists have strongly recognized it as a viable way of internalizing externalities.

In India, United Kingdom and France, river basin is taken for all phases of water management. Elsewhere, such concept is first catching up.

Environmental Impacts

In the previous section, we have observed how water development project can have some serious impacts on other facets of the ecosystems. Under UN initiatives, water management agencies are obliged to identify and evaluate the impact of water projects on the environment and the overall ecosystems. In fact, the environmental impacts are now an elaborate process and in India, it is statutory to undertake environmental Impact study and due clearance before a Project can be implemented.

Social Impacts

Over the recent decades, the project planning for the major water projects ha extended its study from the purely technical evaluation to wider social, environmental and other public concerns. In fact, the public and other social groups are being involved with the formulation of the project to maximize social and economic benefit.

In India, we have seen how much social turmoil was created by the very large Narmada Valley project, causing widespread displacement of population. Similar situation was also experienced in China. In New Zealand and Australia, the right of the aborigines of the affected valleys had to be duly recognized. Box B4.4 shows schematically the mechanism of Public Involvement.

Box B20.4 Mechanism for Public Involvement

1. Public opinion polls and other surveys;
2. Referendum;
3. Ballot Box;
4. Letter to editors or public authority;
5. Public hearing of Inquiries;
6. Representation by Pressure groups;
7. Public Meeting /Demonstration;
8. Organizing Workshops;
9. Formal Protest & Application;
10. Court Action

❑❑❑

21
CHAPTER

Challenges Of India's Water Resources

21.1 Overview

Water Issues in Ancient and medieval India (Prasad et al 1987)
Some of the great ancient river valley civilizations are the Egyptian civilization on the Nile valley, the Mesopotamian civilization in the Euphrates and Tigris valley, Indian Harappan civilization in the Indus valley and the Chinese civilization in the Huang He river valley. All these civilizations, which flourished from 3500 to 2000 BC had the common feature that their basic economic activity was agriculture supported by the water resources of the respective rivers..

The later Aryan civilization, during which the prime religious anthology "Rig Veda' was written, dating around 2000 BC, also flourished along river Ganga. The Rig Veda and the other Vedas refer to the damming and control of river flows, water wheels and canals (Prasad et al 1987) for purposes of irrigation.

From the 6th century for about six centuries, there was considerable political instability and turmoil. Even then, dug-well irrigation and prosperous agriculture along Ganga continued. In the 14th century, two reservoirs and a network of canals were constructed in the Gangetic plain by the Tuglaq rulers.

In the16th century, with the establishment of the Mughal empire, there was considerable political stability for the next three centuries. Irrigation by means of tanks, canals and wells was widely practiced by the farmers to lessen the vagaries of the monsoon and to boost the agricultural production. There was nevertheless no major irrigation scheme undertaken.

India again enjoyed two hundred years of peace and stability during the British rule. In the beginning, not much attention was paid to the irrigation or agriculture by the trade-minded British rulers. But in the first half of the 19th century, various parts of the country came in the grip of famine and which forced the British Government to constitute two commissions—

— the Famine Commission 1880 followed by,

— the Irrigation Commission 1891,

which recommended renovation of several defunct or dilapidated irrigation works and also proposed new irrigation schemes.

However modern irrigation development came to India trough the British along with the relevant technology and the organization. The Thomson College of Engineering at Roorkee was established around 1850 and a research organization named Central Water and Power Research Station at Pune was set up in 1916.

Water Issues in India Today (W.B. Report 1991)

India is blessed with a large number of rivers and also gets reasonably good amount of rainfall of 1100 mm on national average. The problem is of space and time distribution of that water resource. In some river basin like Brahmaputra-Barak, there is high excess of water unutilized and flowing out to the sea. In some parts of India, typically in parts of Rajasthan in Western India,the annual rainfall is as low as 100mm. Further, the major rainfall occurs for only three to four months in a year leading to low utilization of the available water. To make the matter worse, the Indian population is steadily increasing from about 400 mm in 1950 to about 1000 millions in 2000.

The World Bank in conjunction with the Ministry of Water Resource of the Indian Government undertook studies of Indian Water Resources (World Bank Report 1999). It observed that the water availability was as high as 5000 cu. m. per person annually about the time of Indian independence in 1950, which was plentiful compared with UN standard of water stress at 1000 cu.m. per capita. It was apprehended by them that the population will grow to 1.4 billion in 2025 and the water availability will come down to about 1500 cu.m. per annum. The adversarial relationship below between population and water availability is illustrated in Fig. F21.1 of World Bank, which indicates that the water availability may come down to as low 1000 cu.m. per capita, unless issues of development and management are tackled rather seriously and urgently.

The most important framework of issues that should be addressed on all India vision:

- — Finding solutions for competing inter-sectoral demands within a region and/or State keeping in view efficient water utilization minimizing wastage;
- — Planning, developing, and managing water on a river basin basis applying the latest technology, economic and social justice;
- — Equitably and prudently allocate water of a river basin under mutual or judicial awards;

To assist the above process of efficient and responsible management of water resources, following infrastructure and institutions are required to be strengthened and developed.

(*i*) National Water Policy for the whole of India;

(*ii*) Effective Legislation and Regulations for handling live issues of water;

(*iii*) Institutions for planning, development and dispute resolution if any;

(*iv*) Economic management including pricing and incentives in the use of water;

(*v*) Update data generation, analysis and public dissemination.

Some of the enabling factors and initiatives for augmenting water resources include:

— Private-Public Partnership in water project management and maintenance;

— Decentralization of water project management;

— Stakeholder participation and involvement;

— Involvement of grassroot organization;

— Water demand management to minimize wastage and maximize utilization;

— Enabling environment for change and improvement of all initiatives;

— Economic Incentive framework.

Universal Access to Water:

Though much work is being undertaken to make available water resources over all inhabited space and over the whole year, there are still many pockets where water does not reach in the developing countries. As a result, many humans do not still have ready and easy access to water.

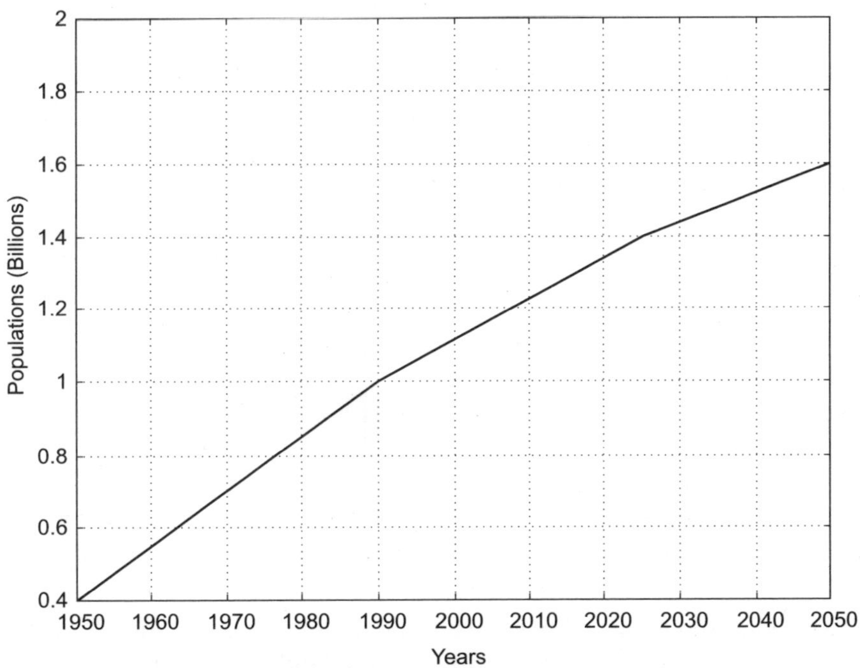

Fig. F21.1A: Growth of Indian Population

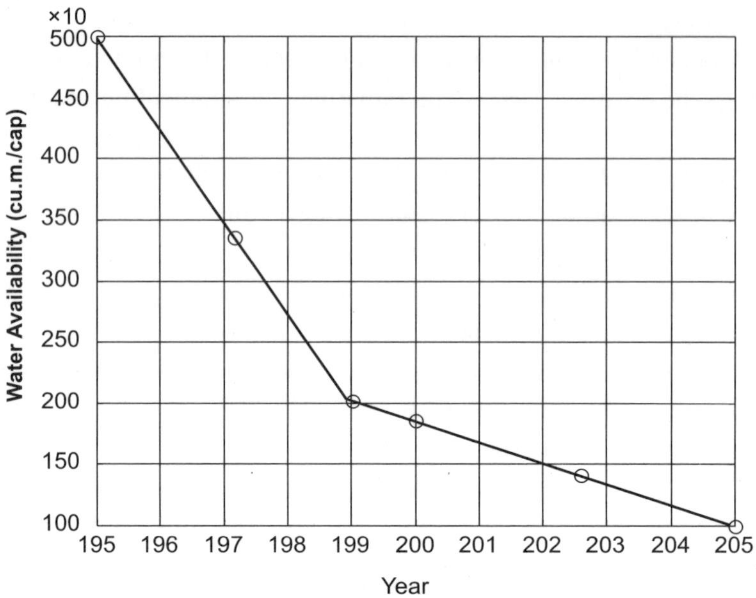

Fig. F21.1B: Trend of Water Availability in India

The Millennium Development Goals (Target 10 of MDG) of the United Nations had ordained in 2000 that the percentage of population not having access to potable water should be halved by 2015. Below are given the trends of water accessibility at the global levels and India:

Region	Year	Population % with access to water
World	1990	77 %
	2002	83 %
India	1990	68 %
	2000	86 %

The Table above shows a high jump in India's progress, beyond the average global progress. Of course, in the developed countries, almost everyone has satisfactory access to water and sanitation.

Two more important water issues in India which need to be duly identified and resolved for the long-term development of Indian water resources pertain to inter-state cooperation in sharing of water and inter-basin transfer of water.

The catchment areas of all major river basins in India are located in more than one administrative States of India and many rivers have a number riparian States. It is therefore in the fitness of the situations that all planning and development of a river basin must be conceived on the basis whole and the states share the water on mutually agreed and equitable basis.

The inter-basin transfer of water from the surplus basin to the deficient basin to raise the levels of all India utilization of water is a live issue which calls for finalization of technical solutions awaiting the political consensus.

A summary of various important Indian water issues is presented in Box B 21.1.

Box B21.1:
Principal Water Issues in India

1. India's water resources are significantly dependant on monsoon rains. This indicates the vulnerability of both agriculture and other water uses to the not so unusual vagaries of the monsoon.
2. The above variations of rainwater run-off need to be stabilized by creation of adequate water storage capacities. India already occupies fourth place globally in the dam capacity at 250 cu. km. The existing capacity could still be further augmented by knowing the higher capacities of the following nations.

U S A	592 cu. km.
China	480 cu. km.
Russia	360 cu. km.

3. Indian agriculture consumes up a lion share of the available water resources in India at 84 %. It is recognized that the efficiency of water use in agriculture is low. Technology and economy of water need to be improved on the lines of Israel and other developed countries.
4. Though the national level of water resource availability at 2000 cu.m. per capita is much higher than the threshold water stress level of 1000 cu.m. per capita, significant part of the North-western India like Rajasthan and Gujrat gets much less below the threshold. The earlier Rajasthan canal system and the Narmada canal system, currently under execution, are completed; it will much alleviate the availability of water. Further, inter-basin transfer of water must be explored.
5. More than 50 % of India's run-off water flows through the northern river systems of Ganga and Brahmaputra basins, while the southern river systems Cauvery and Krishna etc are significantly water deficient. Inter-basin water transfer is therefore important for India's long-term development and shall be pursued towards political, economic and technological solutions.
6. India's sizeable rechargeable ground water resource should be used prudently. Particularly in the semi-arid areas of the western India, it can provide much needed water. No doubt, the Indian Ground Water Board is doing commendable work in assessing the quantity distribution of the groundwater and also the data on the rate of recharge. Perhaps, more mappings and network informations need to be developed

21.2 Outlne of Water Resource Situation

Supply & Demand

We have already mentioned that India has a large number of rivers and also gets reasonably good monsoon rains. But at the same time, the country has a big population, large area of irrigated agriculture land and considerable industrial operation, which generates high demand for water. Thus, the water resource remains critical for India and her development plans. We present here a short outline of India's water resources.

Water Supply Perspectives

Rainfall: India's water resources are predominantly based on rainfall. The average rainfall is 1170 mm annually. But there is considerable variation in rainfalls across the country and again from year to year. The rain occurs mostly in the southwest monsoon season over three to four months, June to September in the year. Thus there arises the necessity of creating large dam storage capacity for maximum utilization of the surface run-off.

Regional rainfall variations are also high, ranging from the low 100 mm to high 11,000 mm in Meghalaya in the Northeast. Generally, there is adequate rainfall in the Eastern and Northeastern India including West Bengal, Eastern Uttar Pradesh, Bihar, Northern Orissa and the Cauvery delta – typically from 1000 mm to 2000 mm annually. In Western India – West Uttar Pradesh, Punjab, Haryana, Rajasthan and the rain-shadow of Deccan plateau, rainfall ranges from 300 to 800 mm.

Surface Water: India's average annual surface water run-off generated by the rainfall and snowmelt is established to 1869 bcm (billion cubic metre). However, it is estimated only about 690 bcm or 37 % of the potential surface water resources can actually be mobilized to effective use, because of the uncertain rainfall during the short period of few months in the year.

Consequently, the water availability is highly variable. While the northern rivers are more or less perennial fed from the rainfall and supplanted by the snowmelt in the summer, the southern rivers get dry during the major part of the year. By contrast, the Brahmaputra-Barak river system due to flood flow nature, is generally utilizable to a low of 24 bcm out of its average annual flow of 586 bcm.

Groundwater: India's rechargeable annual groundwater potential has been estimated to about 452 bcm in the aggregate terms. About half of the amount is located in the Ganga-Brahmaputra-Barak basins system. On an all India basis, using aggregate estimate, only about 30 % of the groundwater potential could be tapped to use for irrigation and domestic purposes.

However, the above aggregate use has been ballasted by the comparatively higher utilization in the surplus Eastern and the Northeastern India. The situation is adverse in other parts of India, where underground reserve has fallen below the recharge level by several metres. In particular, Punjab and Haryana have already exploited most of the shallow water rechargeable reserve. In other areas, in Rajasthan, Gujrat, most of the western UP and the Deccan States, the water tables are fast depleting at the unsustainable rate.

In fact, in the Tenth Five Year Plan (2002-07) document (GOI,2002) notes that in 310 out of 5711 blocks in the country, groundwater was considered to be over-exploited *i.e.,* average withdrawal is in excess of replenishment. Consequently, the groundwater level has gone down by as much as 4 metres in the affected districts between 1981 and 2000.

Per Capita Water Availability

India vs. World

India has about 16 per cent of world's population but has only about 4 per cent of world's total renewable water resources (Indian 10th Plan, GOI, 2002). This highlights the pressure of population on available water resource, which would have been otherwise considered adequate at 1170 mm average rainfall.

Fig. F21.1B, derived from the World Bank Report, also shows the downward future water availability scenario from 5000 cu. m. per capita in 1950 to 1000 cu. m. per capita in 2050, which is the UNDP threshold of water scarcity. However, many nations are making good living with much lower level of water availability. Table T21.1 shows Israel managing with per capita water of 382 cum and Singapore with 211 cum, two fairly developed countries. These examples underline the importance of improvement in water use in India.

Table T21.1: National per capita Water Availability

Country	Per capita Water Availability (Cum per annum)
AFRICA	
Algeria	528
Egypt	794
Kenya	1069
Libya	111
Morocco	1110
South Africa	1106
ASIA	
Singapore	139
EUROPE	
Denmark	416
United Kingdom	1219
MIDDLE EAST	
Israel	255
Jordan	157
Kuwait	8
Lebanon	1189
ASIA	
China	2276
India	1754

Source: World Resource Institute,2005

India ranks at the top of world in drawing 558 bcm per year for irrigation (FAO) of 72 million hectares. In fact, the irrigation water amounts to 84 % of total water usage. This underpins the necessity of improving the irrigation technology and makes do with much less water without adversely affecting agricultural productivity.

Water Potential Perspectives By River Basin

The water availability potential of various river basins of the country can be seen from the Table T21.2. The Table shows both the river run-off and the rechargeable groundwater. It is seen that in many river basins – Cauvery, Pennar, Sabarmati, West-flowing rivers in Kutch and Sourashtra (including Luni), east-flowing rives in between Mahanadi and Godavari, and Pennar to Kannya Kumari, the per capita water availability is below 1000 cum annually, even today. This is a forewarnng of general water stress and deserves special attention of water developers and managers. Secondly, Ganga and Brahmaputra river systems with 1110.62 bcm water account for 59.4 percentage of total Indian surface water for 44.7 percent of total land area. This also indicates an overall water scarcity in the rest of India outside the Ganga-Brahmaputra basin areas.

Water Demand Perspective

Irrigation has played a major role in converting India from a highly food deficient country to a surplus country. However, Indian agriculture uses water most extravagantly. India ranks at the top of all countries in withdrawing 558 bcm of water for irrigation (FAO 2003). In fact, the irrigation water constitutes nearly 84 % of total water usage. That underpins the necessity of improving the irrigation technology to higher efficiency and economise water consumption without adversely affecting agricultural productivity.

In any case, this historical situation of very high agricultural water consumption, while the industrial and domestic consumptions are minor players is definitely to change. As the country gets increasingly industrialized, demand for industry and power will multiply to much higher share of water and the domestic demand will conspicuously go up for higher quantity and quality of water. In 1995 in Europe, the demand for agriculture, industry and domestic were respectively in percentages 38, 38 and 24 (See Table T20.10). However, compared with Europe, much areas of Indian agriculture are semi-arid necessitating irrigation and hence requiring higher allocation of water.

The inter-sectoral allocations of water for the three major demand sectors for 1990 (actual) and for 2025 (projected) are estimated in the World Bank Report 1990 and presented in Table T21.3.

Table T21.3: Inter-Sectoral Water Allocations

Demand Types	Annual Demands		
	1990 (bcm)	2025 (bcm)	Annual Increase(%)
Domestic	25	52	6
Industry	67	228	10
Agriculture	460	770	5

Source: World Bank Report 1990; bcm – billion cubic meters

Table T21.2: River Basin and Water Availability

River Basins	Average Annual Surface Water (bcm)	Replenishable Ground Water (bcm)	Per Capita Surface Water (cm)	Per Capita Surface + ground Water (cm)
Indus	73.31	26.55	1749	2383.30
Brahmaputra -Barak	585.60	206.07	16,589	17,612.66
Ganga	525.02	171.00	1471	1950.73
Godavari	110.54	40.60	2048	2799.93
Krishna	78.12	26.40	1285	1719.64
Cauvery	21.36	12.30	728	1147.63
Subernarekha	12.37	1.8	1307	1497.89
Brahmani -Baitarini	28.48	4.05	2915	3329.58
Mahanadi	66.88	16.50	2513	3134.59
Pennar	6.32	4.39	651	1159.79
Mahi	11.02	-	1052	1052
Sabarmati	3.81	-	360	360
Narmada	45.64	10.80	3109	3869.46
Tapi	14.88	8.27	1007	1564.19
West-Flowing Rivers				
Tapi to Tadri	87.41	28.90	3383	3387.98
Tadri–Kannya Kumari	113.53	-	3480	3480
Kutch,Saurastra (Luni)	15.10	-	683	683
East-Flowing Rivers				
Mahanadi to Godavari	22.52	-	953	953
Pennar-Kannyakumari	16.46	-	366	366
Area of inland drainage in Rajasthan				
Minor Rivers drain to Bangladesh /Myanmar	31.00	-	14623	14623

Source: Indian Water Resource Society 1997

21.3 Indian Legal Regime and Institutional Overview

Basically, the air a man breathes and the water he drinks are the inalienable right of the humankind. In this sense, water is a private goods. However, because of the very wide influence of water on life, climate, transportation, industry and agriculture, the water resource is coming under the national law and regulation. In fact, we have already seen in the earlier chapter that the quality and scarcity of water resource have now taken a global perspective.

Constitutional Provision & Law

India is a union of many states governed by a written Constitution. The Seventh Schedule of the Constitution lays down the respective power and the responsibilities of the Union and the States. The Indian parliament has extensive power to make laws with respect to any matters enumerated in the List I in the Seventh Schedule, referred to as the "Union List". The Legislature of any State has exclusive power to make laws for such State or any part thereof with respect to any matters enumerated in the List II in the Seventh Schedule (referred to as the "State List"). The Parliament and the Legislature of any State have also the power to make laws with respect to any matters enumerated in List III in the Seventh Schedule (referred to as the "Concurrent List").

Article 249 of the Constitution empowers the Parliament to make laws in the Stae list, if the Council of the States declares by resolution empowered by not less than two-third of members present and voting that it is necessary or expedient in national interest that the Parliament should make laws with any matter enumerated in the State List specified in the Resolution. It shall then be lawful for the Parliament to make laws for the whole or any part of the territory of India with respect to that matter, while the resolution remains in force.

Now regarding water, under the Entry 56 of the Union List, the regulation and development of inter-state rivers and river valleys are the primary responsibility of the Union.

But under Entry 17 of the State List, water, that is, water supply, irrigation, and canals, drainage, and embankments, water storage and water power are the responsibility of the State.

In the defacto interpretation of the responsibilities, the power of the States has emerged as pre-eminent. Thus, notwithstanding the power conferred on the Union Government in Entry 56, water has come to be perceived as a State subject. This Union-State arrangement is a legacy of the British times, for example, from the Government of India Act of 1919 in which irrigation was a subject matter of provinces, with the Government of India playing a role of coordinator, for advice and dispute resolution.

In practice, the Government of India has so far dominated, financed, and administered all major River Valley projects (see Box B21.1). Besides, water being a national resource affecting the whole country, the Ministry of Water

Resource at the Union level has naturally played the role of an Apex Adviser, also enunciating National Water Policy by deliberation in the national parliament. The Central Government could also exercise its directives in the water matter through a number of mechanisms as:

(*i*) The National Planning Commission chaired by the Prime Minister sanctions financial allocations to the States;

(*ii*) Through the Inter-Sate Water Disputes Act of 1956, National Policy or otherwise several pan-Indian National Institutions like the longstanding Central Water Power Commission and the Central Ground Water Board and the finally the Ministry of Water Resources, Union effectively monitors and influences all major water decisions of the States;

(*iii*) Generally, all major water projects require Environmental clearance, which is controlled and granted by the central Ministry of environment and Forest.

National Water Policies (NWP)

Forty years after the Indian independence came the first National Water Policy in 1987 during the Prime Ministership of Rajiv Gandhi of the Congress Party, enunciating inter alia (Anand 2007):

(*a*) Water resources available should be turned into 'utilizable' resources to the maximum extent possible;

(*b*) The drinking water should be given the first priority, followed by irrigation and the hydropower;

(*c*) Planning should be done on the basis of hydrological basins and appropriate organizations should be set up for the planned development and management of such basins;

(*d*) Water should be made available to water-short areas including inter-basin transfers;

(*e*) Water rates should be used to convey the scarcity value and encourage the economical use of water;

However, the main specific commitment made in the Policy document that "adequate drinking water facilities should be provided to the entire population both in the urban and the rural areas by 1991" remains unfulfilled even today, albeit considerable improvements achieved in the water resource development.

The Second National Water Policy was announced in 2002 during the prime Ministership of Atal Behari Bajpayee of BJP. The 2002 NWP retained the basis of the earlier NWP with some additional elaborations (Anand 2007):

(*a*) The need to manage carefully the resettlement and rehabilitation of persons affected by the projects;

(*b*) Considerations of financial and physical sustainability replacing the water rate concept (which was perhaps considered inappropriate for the large poor population);

(*c*) Welcoming private sector participation;

(*d*) Equitable water sharing and distribution among the States and the need to amend the Inter-State Water disputes Act 1956 (seeking better implementation imperatives),

(*e*) The need for a paradigm shift in the management ethos and to improving the performance of existing water resources infrastructure.

However, because of change of Government back to the Congress party, NWP 2002 was not put into effect. I any case, the policy statements are not legally binding but symbolic of good intention. However, as broad guidelines, they help to focus the approach, planning and budgeting prior to implementation.

21.4 Water Institutional Framework

The Institutional frameworks, both formal and informal, for the development and management of water resources have grown considerably in India today. But there are structural weaknesses and lack of clarity in defining roles, responsibilities, and authority and finally the preponderance of bureaucracy, which all combine seriously hampering the operational efficiency. But, hopefully the wind of change is already blowing.

Now that the State has the primary responsibility for water, we start institutional survey with the State organ.

State Level Organization

Albeit water being primarily a State responsibility, the State level institutions governing multi-sectoral water use and allocation is not strongly developed in all States. Because the irrigation is the largest user of water, State Irrigation Departments have till recently been responsible for construction, maintenance and management of water resource development schemes. Thus irrigation enjoyed the priority of water use in violation of the national Water policy.

Recognising the need to rectify such untenable arrangement, States like Tamilnadu and Orissa, under the inspiration of the World Bank, were the first States to create independent management for water, e g State Water Council in Tamilnadu and State Water Resource Board in Orissa. Other States have since taken to the model as well.

Local Level

Under the State level Water Board or Council, there is need for the local specialist institutions to actually run a particular water project or river valley. Some of the authority of the State Board should devolve on these specialist institutions.

Because the local Governments, like the Zilla Parishad or a Panchayet, developed after the 73rd and 74th Constitutional amendments (1992), that build, operate and maintain the local water supply should be further trained and strengthened technically, professionally and managerially for upgradation of all water services.

Grassroot Level

It might be necessary to identify specialist organization below the local level or Zilla Parishad but possibly above or at Panchayet level for the grassroot management of water issues and problems. For many water management issues have their origin and solution at the very location level, where the grassroot team can identify and solve the issue mere consensually and effectively. The availability of water is becoming critical issue for rural water supply system throughout India, especially in the Deccan plateau and arid western regions. Further, because of overuse by the irrigation, local ground water levels are depleting; contamination of drinking water is also a problem in cities with inadequate effluent treatment or large villages, which typically do not have sewage disposal network. Very often, these issues are entirely resolvable if the local community got together and took pragmatic ameliorative actions.

Inter-State Institutions

All major Indian rivers and their tributaries flow over several States. Inter-State Institutions, under the aegis of the Union Government, are a necessity for allocation, planning and scientific management of water of such basins for the optimum and equitable development.

Among the successful inter-state institutions, mention may be made of the Damodar valley Corporation, Bakra-Beas management Board, Upper Yamuna River Board and flood control institutions like Ganga Flood Control Commission and Brahmaputra Flood Control Board, etc. details of these institutions may be seen in Box B 21.2

Union-Level Institutions

At the Central level, the Ministry of Water Resources (MOWR) is recognized as the apex nodal Ministry responsible for water. Under the MOR, there are a number of technical agencies, *e.g.*,

- Central Water Commission (CWC), concerned with surface water;
- Central Ground Water Board (CGWB), concerned with groundwater matters;
- National Water Development Agency (NWDA) , primarily assessing inter-basin water transfer options.

Other agencies and Ministries are also involved with water, *e.g.,* Ministry of Environment and Forest (MOEF), the Central Pollution Control Board (CPCB), Rajiv Gandhi National Drinking Water Mission in the Ministry of Rural Affairs dealing with rural water supply and sanitation.

Civil Society and the Public-Private-Partnership

No large water utility has yet been fully privatized. The NWP 2002 did conceive of the Public-Private-partnership (P-P-P) model of developing water projects. But with the development of the Capital Market on one hand and the enormous need to mobilize capital investment, the Governments today are seeking and experimenting with idea of private participation in water infrastructure in some form. There is also realization that the grassroot waterworks or even larger water projects will work more effectively with the community involvement or by partnership with private owner. For such water utilities, following models are being tried also (Anand 2007).

(*i*) **Performance contracts:** These contracts are for medium terms of Service Contract: where the ownership and basic operations remain with the State sector, various peripheral services are being contracted out to private parties on competitive tender basis, commonly on yearly contract.

(*ii*) **Management Control:** In Navi Mumbai, four private sector contracts have been awarded time of to manage the water supplies in 15 areas with specified about 5 years.

(*iii*) **Leases and Concessions:** These are longer-term contracts. A Water treatment plant in Delhi has been given to a multi-national company to run. In Tripur, a private sector has been engaged to build, operate, and maintain a water supply project of Tripur Water Development Authority through a special purpose vehicle (SPV) supported by the U S Agency for the International development.

Fig. F21.2 shows a very revealing chart of various types of water projects and their potential implementation agencies.

Box B21.2 River Basin Management; Indian Experience

Damodar Valley Corporation (DVC):

This is the first endeavour at the national level of establishing a Basic Planning and Development, modeled after the Tennessee Valley Authority (TVA, created by an Act of the Indian Parliament, 1948. The original mandate of the DVC included the construction and operation of irrigation facilities, water supply, drainage, hydroelectric and thermal power generation, flood control, navigation, aforestation, control of soil erosion, public health, agricultural, industrial, economic and general well-being in the Damodar Valley. The Damodar Valley Project, as envisaged by the planners, was a multi-purpose project comprising eight dams, and one diversion barrage to provide flow control potential, generation of 200MW hydroelectric power and provision of irrigation facilities for a command area of 308,000 hectares. The Chairman and two members were to be appointed by the Central Government.

Ganga Flood Control Board (GFCB) &
Ganga Flood Control Commission (GFCC):

GFCB was set up in 1972 by a GOI Resolution; the GFCC was set up per clause 5 of the same Resolution to assist GFCB and to undertake any specific work in the Ganga basin.

Brahmaputra Board:

The Brahmaputra Board was set up in 1980 to prepare a Master Plan for the flood control in the Brahmaputra Valley, taking into consideration the overall development and the utilization of water resources of the valley for irrigation, hydropower, navigation, and other beneficial purposes. The Chairman of the Board was to be appointed by the Government of India.

Bhakra-Beas Management Board (BBMB):

The Bhakra-Beas Management Board was constituted by an Executive Order per Sec.79 of the Punjab Reorganization Act, 1966 to regulate supply of the Sutlej, Ravi, and Beas rivers to the States of Punjab, Haryana, Rajasthan and the National Capital Region of Delhi. The Chairman was again to be appointed by the Government of India with the members to be drawn from the coObasin States. The BBMB was to be also responsible for distribution of hydropower to the States.

Upper Yammuna River Board (UYRB):

UYRB was constituted for allocating available flows among co-basin States within the framework of MOU signed by the Chief Ministers of the co-basin States.

Source: Burchi & Singh 1997 (India – WRM Review, WB 1998)

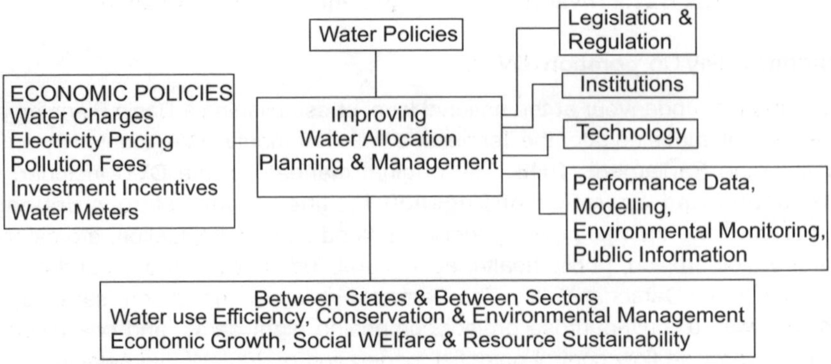

Fig. F21.2 Water Resource Management Enabling Factors

21.5 Reform, Strategy and Action

Reform Initiatives

The first task is to improve and modernize the water infrastructure and Institutions. The Fig. F21.2 shows the various enabling factors, which should be reformed or upgraded on the following lines:

(*a*) Policy: Improving Policy framework, for example, National Water Policy, 1987 or 2002 version should also clearly name implementing agencies for various policies enunciated therein;

(*b*) Legislation: Strengthening the legislative and regulative framework providing for explicit empowerment of the Central Government explained in later parts;

(*c*) Institutions: Institutions are very vital for water development, particularly the public-private-participatory models for multiplying manifold investment, execution and operation management;

(*c*) Economic Return: Time is soon coming to think about some economic pricing or return to provide, if not for investment, for efficient maintenance and preventive wastage;

(*d*) Technology: Upgrading technology to enhance water productivity, particularly for agriculture and domestic distribution;

(*e*) Information & Data: Collection of data and gathering of information require to be strengthened to develop project models, to improve performance and to draw up long term plan.

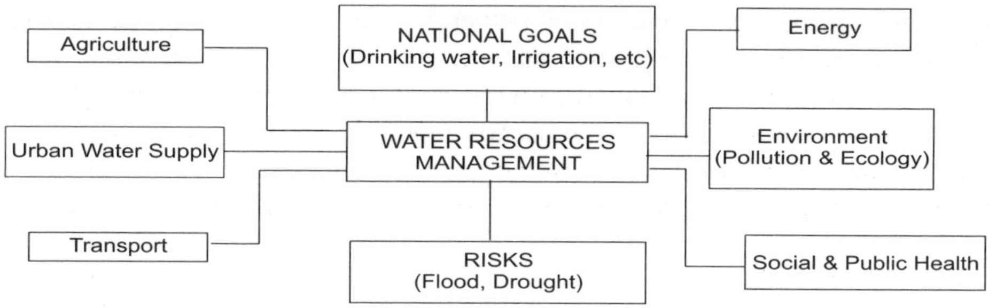

Fig. F21.3: Water Use Management : Sectoral Network

Strengthening the Laws

The present Constitutional provisions and the laws enacted thereunder look upon water resource primarily as State responsibility and restrict the development of water as an all India resource and also hinders the resolution of interstate disputes. The following amendments of Acts are proposed on the basis of World Bank recommendations (WBR 1990):

1. River Board Act,1956: The widespread view is that there is great need for improved collaboration between States for inter-State Rivers. The present Act requires prior consent of the concerned States. A simple amendment would be to bestow power to the Central Government to establish RBO suo moto. This change would be permissible as the Act comes under the Union List (Entry 56) of the Constitution.
2. Inter-State Water Disputes Act, 1956: The present Act also comes under the Union List. The implementation of the Tribunal Awards under the present Act is proving unsatisfactory and again the formation of the very same Tribunal requires, in the first instance, the consents of the co-basin States, failing which the Central Government can constitute a tribunal. The following amendments are proposed:

— Empowering the Central Government to institute a Tribunal suo moto, if considered in the national interest;

— Introducing mandatory time limit of implementing the Awards of a Tribunal with some safeguards to the States;

— Providing for legally enforceable periodical review of States implementation.

Public vs. Private Water Rights

With the goal of more equitable and effective use of the precious water resources, there is a strong logic of regulating the extravagant ad wasteful withdrawal of water by individuals and agencies. Water in future should be treated as public goods and the concept of usufractuary or absolute rights to available water needs to be revisited and limited. Local Governments, Panchayets or the community institutions should be empowered to set limits to water use through quantified permits with some legal safeguards to users. For surface water, automatic riparian rights of unlimited withdrawal of water should be derecognized. Similarly, for ground water, the presently practiced absolute discretion of the user to pump out water should be subject to user's quantity to be issued by the local empowered Institution with some judicial protection to the landowners.

Strategy for Action

To meet the Water Challenges of India: There are two broad strategies; one is to improve and augment the water resource capacities of necessary quality. In Fig. F21.2, we have shown in seven boxes the important enabling factors of supply management. They should constitute the strategic arms of our management of water resources.

In Fig. F21.3, we present the various users and consumption sectors, where the water goes. Agriculture, Energy (Hydro-energy & Thermal energy), Domestic and Drinking water are the three largest sectors of water consumers. Monitoring the use of water in three sectors, regulating the wastage, improving the technologies in irrigation, energy and domestic water distribution and supplies etc should constitute the main planks of demand management strategies.

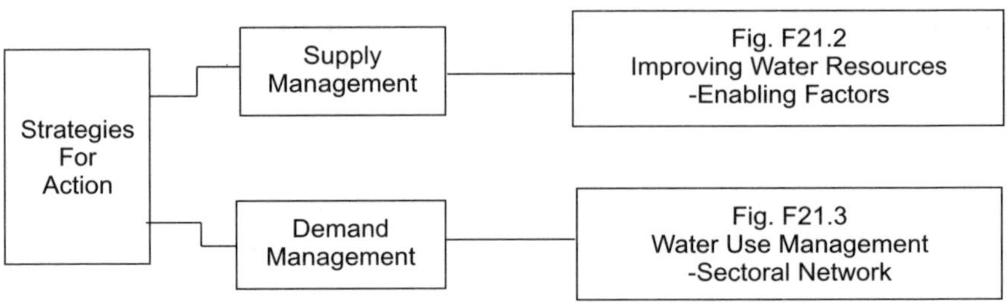

Fig. F 21.4: Overall Water Resource Strategies

The overall strategies for action do therefore cover both the Supply and Demand management as illustrated in Fig. F21.4 below.

The broad strategies outlined through the Three Figs. F21.2, F21.3 and F21.4 and Box B21.3 call for local actions, which will naturally vary from region to region over the Indian sub-continent. India is a big country, which may be subdivided into five regions with distinctive water perspectives (Chitale 2000):

1. The snow-fed Himalayan area;
2. Brahmaputra-Ganga Eastern Rivers with high South-West Monsoon rainfall;
3. Semi-arid and arid lands of Northwest India;

4. Undulating Central and Southern Peninsula;

5. South Area of two rainfall seasons.

Because of the basic differences in the climatic and geomorphologic characteristics, it will be necessary to evolve separate appropriate strategies for the management of water resources in these five regions.

There is another important strategy of inter-basin transfers of surplus water of the Ganga-Brahmaputra basins to the southern low-flow basins of Krishna and Cauvery. This proposal has been mulled over a long time but actions on them had not been initiated for the innumerable and huge financial and technical problems. This subject will be dealt with in a later Chapter.

Box B21.3 Recommended Matrix of Implementation Agencies (Anand 2007)

	Formulating Water Policies, Laws etc	Water Resource Management, Inter-state River, Aquifers, Large-scale Water Projects	Water Resource Management, Inter- state Rivers, Aquifers, Small-scale Water Projects	Drinking Water Supply Infrastructure Investments (Treatment Plants Mains)	1 Operation and Maintenance of Drinking water Supply	1 Finances, Accounts Billing
Federal Government	*	*	*	*	*	*
State Government	*	*	*	*	*	*
Local Government				*	*	*
Private Sector-Large				*	*	*
Firms						
Private Sector-Small				*	*	*
Individual Citizens				*	*	

❑❑❑

(Continued) [illegible]

5. Southeastern [illegible]

Because of the basic differences in the [illegible] and [illegible] situations, it will be necessary to develop separate appropriate [illegible] for the [illegible] magnitude of [illegible] resources in [illegible] section.

There are [illegible] of [illegible] Group [illegible] to the south [illegible] and [illegible] had not been [illegible] obvious [illegible]

22

CHAPTER

Indian Water Resources II* Rainfall, Rivers, And Inter-Basin Transfer Of Water

22.1 Indian Rainfall

Monsoon Winds

The rains in India are mostly caused by South-West Monsoon winds. The Monsoon is believed to be an Arabic word meaning winds changing direction. Around end May and early June, hen the water gets very warm over India, the South-Eastern Trade winds blowing over the Indian oceans gets deflected eastwards, while crossing over the equator due to the rotation of the earth. This trade wind now becomes South-West monsoon and blows over India carrying high moistures, as indicated in Fig. F22.1. A part of the South-West monsoon coming from the Arabian sea sweeps over the west of South India bringing rains. The other part of the same southwest monsoon coming from the Bay of Bengal sweeps over the Eastern India causing rain. Gradually, the Monsoon covers the whole of India by about the second half of June. The S-W monsoon prevails over India June to September.

Indian Geography and Pattern of Rainfall

The long Himalayan range extends 2500 km from the east to the west like a garland over the whole of India (see Fig.F22.2). The range is believed to be geologically young, having being formed due to tectonic drift of the continent northward. In the South Peninsula, there are number of very old hill ranges –

- Western Ghat extending 1600 km along the west coast, ranging in height upto 2500 m and the hills are as old as 2500 million years;
- Eastern Ghat extending along the east coast from Orissa in the east to Nilgiri in the south, aged about 1300 million years, height ranging from 500 m to 2500 m;

- The Aravalli Range in Rajasthan in the north-western India, about 800 km long, height ranging to 1000 m;
- In the central India, the Deccan plateau consists of Vindya Hills (1400 million years old) and the Satpura range (1000 million years old);
- In the northeast India, there are several minor ranges including Garo Hills, Jaintea Hills and the Khasi Hills.

The alignment of the above hill ranges are shown in Fig. F.22.2 and also shown the amount of rainfall in millimeters per annum in both the Lee side and the downside of the hills. It will be seen how profoundly these hills interfere with the flow of the Monsoon winds, causing high rainfall in the lee side and scanty rainfall on the downside.

The highest rainfall of 11,400 mm occurs at Cherapunjee at an elevation of 1330m in the northeast India. At the other places at Khasi and Jaintea Hills of the Northeast, the rainfall averages 10,000 mm.

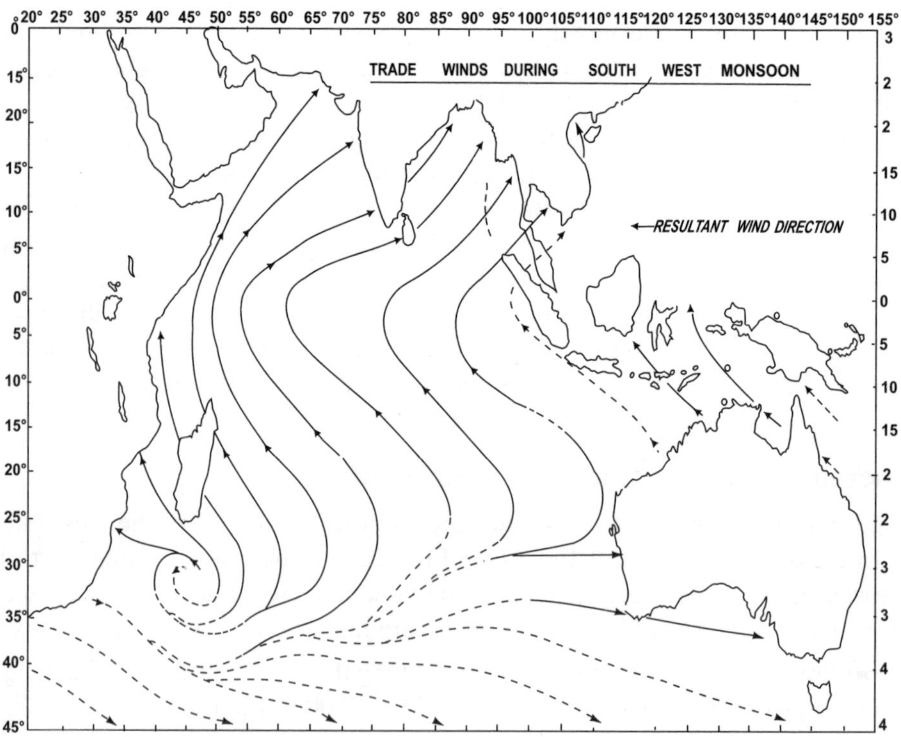

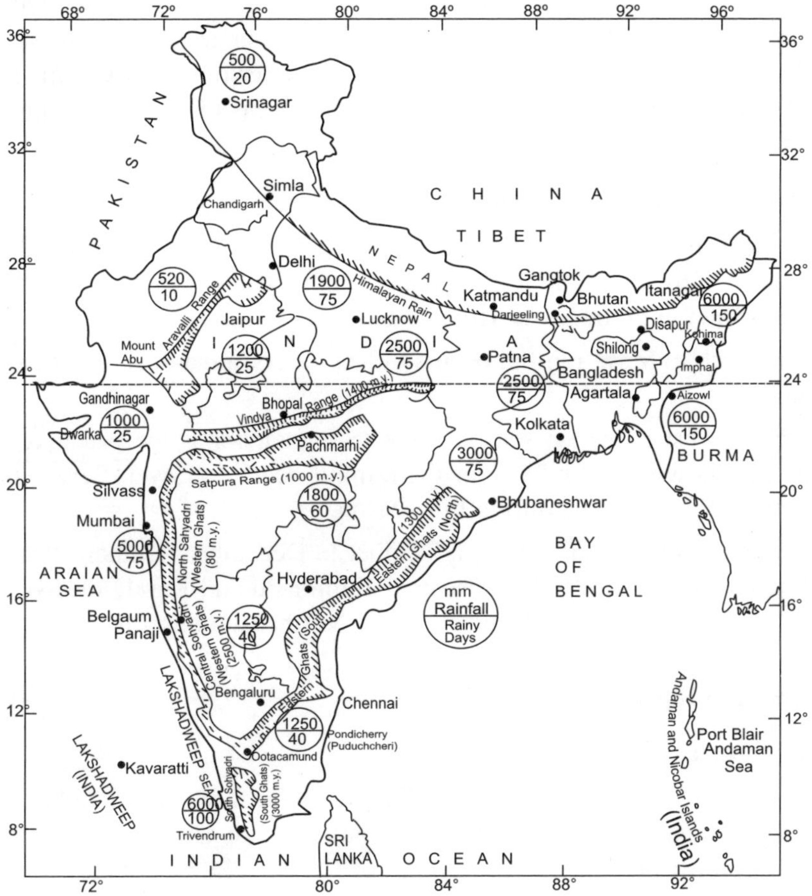

Fig. F22.2: Mountain Ranges, Rainfall and Rainy Days
(Source: Rao)

Along the west coast of India on the windward side there occurs quite high rainfall, at places around 6000 mm but on the leeside, the monsoon cloud is obstructed and the rains become scanty, as low as 500 to 600 mm. Such conspicuous differences of rain, of course, do not occur on the plains of India.

Along the Himalaya foothills, the rainfall varies from 1000 mm to 2000 mm around the eastern India.

In the eastern India, the S-W monsoon hits the coast hardest and wanes as the monsoon moves inland. In West Bengal and Orissa, the rain averages 1500 mm annually but in Rajasthan, deep in the West, the rain gradually subsides to 500 mm or even lower in the some region.

Down South along the Eastern Ghat, the rainfall averages 1500 mm on the seaside but on the lee side it reduces to below 1000 mm.

In Central India above the Vindya Hill range, the rainfall hovers around 900 mm. The various rainfall figures over India are indicated encircled in Fig. F22.2

The rain does not fall continuously even during the monsoon, but only intermittently. In Fig.F22.2 are also shown the number of rainy days, again encircled (lower part) for identification. The number of actual wet days vary from 150 days in the North-East, 75 days in the plains of West Bengal to even 10 days in the North-Western deserts.

Seasonal Variation of Rains

During the Winter (January-February), there rains in the Kashmir and the neighbourhood to about 200 mm which helps the Winter crops in the northern India and snowfall in the lower Himalayas.

From March to May, rainfall occurs in Kerala, Assam and also in Jammu & Kashmir.

The Southwest monsoon season, June to September is the principal rainy season in India when over 75 % of the rainfall occurs all over. Severe floods also occur in Eastern India during this season.

During the post-monsoon period, October to December, southern Peninsula including Andhra Pradesh, Rayalaseema, Tamilnadu and Kerala receive some rain. In Tamilnadu, this rain is known as the Northeast monsoon.

22.2 Indian Rivers

Major Rivers

There are fourteen major rivers in India. Of these, three rivers Indus (west-flowing to Pakistan), Ganga and Brahmaputra are of Himalayan origin and situated in North India above the Tropic of Cancer (23.5 Latitude). Between the Tropic of Cancer and the 20 Latitude, there are seven rivers including three east-flowing Subarnarekha, Brahmani and Mahanadi and four west-flowing Sabarmati, Mahi, Narmada and Tapi. In peninsular India below 20 latitude, there are again four east-flowing rivers Godavari, Krishna, Pennar and Cauvery. The geogarphical distribution of rivers is shown on the map of India (Fig.F22.3).The catchment areas and the discharges of these rivers are also presented in Table T22.1.

From this Table T22.1, it may noted that 14 major rivers have each catchment areas of generally of 20,000 sq. m. or more, which is the reason of leveling them as major rivers. The other point to note that rivers, which flow through areas of higher rainfall, have usually higher discharge, while the others have lower discharges. This situation is diagrammatically in Fig. F22.4 to a base of the catchment areas.

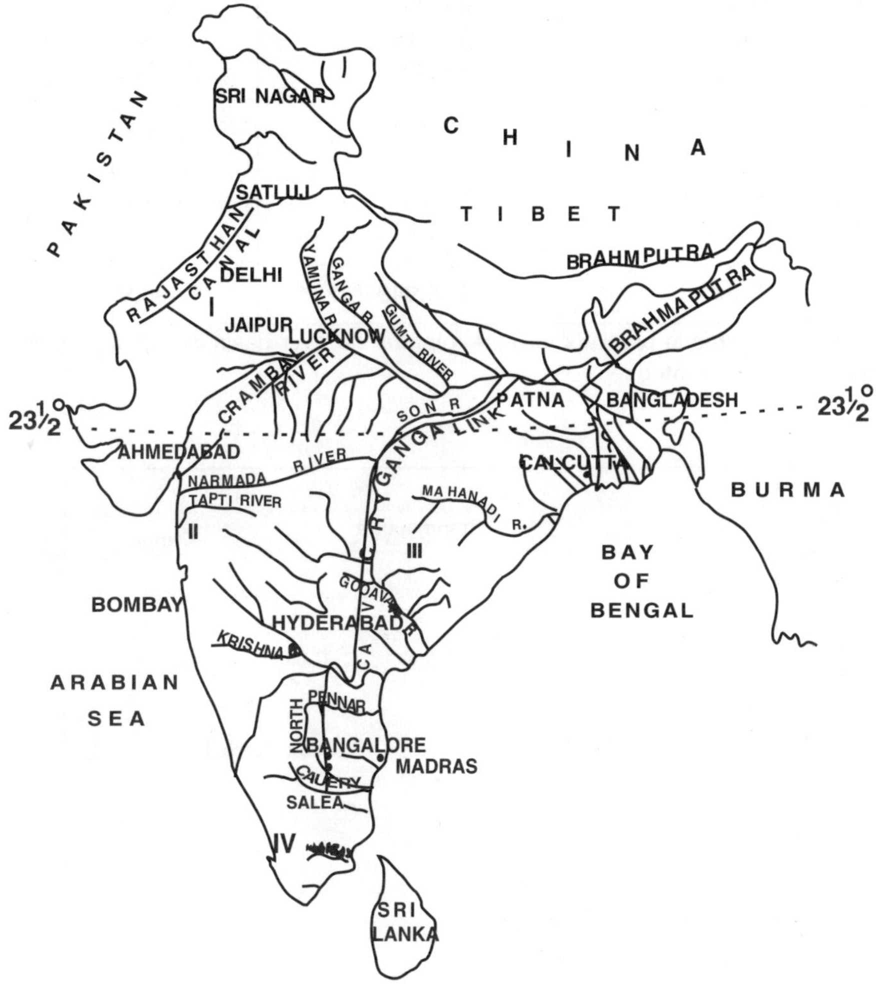

Table 22.1: Discharges of Major River Basins

S. NO.	Name of Rivers	Area (sq.km.)	Annual Discharge Million cu.m.
1	The Indus	(468,068) 321,289	(207,800) 41,955
2	The Ganga	(1,050,000) 861,404	 493,400
3	Brahmaputra	()580,000) 187,113 +70,895	510,450
4	Sabarmati	21,674	3,200

Contd...

5	Mahi	34,842	8,500
6	Narmada	98,796	40,705
7	Tapi	65,145	17,982
8	Subarnarekha	19,296	7,940
9	Brahmani	39,033	18,310
10	Mahanadi	141,589	66,640
11	Godavari	312,812	105,000
12	Krishna	258,948	67,675
13	Pennar	55,213	3,238
14	Cauvery	87,900	20,950
	Total	2,575,949	1,405,945

Note: The figures in brackets indicate total for the river basin including parts in neighbouring countries.

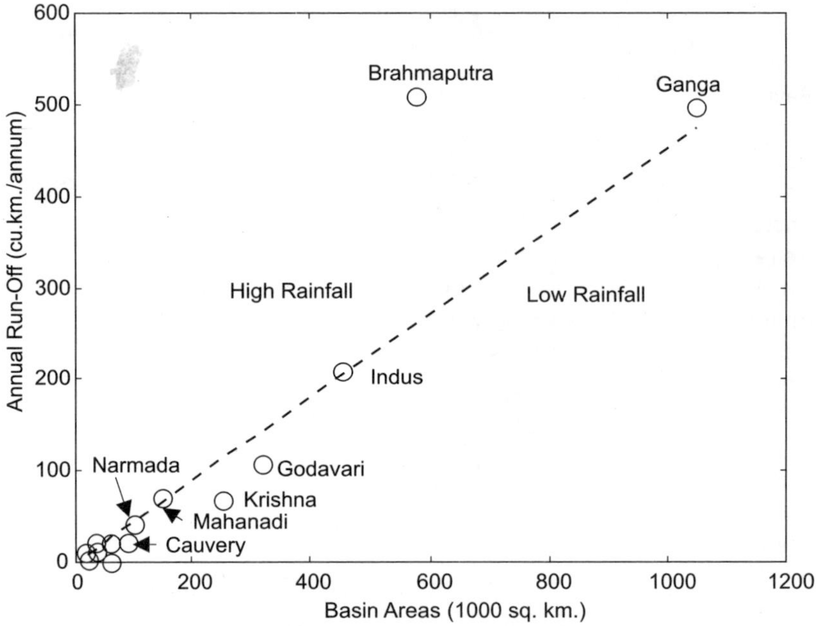

Fig. F22.4: Catchment Areas & Flow of Major Rivers

Table T22.2: Discharges of Medium Rivers

S No	Name of River State	Source Indian	Length	Area sq km charge	Annual Dis (Million cum)
		West Flowing Rivers			
1	Shetrunji	Village Dalkania	182	5,5514	280
2	Bhadra	Rajkot Dist.	198	7,094	350
3	Dhadhar	Panchmaqhal Dist.	135	2,770	690
4	Burhabalang	Mayurbhanj Dist.	164	4,837	2,177
5	Baitarni	Keonjhar Dist.	365	12,789	5,755
6	Purna	Dhulia Dist.	142	2,431	1,093
7	Ambika	Dhangs District	142	2,715	1,247
8	Vaitarna	Nasik Dist.	126	2,572	4,350
9	Ulhas	Pune Dist.	122	4,637	3,014
10	Savitri	Kolaba Dist.	80	2,257	1,467
11	Mandavi.	Belgaum Dist.	87	2,0332	1,320
12	Kalindi	Belgaum Dist.	153	5,179	6,537
13	Gangavati/Bedti	Dharwar	152	3,902	4,925
14	Sharavati	Shimoga	122	2,209	4,545
15	Netravati	Canara	103	3,657	4,615
16	Ghalia/Beypore	Elamtalvi Hills	169	2,788	5,200
17	Bharatpuzha	Anamali Hills	251	5,397	8,800
18	Periyar	Sivsjini Hills	228	5,243	12,300
19	Pamba	Sivsjini Hills	177	1,961	6,300
		East Flowing Rivers			
20	Rushikulya	Phulabani Dist.	146	7,753	1,800
21	Yamasadhara	Phulbani	221	10,830	3,500
22	Nagavali	Kalahandi	217	9,410	2,400
23	Sarda	Vishakhapat. Dist.	104	2,725	700
24	Yeleru	Vizag Dist.	125	3,809	9,000
25	Gundlakamma	Kurnool Dist.	220	8,494	1,000
26	Musi	Nellore Dist.	112	2,219	260
27	Paleru	Nellore Dist.	104	2,480	300
28	Munru	Nellore Dist.	122	3,734	450
29	Kunleru	Velikonda Dist.	73	3,534	420
30	Swarnamukhi	Palka	130	3,225	500
31	Kortalyaiar	Chingelpet	131	3,521	350
32	Palar	Kolar Dist.	348	17,871	1,780
33	Gingee	South Arcot	94	3,044	300
34	Ponnaiyar	Kolar Dist.	396	16,019	1,600
35	Vellar	Chithri Hills	193	8,558	850
36	Vaigai	Madurai District	258	7,741	770

Contd...

37 Varshalli	Madurai Dist.	125	2,104	310
38 Cundar	Madurai Dist.	146	4,838	480
39 Vaippar	Tirunelveli Dist.	130	5,288	530
40 Tamarpani	Tirunelveli Dist.	130	5,482	1.370
Rivers flowing into other countries				
41 Karnaphuli	Mizoram	144	3,999	2,600
42 Kaldan	Mizoram	290	7,933	5,160
43 Imphal	Manipur	-	7,255	4,716
44 Tixu Nanitaluk	Nagaland	148	6,449	4,200
	Total		**239,302**	**112,211**

Medium & Minor Rivers

Rivers, which have individual catchment area generally below 20,000 sq.m., but above 2,000 sq.m., are identified as medium rivers. In Table 22.2 are listed 44 such medium rivers flowing out to the east coast or the west coast, excepting the rivers in the North-East flowing into the neighbouring countries. The medium rivers generally flow within a coastal State; hence they are grouped by States. Their total discharges constitute less than 10 % of the total countrywide run-off.

A third category of rivers, identified as minor rivers have very limited catchment areas of 2000 sq m or below each flows relatively short distances from the coastal hills (Eastern Ghat or Western Ghat as the case may be) to the sea. Table T22.3 shows the estimated Statewise total flows emanating from the minor rivers in each State.

There are some minor desert rivers (Rao, 1975), which are also listed along with the minor rivers in the Table T22.3.

In Table T22.4 are presented the totals of all discharges and catchment areas of all Indian rivers.

Indian River Basins

Against each catchment area of respective Indian Rivers are shown the cultivable area, sown area, and the irrigated area in Table T4.19. The total river basins constitute 83 % of the total area of the country. Secondly, the total cultivable land at 193.5 million hectares constitutes 62 % of the total basin area of 3.12 million sq. km.. In comparison with the other international data, the Indian percentages of utilization are high.

Shortage of Water in River Basins

Conservation of water by storages is important as the rains, in the average, last about 4 months in a year, when the rivers achieve their high discharges. Thus, storage of water in large reservoirs is very important for agriculture. Fig. F22.5 shows the storage capacities of the important rivers in Bar chart forms.

Table T22.3: Discharges of Minor Rivers. [Abstract]

S. Nos.	Indian States	Catchment AreaSq. km.	Run-off Million cum
1	Gujrat	4,634	1,160
2	Karnataka, Goa, Maharashtra	2,146	966
3	Kerala	19,489	40,073
4	West Bengal, Orissa	10,000	6,500
5	Andhra Pradesh	44,142	9,746
6	Tamil Nadu	16,162	1,625
	Total	96,537	60,088
	Other Minor Rivers	109,000	57,000
	Total	205,537	117,088
Desert rivers:			
7.	Luni & Other rivers	80,000	
8.	The Macha	2,515	
9.	The Rupen	4,000	
10.	The Saraswati	2,934	
11.	The Banas	8,674	
12.	The Ghaggar	1309	
13.	Others	1,309	
Total (Desert Rivers)		99,432	
Grand Total of Run-off for Minor and Desert rivers			127,000 million cum

Table T22.4: Details of Catchment Areas of 4 Groups of Rivers

River Basins	Catchment Million sq km	Total area %	Run-off Cu km	Run-off %	Population %
Group I: Major River Basins	North 1.44 Central 0.42 South 0.72 Total 2.58	83%	1,406	85%	80%
Group II: Medium	0.24	8%	112	7%	
Group III: Minor	0.20				20%
Group IV:	0.10	9%	127	8%	
Desert Rivers					
Total	**3.12**	**100%**	1,645	**100%**	**100%**

Table T22.5: Uses of Water Supplies for Different Water Basins

S. No	Basins	Govt. & Private	Tanks	Other sources	Total sources water	Tubewells	Other Wells	Total Ground water	Grand Totals
1.	Indus	3,462	1	464	3,927	1,545	862	2,407	6,334
2.	Ganga	7,618	1,205	1,767	10,590	4,244	4,662	8,906	19,496
3.	Brahmaputra - Barak	328	3	305	636	-	130	130	766
4.	Sabarmati	89	15	3	107	11	200	211	318
5.	Mahi	98	31	4	133	12	125	137	270
6.	Narmada	11	53	9	73	11	200	211	284
7.	Tapi	172	4	5	181	4	190	194	375
8.	Subarnarekha	92	10	24	126	-	11	11	137
9.	Brahmani	125	79	74	278	-	30	30	308
10.	Mahanadi	1,117	216	167	1,500	-	60	60	1560
11.	Godavari	1,137	501	64	1,702	28	609	637	2339
12.	Krishna	2143	393	112	2,648	28	775	803	3451
13.	Pennar	97	120	17	234	12	230	242	476
14.	Cauvery	1,110	270	120	1,500	5	330	335	1835
	Total	17,599	2,901	3,135	23,635	5,900	8414	14,314	37,949

Medium and Minor Rivers of India

1.	Rivers of N.E. Regions	-	-	36	36	-	-	-	36
2.	Luni and others in Saurashtra / Kutch	177	77	5	259	81	1,029	1,110	1,369
3.	Between Tapi and Mahi	10	10	3	23	6	6	12	35
4.	Between Tapi/ Kanya Kumari	543	167	204	914	2	120	122	1,036
5.	Between Ganga/ Vaitarni	47	23	20	90	33	3	36	126
6.	Vaitarni	60	40	35	135	-	13	13	148
7.	Betw. Mahanadi/ Godavari	247	180	53	480	12	28	40	520
8.	Betw. Godavari/ Krishna	-	30	5	35	12	10	22	57

Contd...

9.	Betw. Krishna/ Pennar	26	75	5	106	12	70	82	188
10.	Betw. Pennar/ Cauvery	164	404	37	605	37	535	572	1,177
11.	Betw. Cauvery/ Kanya Kumari	112	243	6	361	10	2631	273	634
	Total for Medium/ Minor Rivers	1,386	1,249	409	3,044	205	2,077	2,282	5,326
	Totals for Major, Medium & Minor Rivers	18,985	4,150	3,544	26,679	6,105	10,491	16,596	43,275

22.3 Interbasin Transfer of Water – National Water Grid

Surface Water Availability – All India Scenarios: (Rao 1975)

From Table T21.2, we see that the per capita availability of surface water is excessively high in the Brahmaputra basin; in many other basins, it is quite sufficient, being well over 1000 cu.m. per capita per annum, and water is scarce or at stress level in many basins, *e.g.,* Cauvery, Pennar, etc. Again we observe from Fig.F22.2 that rainfall is highly uneven on all India bases. The task is therefore to identify the scarcity areas on regional bases and to examine if it is technically feasible to transfer water from the nearest surplus basin rivers to balance out the scarcity over the geographical spread of India. The second task is to store water in sufficient quantities from existing ten percent to twentyfive percent of the annual run-off. These measures will demand heavy investments in fund, organization and time of few decades, but once implemented can relieve India of potential scarcity for much of twentyfirst century.

Late Dr. K L Rao, an eminent water engineer (Rao 1975) had proposed to divide the country into four water zones (see Fig. F22.6A).

Zone I – Himalayan Rivers consisting the Indus, Ganga and Brahmaputra;

Zone II – West-flowing river basins excluding Kerala State;

Zone III – East-flowing rivers upto Cauvery basin;

Zone IV – Balance country (including Kerala).

Now Fig. F22.6A shows the cultivable land area *vis-à-vis* water availability in percentage terms. On apparent reading, it would appear that, except the Himalaya Zone I, all other zones are water deficient. This is really not the exact position, for Zones II and III are water deficient compared to Zone I, which is highly water-surplus. In absolute term, Mahanadi Basin in zone II is quite water surplus as having surface water of 2513 cu m per capita per annum and combined with groundwater 3135 cu m per annum (See Table T21.2), we have to remember that the recommended zoning is for the purpose of evolving an inter-basin water transfer scheme.

Regarding the feasibility of inter-basin transfers, there are several old systems already in operation, *e.g.*, Periyar Diversion, Kurnool-Cuddapah canal, Perambikulam-Aliyar project, Rajasthan canals (connecting Chambal river), Beas-Sutlej Link, Narmada dams and canals project.

In USA, there are several working inter-basin water transfer schemes. In California, 5200 million cu m is transferred to the Southern parts over a distance of 720 km. In the Artysh Karganda scheme in Central Kazakhstan, 75 cumecs water is transferred over a distance of 450 km by stage pumping upto 22 m head.

We will now briefly outline the National Water Grid system, which was conceived by the National Water Power Commission several decades ago but as yet remains unexecuted because of its gigantic scale and lack of political consensus; but also to a lesser extent on issues of availability of fund and doubt of technical feasibility.

Proposed Network of National Water Grid

Considering the locations of the surplus rivers systems in Zone I in the North to the deficient river systems in the South, it could be inferred that the general direction of the main arteries of the Grid should be north to south. *Secondly,* most of the rivers in India either flow to the East or the West; the North-South alignment of the major grid artery will be thus able to connect a number of rivers. With the above broad concepts of the grid, the inter-river linkages can be visualized as follows:

1. Ganga-Cauvery Link connecting the Ganga in the North with the Cauvery in the South and passing through the basins of the Son, Narmada, Tapi, Godavari, Krishna and Pennar.
2. To further boost the flow in Ganga, a link from Brahmaputra to Ganga through Bangladesh.
3. Canal from Narmada to Gujrat and West Rajasthan, inter-connecting rivers Mahi, Sabarmati and Luni and which are already under construction.
4. Canal from the Chambal river to Central Rajasthan, which has been also implemented in some form.
5. A number of network link canals from the surging rivers of the western Ghat rivers to the East.

The above schemes of inter-linking canal network has been superimposed in Fig. F22.6B. Another important aspect of inter-basin transfer will be to build a series of storages and reservoirs along the arterial routes to feed the southern basins during the dry seasons. In any case, such reservoirs may be necessary to install major pumping stations to overcome geodetic elevations.

Ganga-Cauvery Link

Ganga has good perennial flow near Patna but downstream of the two tributaries Ghagra and Gandak. Taking the average annual discharge of Ganga from Table T22.1 at 493,400 million cu m, which amounts to a flow of 15,646 cumecs (cubic metre per second). The National Grid of the CWPC planned to undertake the following inter-connections:

(*i*) To draw 1700 cumecs of water from Ganges downtown of Ghagra and Gandak for 150 days.

(*ii*) This drawal of 1700 cumecs amounted to 10.9 % of the Ganga average flow of 15,646 cumecs. Further, this drawal of 1700 cumecs for 150 days amounted to a diversion of 22,032 million cu m annually, which amounts to 4.47 % of the Ganga discharges of 493,400 million cu m.

(*iii*) Out of this withdrawal of 1700 cumecs from Ganga, 290 cumecs will be pumped to draught-prone areas of South Uttar Pradesh and South Bihar, which are, of course, situated within the Ganga basin itself.

(*iv*) There is a further parallel proposal of diverting 1150 cumecs from Brahmaputra near Dhubri through Bangladesh to upstream off Farakka barrage on the Ganga. That means the Ganga basin will loose very little water by diversion to the southern grid.

(*v*) The diversion route of the grid will be such that water from Ganga will be lifted along the Son river to the proposed Bagri reservoir on Narmada to an elevation of + 423 m. With a total lift of +400 m to overcome the Vindya mountain range, the grid canal will flow along Son, Narmada, Wainganga and Godavari over a distance of 960 km. From here the canal has to traverse another 960 to the Cauvery River. The last leg of the canal can be aligned after detailed survey.

Table T22.6: Storages existing or under construction in various Basins

S. No.	Basins	Storage Million cu m	Percentage of Basin Run-off
1	Indus	14,419	34
2	Ganga	33,476	7
3	Brahmaputra	142	-
4	Sabarmati	1,017	33
5	Mahi	4,286	56
6	Narmada	2,550	6
7	Tapi	8,148	45
8	Subarnarekha	283	4
9	Brahmani, Baitarni, Rishikulya etc	3,953	22
10	Mahanadi	7,926	12

11	Godavari	14,858	15
12	Krishna	29,860	44
13	Pennar	1,978	61
14	Cauvary	5,478	26
15	**West Flowing Rivers**	14,121	18,452
(	Kanyakumari to Tapi, Medium & Minor rivers)	8	
16	**East Flowing Rivers of AP. & Tamilnadu**		
	(medium & Minor Rivers)		
17	**Rivers of Saurashtra & Kutch**		
	(Luni & other minor rivers)	1,810	
	Total	146,826	
	Storage in Tanks and small ponds	15,400	
	Total	**162,226**	

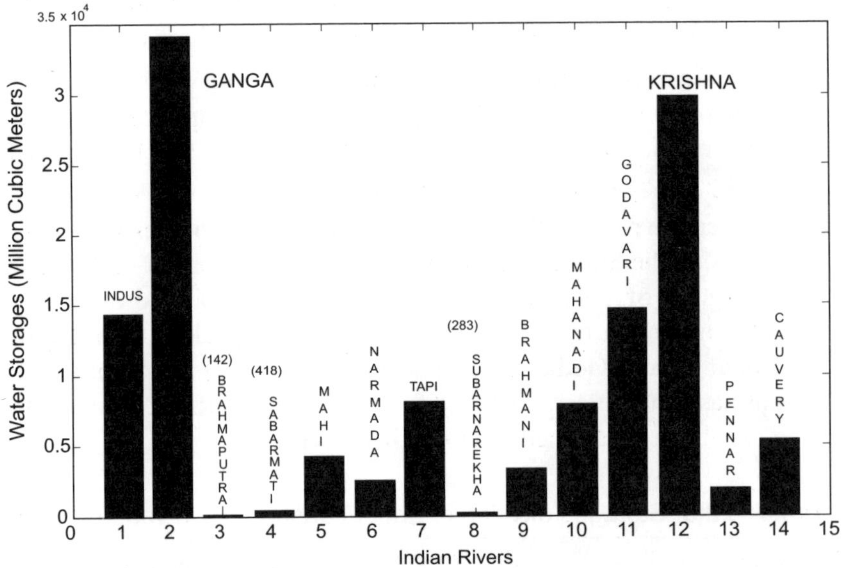

Fig. F22.5: Water Storages In Major Indian River Basins

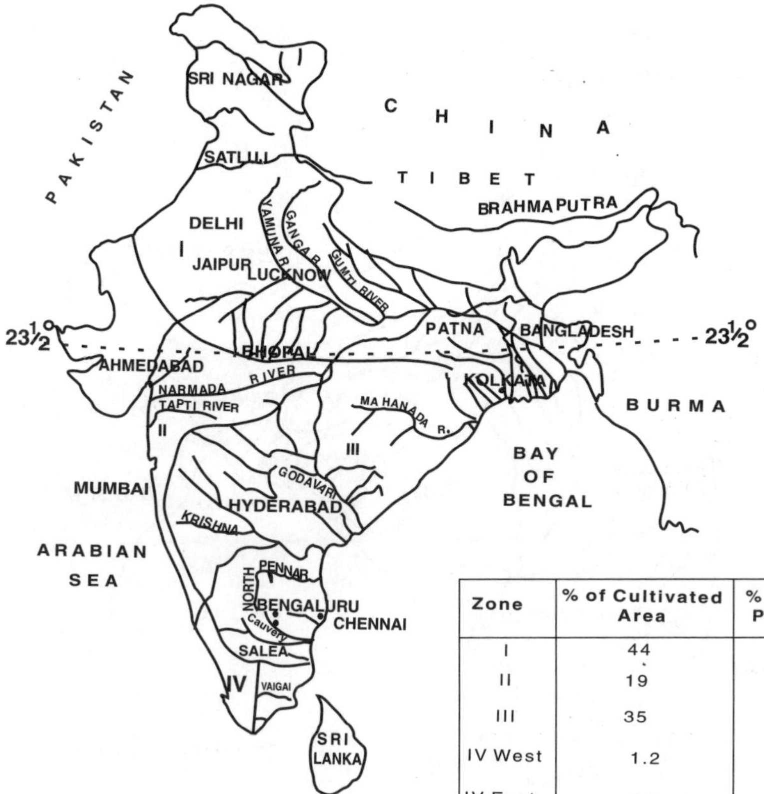

Zone	% of Cultivated Area	% of Water Potential
I	44	64
II	19	9
III	35	19
IV West	1.2	7
IV East	0.8	1

Fig. F22.6A: Different water Zones of India
(Source : Rao)

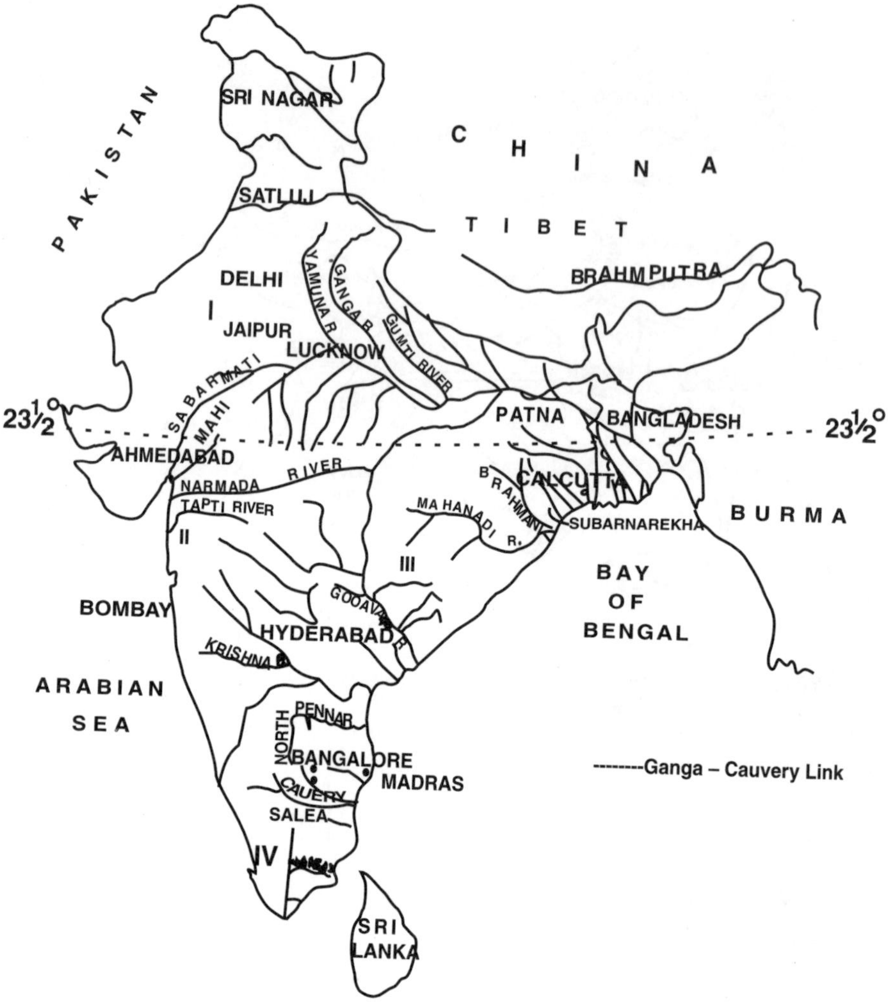
SRI NAGAR
PAKISTAN
CHINA
SATLUJ
TIBET
BRAHMPUTRA
DELHI
I
JAIPUR
YAMUNA R
GANGA R
GUMTI RIVER
LUCKNOW
SABARMATI
MAHI
$23\frac{1}{2}^{\circ}$
PATNA
BANGLADESH
$23\frac{1}{2}^{\circ}$
AHMEDABAD
NARMADA RIVER
BRAHMANI R.
CALCUTTA
TAPTI RIVER
MAHANADI R.
SUBARNAREKHA
BURMA
II
III
BAY
OF
BENGAL
BOMBAY
GODAVARI R
HYDERABAD
KRISHNA R
ARABIAN
SEA
PENNAR
NORTH
BANGALORE
MADRAS
CAUVERY
SALEA
IV
SRI
LANKA
--------Ganga – Cauvery Link

❑❑❑

23
CHAPTER

Urban Water Supply A Novel Dual Water System

The discourse in this Chapter is based on an innovative Paper by Professor Arun K Deb Roy (Deb Roy 1987). The vital importance of good quality water for human consumption, agriculture and industry is well recognized universally.

Dual Water System

In the dual system, two grades of water – one potable and the other semi-potable or non-potable – are supplied to the consumers through separate distribution system matching with quality and quantity requirements. Since only a small faction of water consumed in household is required to be potable, considerable saving in processing cost can be saved by dual water system. The semi-potable or non-potable water can be supplied to the household after ordinary processing. Even the sub-potable water can be supplied disinfected and made safe for occasional drinking but may not meet the primary water regulation for trace heavy metals and organic matter.

The non-potable water in a dual supply will be recycled water from secondary effluents after filtration and disinfection. This water will be used for irrigation of public parks and golf courses, air conditioning, industrial cooling, ground water recharging and other low uses. The quality of non-potable water shall be free of bacteria and viruses at the detection level and relatively clear.

Some typical average urban water usages for the United States are given in Table 23.1.

Table 23.1: Urban Water Use in USA

Use	Percentage
Residential	40
Commercial	15
Industrial	25
Unaccounted	20

Note: Quantitatively, US residence uses water at 230 litres per day, 40% for toilet flushing, 30% for bathing, 15 % for laundering, 6 % for dish washing, 5% for drinking and 4% for miscellaneous uses.

Watman Water Management Model

To achieve the above objectives of high economy and high quality, a water management model is proposed with more than one source of water supplies and with more than one quality of water supply. This model concept will produce alternative plans for supplying water to meet the city and the regional demands.

A general conceptual of the Watman Urban water system model is shown in Fig. F23.1. This model provides the user with a tool to analyze various water supply system alternatives, including multiple distribution systems. Water reuse is essential in developing future long-term water management plans for cities, industries or regions.

The model is capable of analyzing conventional and multiple water distribution systems for three different qualities of water, such as, potable, semi-potable and non-potable.

Box B23.1A: List of Small Water Processing Units
Small System Collection

S2-1 Aeration	S2-16 POUT (GAC)
S2-2 Rapid Mix and Coagulation	S2-17 POUT (Cation)
S2-3 Clarification	S2-18 POUT (Anion)
S2-4 Chlorination	S2-19 POUT (R.O.)
S2-5 Gravity Filtration	S2-20 POUT (Activated Alum)
S2-6 Pressure Filtration	Small System Collection
S2-7 Ultra-Filtration	S1-1 Improved Reservoir
S2-8 Reverse Osmosis	S1-2 Raw Water Pumping
S2-9 Package Complete Filter Plant	**Small System Residual**
S2-10 Ozone Generation & Contact	**Waste Handling & Disposal**
S2-11 GAC Column	S3-1 STP Sludge Handling
S2-12 PAC Feed System	Small System Distribution
S2-13 Cation Exchange	S4-1 High Service Pumping
S2-14 Anion Exchange	S4-2 Water storage Tanks
S2-15 Activated Alum	S4-3 Bottled Water

Box B23.1B: List of Large Water Processing Units

Large Water System Treatment Units

T1 Aerator
T2 Flash Mix & Coagulator
T3 Sedimentation
T4 Filtration
T5 Chlorination Equipment
T6 Chlorination Contact Tank
T7 Ozonator
T8 Reverse Osmosis
T9 Granulator Carbon Adsorption (without regeneration)
T10 Granulator Carbon Adsorption (with regeneration)
T11 Micro screening
T12 Nitrification (with clarifier)
T13 Denitrification (with clarifier)
T14 Breakpoint Chlorination
T15 Dechlorination
T16 Phosphorous Removal by Alum
T17 Phosphorous Removal by FeCl3
T18 Two State Lime Treatment

Large Water System Collection & Transmission

C1 Impounded Resrvoir
C2 Wells
C3 Vertical Turbine Pumps
C4 Submersible Turbine Pumps
C5 Low-Lift Pump Station
C6 Transmission Pump Line & Pumping Station
C7 Inspection Tower

Large Water System Distribution

D1 High Lift Pumping Station & Pumping Main
D2 Water Storage Tank (On Ground)
D3 water Storage Tank (Elevated)
D4 Distribution Main
D5 Interior Plumbing
D6 Well / Land recharge

Large Water System Waste Handling

W1 Alum Sludge
W2 Two Stage Lime Sludge
W3 Nitrification sludge
W4 Denitrification Sludge
W5 Alum Phosphate Sludge
W6 Iron Phosphate Sludge

There are 63 unit processes included in the Watman model. The unit processes are listed in the Box B23.1A & 23.1B. These processes represent a comprehensive list of water supply collection, transmission, treatment, distribution and residential disposal practice available to Utilities. By properly selecting the desired combinations of unit processes, alternative water systems can be generated for evaluations.

Some Typical Dual Water System Solutions

Some diagrams are presented proposing several typical dual water systems.

Fig. F23.2A is a common example of water resuse for conservation of good quality sources.

Fig. F23.2B is an example of an industrial city, where a large portion of municipal water demand is for industrial needs; a dual supply can be effectively used with one non-potable supply for industries, and a potable supply for other uses. The non-potable supply can be derived from the same source with only existing treatment, or from reclaimed municipal wastewater with further treatment.

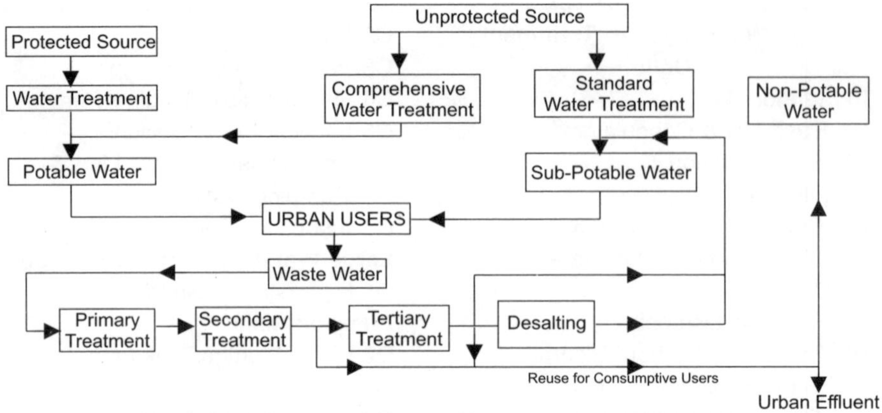

Fig. F23.1: Urban Dual Water System (Watman Model)

In Fig. F23.2 granular activated carbon or reverse osmosis treatment is essential with standard treatment; a dual water system may be more economical than a conventional system.

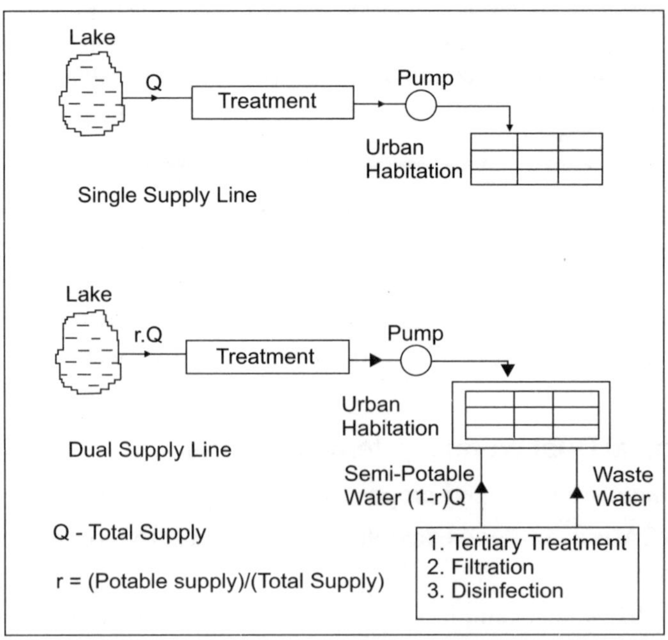

Fig. F23.2A

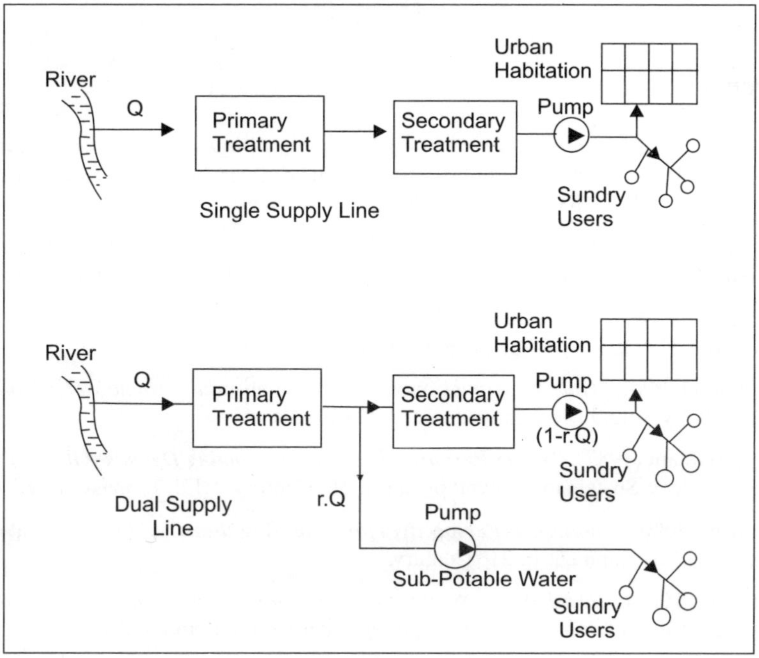

Fig. F23.2B: Urban Water Supply Proposals

Fig. F23.3 shows a regional Water Supply Application with multiple water uses. In the regional management of water, good quality limited sources can be conserved and used for potable supply through a regional water treatment and distribution system. The water for sub-potable uses can be obtained from local polluted sources.

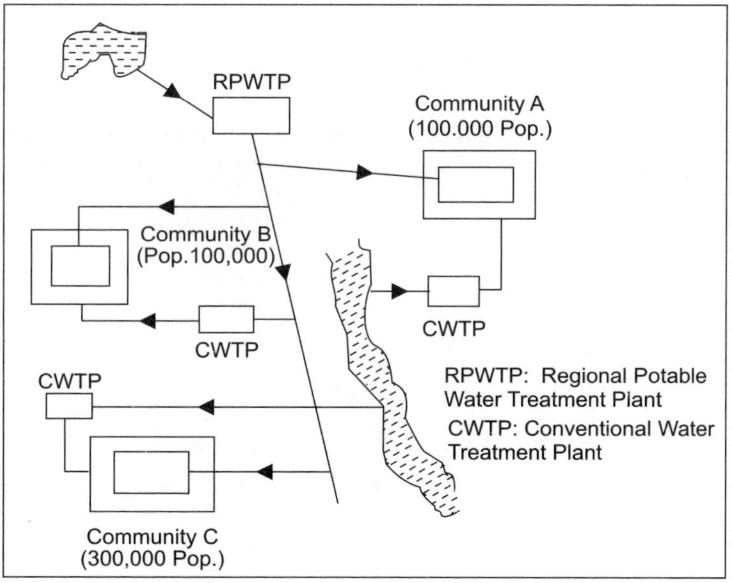

Fig. F23.3: Regional Dual Water Supply System

❑❑❑

Part IV

References

Anand P.B. (2007), *Semantic of Success or pragmatics of Progress*? (An Assessment of India's Progress with Drinking Water Supply), The Journal of Environment & Development, vol. 16 Number 1, March 2007, p32-57, Sage Publication.

Biswas Asit K (2007 January), *Lecture on Water Management in the 21st Century for the Developing World at the Indian Institute of Technology*, Kharagpur, India, c/o.Thirld World Centre for Water Management, Atizapan, Mexico.

Biswas Asit K (19..), *System Approach to Water Management*

Biswas Asit K (1990), *"Objectives and Concepts of Environmentally-sound Water Management"*, Environmentally Sound Management,(Ed. A.K.Biswas)

Cosgrove William (1987), *"Water Resource Sector in the Coming Decade: Global Perspective*, Water Resource & Sustainable Development in 21st Century", (Ed. Kamala Prasad.)

Chitale M.A. (2000), *South Asian Perspectives for Water Resource Development*, Water Resource & Sustainable Development in 21st Century.

Deb Arun K (1987), *"Dual Water Supply for Future for Future Urban Water Management*, Proceedings of the International Syposium on Water for the Future", Rome.

Furan R (1967), *"The problem of Water – a World Study"*

Masters Gilbert M (2000/1991), *"Introduction to Environmental engineering and Science"*, Prentice Hall, New Delhi.

Prasad T, Kumar B S 1987, *"Water Resource Development in India – its Central Role in the Past and Crucial Significance for the Future*, Rome.

Rao Dr.K L (1979), *"India's Water Wealth"*, Orient Longman, New Delhi.

Than N C & Tam D M 1990, *"Water Syst4ems and the Environment"*, Environmentally Sound Management, Asit K Biswas.

UN Water Conference 1977, *Mardel Plata, Argentina*, Part I, "Water Development & Management, Ed. Asit K Biswas.

World Bank Report 1999, *"India: Water Resource Management, Inter-Sectoral Water Allocation*, Planning & Management" in collaboration with the Government of India, Ministry of Water Resources.

Index